VOLUME SEVENTY FIVE

Advances in
ORGANOMETALLIC
CHEMISTRY

VOLUME SEVENTY FIVE

Advances in ORGANOMETALLIC CHEMISTRY

Edited by

PEDRO J. PÉREZ
Laboratorio de Catálisis Homogénea
CIQSO-Centro de Investigación en Química Sostenible and
Departamento de Química
Universidad de Huelva - Huelva
Spain

Founding Editors

F. GORDON A. STONE

ROBERT WEST

ELSEVIER

ACADEMIC PRESS
An imprint of Elsevier

Academic Press is an imprint of Elsevier
50 Hampshire Street, 5th Floor, Cambridge, MA 02139, United States
525 B Street, Suite 1650, San Diego, CA 92101, United States
The Boulevard, Langford Lane, Kidlington, Oxford OX5 1GB, United Kingdom
125 London Wall, London, EC2Y 5AS, United Kingdom

First edition 2021

Copyright © 2021 Elsevier Inc. All rights reserved

No part of this publication may be reproduced or transmitted in any form or by any means, electronic or mechanical, including photocopying, recording, or any information storage and retrieval system, without permission in writing from the publisher. Details on how to seek permission, further information about the Publisher's permissions policies and our arrangements with organizations such as the Copyright Clearance Center and the Copyright Licensing Agency, can be found at our website: www.elsevier.com/permissions.

This book and the individual contributions contained in it are protected under copyright by the Publisher (other than as may be noted herein).

Notices
Knowledge and best practice in this field are constantly changing. As new research and experience broaden our understanding, changes in research methods, professional practices, or medical treatment may become necessary.

Practitioners and researchers must always rely on their own experience and knowledge in evaluating and using any information, methods, compounds, or experiments described herein. In using such information or methods they should be mindful of their own safety and the safety of others, including parties for whom they have a professional responsibility.

To the fullest extent of the law, neither the Publisher nor the authors, contributors, or editors, assume any liability for any injury and/or damage to persons or property as a matter of products liability, negligence or otherwise, or from any use or operation of any methods, products, instructions, or ideas contained in the material herein.

ISBN: 978-0-12-824581-1
ISSN: 0065-3055

For information on all Academic Press publications
visit our website at https://www.elsevier.com/books-and-journals

Publisher: Zoe Kruze
Acquisitions Editor: Sam Mahfoudh
Developmental Editor: Tara Nadera
Production Project Manager: Denny Mansingh
Cover Designer: Alan Studholme

Typeset by SPi Global, India

Contents

Contributors ix
Preface xi

1. **Hydrofunctionalization reactions of heterocumulenes: Formation of C–X (X=B, N, O, P, S and Si) bonds by homogeneous metal catalysts** 1
   Cameron D. Huke and Deborah L. Kays

   1. Introduction 2
   2. Hydroboration 3
   3. Hydroamination 13
   4. Hydroalkoxylation 18
   5. Hydrosilylation 21
   6. P—H bond reactions 23
   7. Hydrothiolation 39
   8. Multi E–H hydrofunctionalization 41
   9. Concluding remarks 44
   Acknowledgments 45
   References 45

2. **Polymerization of terpenes and terpenoids using metal catalysts** 55
   Miguel Palenzuela, David Sánchez-Roa, Jesús Damián, Valentina Sessini, and Marta E.G. Mosquera

   1. Introduction: Terpenes and terpenoids, origin and classification 55
   2. Polymerization of terpenes 57
   3. Polymerization of terpenoids 75
   4. Copolymerization of terpenoids 81
   5. Concluding remarks 85
   Acknowledgments 86
   References 86

3. **Bimetallic frustrated Lewis pairs** 95
   Miquel Navarro and Jesús Campos

   1. Introduction 95
   2. Frustrated Lewis pairs 97
   3. Bimetallic complexes for bond activation 105

	4. Genuine metal-only FLP activation	122
	5. Miscellanea	131
	6. Conclusions	136
	Acknowledgments	136
	References	137

4. Metallic-based magnetic switches under confinement — 149
Alejandro López-Moreno and Maria del Carmen Giménez-López

	1. Introduction	149
	2. Endohedral fullerenes: Zero-dimensional host cages	153
	3. Carbon nanotubes: One-dimensional host nanocontainers	162
	4. Metal-organic frameworks: As three-dimensional platforms	167
	5. Conclusions	181
	Acknowledgments	183
	References	183

5. Recent advances in the synthesis and application of tris(pyridyl) ligands containing metallic and semimetallic p-block bridgeheads — 193
Andrew J. Peel, Jessica E. Waters, Alex J. Plajer, Raúl García-Rodríguez, and Dominic S. Wright

	1. Introduction	193
	2. Group 13 tris(pyridyl) ligands	194
	3. Group 14 tris(pyridyl) ligands	217
	4. Group 15 tris(pyridyl) ligands	234
	5. Outlook	240
	Acknowledgments	241
	References	241

6. Reactivities of N-heterocyclic carbenes at metal centers — 245
Thomas P. Nicholls, James R. Williams, and Charlotte E. Willans

	1. C—H, C—C, and C—N bond activation	245
	2. Reductive elimination	271
	3. C—C reductive elimination	272
	4. C-X reductive elimination	281
	5. C—H reductive elimination	286
	6. Insertion into a metal-C_{NHC} bond	289
	7. Hydrolytic ring opening	294

8. Ring expansion reactions	298
9. Miscellaneous ring openings	306
10. Miscellaneous NHC reactivity	313
11. Dissociation reactions	316
12. Summary	318
Acknowledgment	319
References	319

Contributors

Jesús Campos
Instituto de Investigaciones Químicas (IIQ), Departamento de Química Inorgánica and Centro de Innovación en Química Avanzada (ORFEO-CINQA), Consejo Superior de Investigaciones Científicas (CSIC) and University of Sevilla, Sevilla, Spain

Jesús Damián
Department of Organic and Inorganic Chemistry, Institute of Chemical Research "Andrés M. del Río" (IQAR), Universidad de Alcalá, Campus Universitario, Alcalá de Henares, Madrid, Spain

Maria del Carmen Giménez-López
Centro Singular de Investigación en Química Biolóxica e Materiais Moleculares (CIQUS), Universidade de Santiago de Compostela, Santiago de Compostela, Spain

Raúl García-Rodríguez
GIR MIOMeT-IU, Cinquima, Química Inorgánica, Facultad de Ciencias, Universidad de Valladolid, Valladolid, Spain

Cameron D. Huke
School of Chemistry, University of Nottingham, University Park, Nottingham, United Kingdom

Deborah L. Kays
School of Chemistry, University of Nottingham, University Park, Nottingham, United Kingdom

Alejandro López-Moreno
Centro Singular de Investigación en Química Biolóxica e Materiais Moleculares (CIQUS), Universidade de Santiago de Compostela, Santiago de Compostela, Spain

Marta E.G. Mosquera
Department of Organic and Inorganic Chemistry, Institute of Chemical Research "Andrés M. del Río" (IQAR), Universidad de Alcalá, Campus Universitario, Alcalá de Henares, Madrid, Spain

Miquel Navarro
Instituto de Investigaciones Químicas (IIQ), Departamento de Química Inorgánica and Centro de Innovación en Química Avanzada (ORFEO-CINQA), Consejo Superior de Investigaciones Científicas (CSIC) and University of Sevilla, Sevilla, Spain

Thomas P. Nicholls
School of Chemistry, University of Leeds, Leeds, United Kingdom

Miguel Palenzuela
Department of Organic and Inorganic Chemistry, Institute of Chemical Research "Andrés M. del Río" (IQAR), Universidad de Alcalá, Campus Universitario, Alcalá de Henares, Madrid, Spain

Andrew J. Peel
Department of Chemistry, Lensfield Road, Cambridge, United Kingdom

Alex J. Plajer
Inorganic Chemistry Research Laboratory, South Parks Road, Oxford, United Kingdom

David Sánchez-Roa
Department of Organic and Inorganic Chemistry, Institute of Chemical Research "Andrés M. del Río" (IQAR), Universidad de Alcalá, Campus Universitario, Alcalá de Henares, Madrid, Spain

Valentina Sessini
Department of Organic and Inorganic Chemistry, Institute of Chemical Research "Andrés M. del Río" (IQAR), Universidad de Alcalá, Campus Universitario, Alcalá de Henares, Madrid, Spain

Jessica E. Waters
Department of Chemistry, Lensfield Road, Cambridge, United Kingdom

Charlotte E. Willans
School of Chemistry, University of Leeds, Leeds, United Kingdom

James R. Williams
School of Chemistry, University of Leeds, Leeds, United Kingdom

Dominic S. Wright
Department of Chemistry, Lensfield Road, Cambridge, United Kingdom

Preface

This is a special volume of *Advances in Organometallic Chemistry* on occasion of the first UK-Spain Organometallic Chemistry Symposium (USOCS2019) which was held in Alcalá de Henares (Madrid, Spain) in September 2019. Contributors to this volume participated as invited speakers at such forum.

The relationship between United Kingdom and Spain in the context of organometallic chemistry started in the 70s–80s of the last century, when a number of postdoctoral associates from Spain landed in several UK laboratories at Imperial College London, Oxford, Sheffield or Bristol, among others. Most of them further become the first generation of organometallic chemists at Spain, with a marked influence of the British organometallic community. Along the years such collaborations were maintained, and this symposium somehow recovers the initial spirit of such joint travel.

The contributions contained in this volume provide a fresh view of the next generation of chemists in both countries. Kays has accounted on the use of homogeneous catalysts for the modification of heterocumulenes through the formation of carbon–heteroatom bonds. Mosquera has focused on the polymerization reactions of terpenes employing metal-based catalysts. The area of frustrated Lewis pairs with bimetallic complexes constitutes the topic selected by Campos. Giménez-López describes the field of confined magnetic switches with metal centers. The chemistry of trispyridyl ligands in the context of metallic and semimetallic complexes has been reviewed by Wright. Finally, Willans has entered in the chemistry of N-heterocyclic carbenes from the point of view of the reactivity at the NHC ligand.

It seems obvious that without the participation and commitment of all the authors, this volume would have not been possible, therefore I wish to thank them all for their contributions. Also, I would like to thank Shellie Bryant for her help along the years in the editorial team, and Tara Nodera, which from now on takes such responsibility, along with Denny Mansingh.

Let us hope that this year 2021 will bring mankind a time surpassing the terrible disease we are facing all over the world.

PEDRO J. PÉREZ

CHAPTER ONE

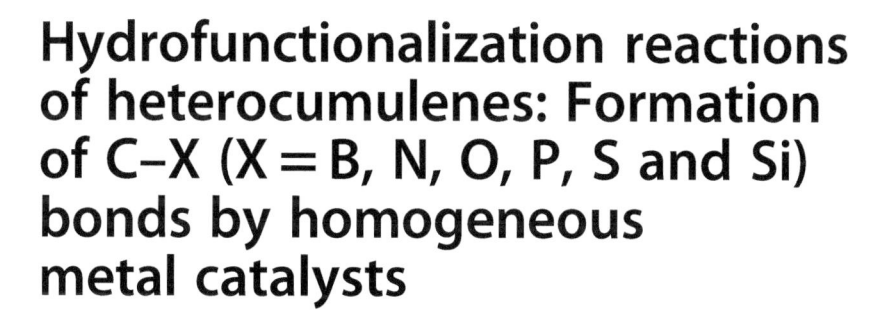

Hydrofunctionalization reactions of heterocumulenes: Formation of C–X (X = B, N, O, P, S and Si) bonds by homogeneous metal catalysts

Cameron D. Huke and Deborah L. Kays*

School of Chemistry, University of Nottingham, University Park, Nottingham, United Kingdom
*Corresponding author: e-mail address: deborah.kays@nottingham.ac.uk

Contents

1. Introduction	2
2. Hydroboration	3
2.1 Group 1 catalysis	4
2.2 Actinide catalysis	6
2.3 Transition metal catalysis	8
2.4 Aluminum catalysis	9
2.5 Catalyst-free	11
3. Hydroamination	13
3.1 Group 1 and 2 catalysis	13
3.2 Transition metal catalysis	15
4. Hydroalkoxylation	18
4.1 Actinide catalysis	19
4.2 Transition metal catalysis	21
5. Hydrosilylation	21
5.1 Transition metal catalysis	21
6. P—H bond reactions	23
6.1 Hydrophosphination	24
6.2 Hydrophosphinylation	35
6.3 Hydrophosphonylation	39
7. Hydrothiolation	39
7.1 Lanthanide catalysis	40
8. Multi E–H hydrofunctionalization	41
8.1 Actinide catalysis	41
8.2 Transition metal catalysis	42
9. Concluding remarks	44
Acknowledgments	45
References	45

1. Introduction

Heteroatom-containing organic molecules have found widespread success and applications in both industry and academia in areas such as pharmaceuticals,[1–3] fine and bulk chemicals, agrochemicals,[4] coordination chemistry[5–7] and catalysis.[8–10] The resulting demand and growing need for more sustainable chemical processes has led the chemical community to develop more efficient and selective processes for the design of these molecules. A popular route among their syntheses involves the addition of a E—H bond (E=O, S, N, P, B, etc.) to unsaturated organic molecules such as an alkene or alkyne, generally named hydrofunctionalization (also known as heterofunctionalization or hydroelementation). These reactions are highly desirable as they have the potential to be 100% atom efficient and utilize easily accessible low-cost starting materials. Additionally, the use of catalysts in these reactions can further improve the worth of this methodology by controlling selectivity and improving yields.

In the last decades hydrofunctionalization has witnessed intensive research, mainly using alkene,[11–15] alkyne[16–20] and allene[21–26] substrates (with some additional examples using carbonyls[27,28] and nitriles[29–32]). In more recent times, the desire of chemists to synthesize more heteroatom-rich moieties have resulted in the use of heterocumulene substrates with E–H substrates. The products of these reactions are heteroatom-rich urea, thiourea and guanidine derivatives that have use as pharmaceutical analogues,[33,34] organic synthons[35,36] and multidentate ligands.[37–40]

Compared to their counterparts with a single multiple bond, heterocumulenes have more synthetic possibilities with potential reactions spreading across, unsaturated bond selectivity, multiple heterocumulene insertions into a single E–H moiety, and double hydrofunctionalizations of a heterocumulene substrate. In the last decade, although tremendous progress has been made on these reactions, this greater synthetic potential is still comparatively lacking; this chapter pays particular attention to highlighting these more novel successes.

This chapter will outline advancements in the hydroboration, hydroamination, hydroalkoxylation, hydrophosphination, hydrophosphinylation, hydrophosphonylation, hydrothiolation and hydrosilylation of the most common heterocumulene substrates, isocyanates, isothiocyanates and carbodiimides by homogeneous catalysts and, where comparable (or useful for the reader), catalyst-free methodologies. The reader should note the hydroamination of carbodiimides is not covered by this chapter as this widely-explored research area has been reviewed in-depth previously.[2,41,42] In this respect, the hydroamination of isocyanates and isothiocyanates is included in this chapter, as to the best of our knowledge they have not been reviewed in detail previously and, to date, have not been as widely investigated.

2. Hydroboration

The addition of a B—H bond across an unsaturated moiety is a well-studied reaction, and typically forms the initial step toward a wide variety transformations such amination, halogenation, carbonylation and cross-couplings that manipulate regio-, stereo- and chemoselectivity.[43,44] The majority of hydroboration chemistry of heterocumulenes is focused on the use of carbodiimides, either due to the valuable nature of N-borylformamidines as useful intermediates for valuable guanidine derivatives or the general incompatibility of iso(thio)cyanates with H–B (R)$_2$ substrates. There are two main selectivity concerns with the hydroboration of carbodiimides. The first of these is that of C—N bond selectivity for unsymmetrical carbodiimide substrates, where HBpin (pin = Me$_4$C$_2$O$_2$) can preferentially reduce one of the two C=N bonds (Scheme 1, left). The second is that of single vs double hydroboration, involving the reduction of one C=N bond by HBpin or double hydroboration which is the reduction of two C=N bonds by two HBpin molecules (Scheme 1, right).

While this reaction can proceed without a catalyst under forcing thermal conditions (160 °C), there is either little no selectivity with a mixture of single and double hydroboration products being formed, or there are limitations in the use of the alkyl or aryl groups; this reactivity is further described in Section 2.5.[45] This lack of selectivity and harsh conditions has driven the development of catalytic hydroboration chemistry over the past decade.

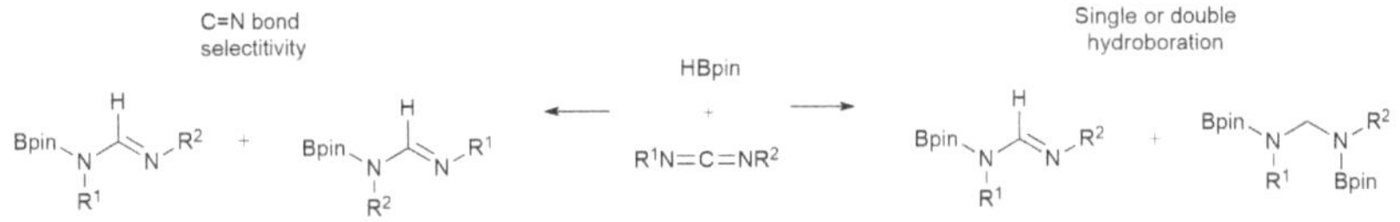

Scheme 1 Left: hydroboration of an unsymmetrical carbodiimide forming a mixture of borylformamidines, and right: hydroboration of a symmetrical carbodiimide forming a mixture of single and double hydroboration products.

Fig. 1 s-Block catalysts for the hydroboration of heterocumulenes, **1–4**.[46–49]

2.1 Group 1 catalysis

Complex **1** (Fig. 1) catalyzes the double hydroboration of ^{i}PrN=C=N^{i}Pr and ^{t}BuN=C=O, whereby two equivalents of HBpin insert across the unsaturated bonds (Scheme 2).[46] The hydridicity of the [HBPh$_3$]$^-$ ion is critical for catalysis, with the perfluorinated analogue [Mg(THF)$_6$][HB(C$_6$F$_5$)$_3$]$_2$ exhibiting inferior catalytic performance in optimization reactions for the hydroboration for other C=X substrates such as pyridine and benzophenone.[46]

The β-diketiminato precatalyst **2** (Fig. 1) has been used in the hydroboration of various alkyl and aryl carbodiimides (Scheme 2).[47] Increase of the steric bulk (to *tert*-butyl) necessitated longer reaction times, from 15 h (for R=Cy, ^{i}Pr) to 60 h, along with elevated temperatures (80 °C). Aryl-substituted carbodiimides also required the use of elevated temperatures to reach high conversions, albeit with similar reaction times to secondary alkyl carbodiimides. All symmetrical carbodiimides selectively underwent hydroboration to afford the *E*-stereoisomer. The use of two equivalents of HBpin at temperatures of 80–100 °C failed to afford double hydroboration products, instead leading to the decomposition of HBpin, as signified by B$_2$pin$_3$ formation. Unsymmetrical EtN=C=N^{t}Bu gave the expected *E*-stereoisomer, whereas EtN=C=N(CH$_2$)$_3$NMe$_2$, with less steric discrepancy between the terminal alkyl groups, gave a 1:1 mixture of the two possible regioisomers, as identified from the ^{1}H NMR spectrum. In a subsequent report, Hill and co-workers demonstrated that under similar conditions the hydroboration of alkyl and sterically encumbered aryl isocyanates catalyzed by **2**, did not afford the expected products, but instead underwent hydrodeoxygenation yielding methylamines, RN(Bpin)CH$_3$ (R = ^{i}Pr, ^{t}Bu, Cy, Et, ^{n}Pr, Ph, Mes and Dipp, where Mes=2,4,6-Me$_3$C$_6$H$_2$ and Dipp=2,6-^{i}Pr$_2$C$_6$H$_3$) and O(Bpin)$_2$.[50] Under less forcing conditions,

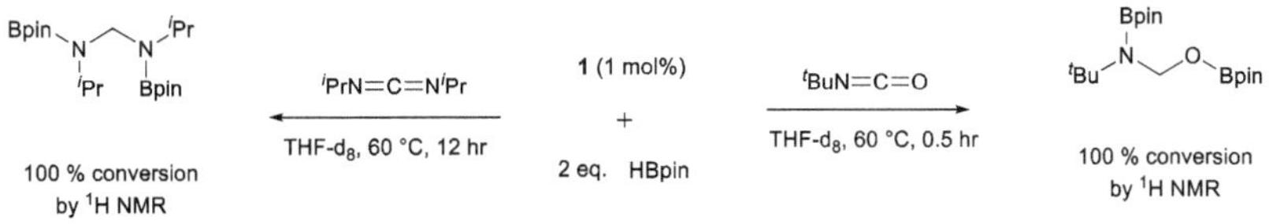

Scheme 2 Left: double hydroboration of diisopropyl carbodiimide, and right: double hydroboration of *tert*-butyl isocyanate.[46]

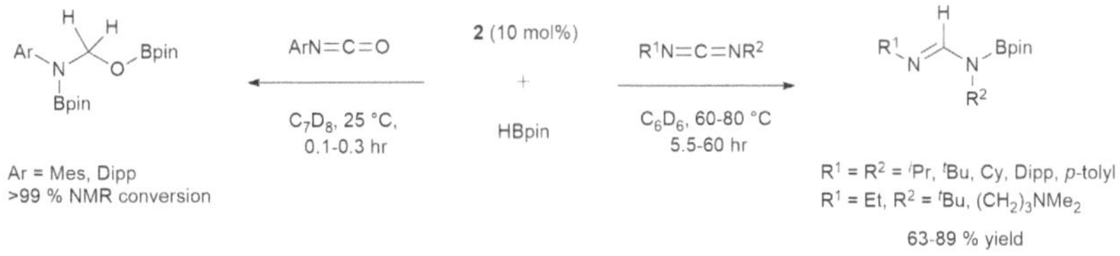

Scheme 3 Left: Dihydroboration of bulky aryl isocyanates by **2** (Mes = 2,4,6-Me$_3$C$_6$H$_3$, Dipp = 2,6-^{i}Pr$_2$C$_6$H$_3$), and right: monohydroboration of symmetrical and unsymmetrical carbodiimides by **2**.[47,50]

bulky MesN=C=O and DippN=C=O rapidly form double hydroboration products *in situ* (Scheme 3, left). Further attempts with these milder conditions using PhN=C=O and a range of aliphatic isocyanates proved unsuccessful, with only initiation of the catalyst being detected and further heating at 60 °C forming methylamines. DFT calculations suggest a mechanism whereby after double hydroboration of the isocyanate, the bis-borylated product undergoes a series of sequential low-energy insertions followed by reductions by HBpin and reinsertion until the completely reduced methylamine is formed.

Stoichiometric reactions of ^{i}PrN=C=N^{i}Pr and **2** yielded C-*n*-butyl amidinate complex **2a** (Scheme 4, left) via insertion of the ^{i}PrN=C=N^{i}Pr into the Mg—C bond, and stoichiometric studies using a similar dimeric hydridic magnesium complex afforded the insertion product **2b** (Scheme 4, right).[47] However, stoichiometric addition of HBpin to this intermediate formed two species in a 1:5 ratio; the minor species was the monohydroboration product and the major species featured a product with coordination to the magnesium center still present. A larger scale reaction between [LMgH]$_2$ L = HC{(Me)CN(2,6-^{i}Pr$_2$C$_6$H$_3$)}$_2$, ^{i}PrN=C=N^{i}Pr and HBpin, afforded crystals of cyclic **2c** (Scheme 4, left) but attempts to induce hydride elimination were unsuccessful. Catalytic behavior was only obtained by stoichiometric addition of HBpin and ^{i}PrN=C=N^{i}Pr. Kinetics studies using CyN=C=NCy revealed an allosteric effect whereby binding of a ligand

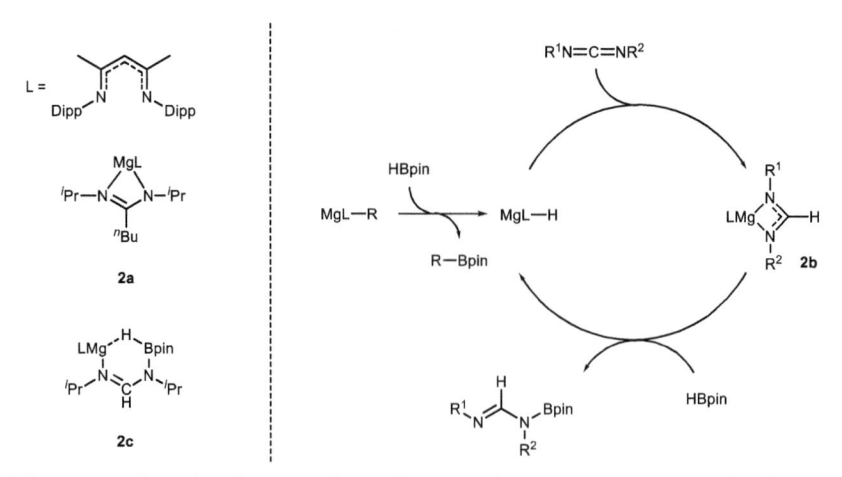

Scheme 4 Left: Isolated intermediates from stoichiometric reactions and right: proposed reaction mechanism by Hill et al.[47]

at one site affects the affinity for other substrates, such that catalytic activity was dependent of additional equivalents of carbodiimide and HBpin.[51]

The reaction between precatalyst **3** (Fig. 1) and HBpin affords the terminal hydride complex [Tism^{PriBenz}]MgH ([Tism^{PriBenz} = tris[(1-isopropylbenzimidazol-2-yl)-dimethylsilyl)]methyl]; this a versatile catalyst for the hydroboration of styrene, pyridine and carbodiimides.[48] For the carbodiimides (R = ^{i}Pr, Cy), precatalyst **3** is more active than **1** and **2**. Complex [Tism^{PriBenz}]MgH exhibits diverse reactivity with a variety of substrates to afford a series of complexes [Tism^{PriBenz}]MgX [X = F, Cl, Br, I, SH, N(H)Ph, CH(Me)Ph].

Hill and co-workers have pursued magnesium β-diketiminato precatalysts further, synthesizing complex **4** (Fig. 1) bonded by two *ortho*-fluorides to the Mg center along with one Mg—H bond. Complex **4** is capable of the double hydroboration of ^{i}PrN=C=N^{i}Pr at 60 °C yielding the respective bis(*N*-boryl)aminal in 70% yield.[49]

2.2 Actinide catalysis

Eisen et al. have prepared a series of actinide complexes **5–12** for the hydroboration of carbodiimides (Fig. 2).[52] Initial optimization reactions showed that the steric demand of the ancillary ligands demonstrated little influence on the catalytic behavior, and that thorium-containing precatalysts in **5–12** gave higher yields than the respective uranium analogues.[52] The best-performing precatalysts were **9–12** all giving rise to the monohydroboration

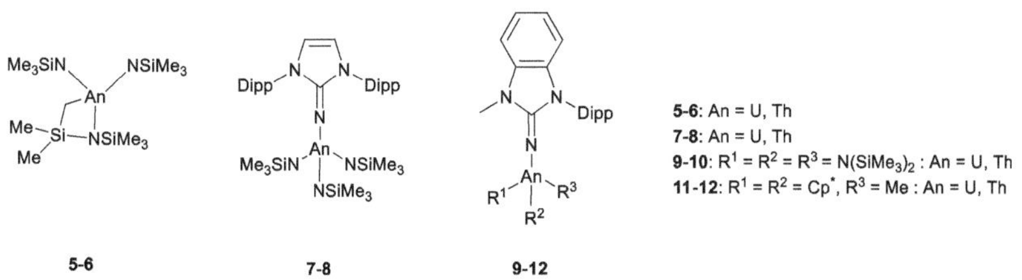

Fig. 2 Actinide complexes **5–12** for the hydroboration of carbodiimides developed by Eisen et al.[52]

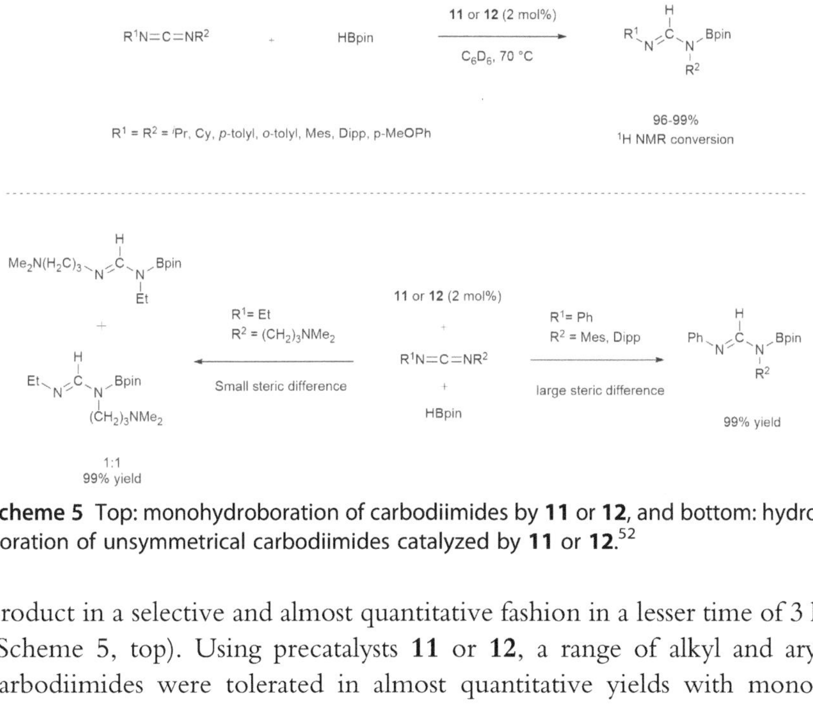

Scheme 5 Top: monohydroboration of carbodiimides by **11** or **12**, and bottom: hydroboration of unsymmetrical carbodiimides catalyzed by **11** or **12**.[52]

product in a selective and almost quantitative fashion in a lesser time of 3 h (Scheme 5, top). Using precatalysts **11** or **12**, a range of alkyl and aryl carbodiimides were tolerated in almost quantitative yields with monohydroboration selectivity, even with two equivalents of HBpin. For unsymmetrical carbodiimides with similar-sized substituents little selectively was observed, but the selectivity improved for substituents with a larger steric discrepancy. For both products, the Bpin group is bound to the nitrogen atom bearing the bulkier substituent (Scheme 5, bottom).

It is postulated that a highly ordered transition state is responsible for the selectivity, with the formamidate ligand adopting a preferential orientation with the larger substituent pointing away from the Cp* group (Cp* = C$_5$Me$_5^-$) and benzimidazolin-2-iminato ligands (Scheme 6, left). Based upon experimental observations, the first step of the mechanism is a rapid

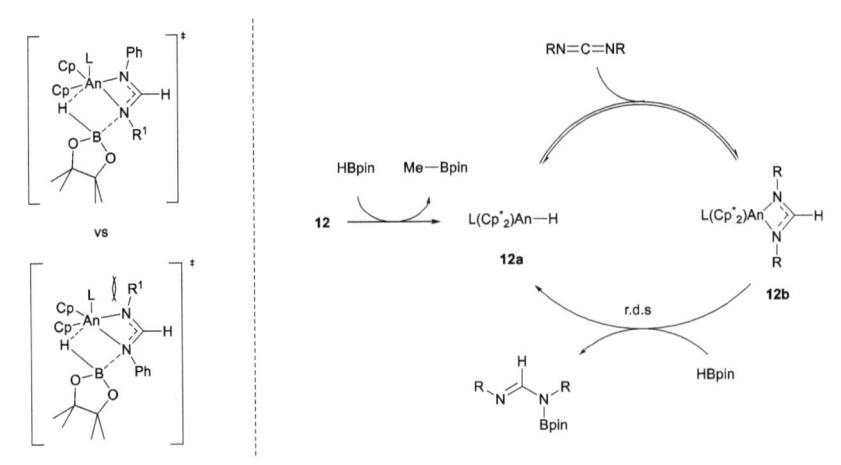

Scheme 6 Left: orientations of highly ordered transition states responsible for R group selectivity for hydroboration of unsymmetrical carbodiimides with the bottom left highlighting the steric clash between the bulky R¹ group and the benzimidazolin-2-iminato ligand, L = 1-methyl, 3-(2,6-diisopropylphenyl)benzimidazolin-2-iminato. Right: mechanism for hydroboration of carbodiimides by precatalyst **11**.[52]

reaction between **12** and HBpin, which releases MeBpin and generates hydridic compound **12a** as the catalytically active species (Scheme 6, right). Insertion of the carbodiimide into the An–H bond of **12a** is in equilibrium, producing the amidinate complex **12b**, and reaction with another molecule of HBpin releases the product and regenerates **12a**, in the rate determining step.

2.3 Transition metal catalysis

In 2020 Eisen and co-workers further expanded the use of imidazolin-2-iminato ligands to develop hafnium-based precatalysts **13** and **14** (Scheme 7).[52] As with the actinide complexes **5–12**, temperatures of 70 °C are required for catalytic turnover of carbodiimide substrates, but for alkyl carbodiimide substrates the hydroboration products were afforded in near quantitative yields in only 1 h (Scheme 7). For aryl carbodiimide substrates yields were highly dependent on the sterics and electronics. As with the actinide precatalysts **5–12**, the Bpin was selectively added to the more sterically encumbered side of the carbodiimide.

A mechanism was proposed (Scheme 8) whereby the first step is a rapid reaction between **13** and HBpin, which releases the three equivalents of PhCH₂Bpin and simultaneously generates the [(ImRN)HfH₃] **13a** as the catalytically active species.[52] Subsequent insertion of the carbodiimide substrate into the Hf—H bond produces the hafnium amidinate complex **13c**,

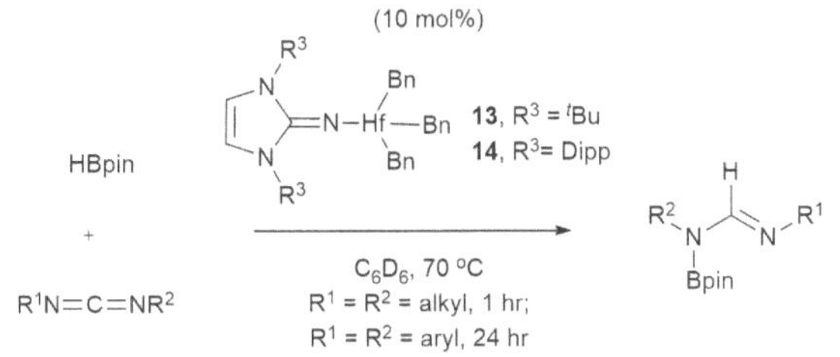

$R^1 = R_2$ = Cy, ^{i}Pr; 99% yield
$R^1 = R^2$ = 2-MeC_6H_4, 4-OMeC_6H_4; 11% (**13**) and 3-5% (**14**)
R^1 = Ph, R^2 = Mes; 38 % (**13**) and 35% (**14**)

Scheme 7 Hydroboration of carbodiimides catalyzed by **13** or **14**.[52]

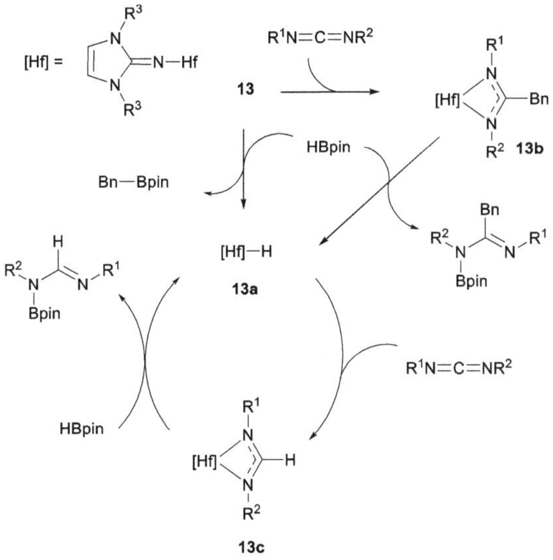

Scheme 8 Mechanism for the hydroboration of carbodiimides by **13**.[53]

which further reacts with another molecule of HBpin, to release the borylformamidine product and regenerate the active species **13a**. Parallels can be drawn to the mechanism of lanthanide precatalyst **12**: after the initial formation of the hydridic active species, both mechanisms are analogous.

2.4 Aluminum catalysis

Roesky et al. have investigated β-diketiminato (**15** and **16**) and hydridic aluminum (**17–19**) precatalysts (Fig. 3) for hydroboration chemistry.[53,54] Precatalyst **15** catalyzes the hydroboration of alkyl carbodiimides

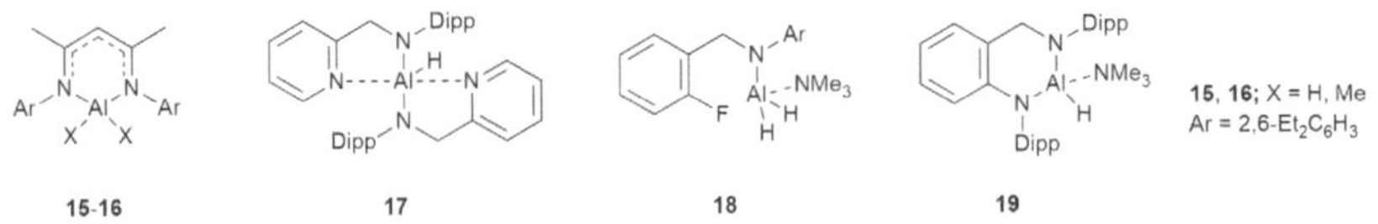

Fig. 3 Aluminum based hydroboration catalysts by Roesky et al.[53,54]

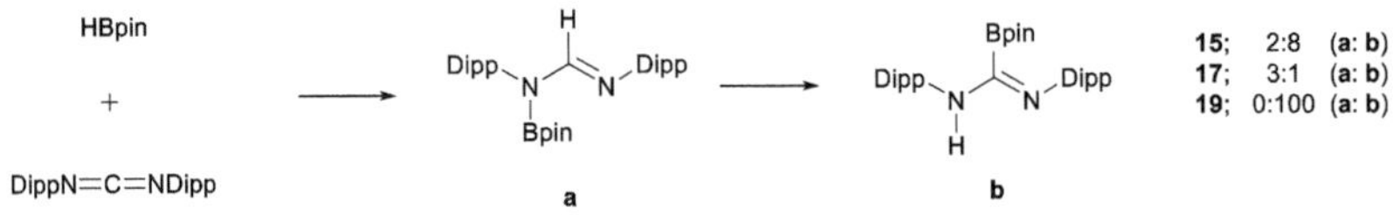

Scheme 9 Monohydroboration of DippN=C=NDipp to N-borylformamidine (**a**) followed by conversion to C-borylformamidine (**b**).[53]

(cyclohexyl and isopropyl) with near quantitative conversion in 12 h at 80 °C; a reduction in temperature (70 °C) led to a reduction in yield to ca. 80%. The bulkier ^{t}BuN=C=N^{t}Bu required significantly longer reaction time, but the even bulkier DippN=C=NDipp substrate was reacted in quantitative conversion in under 2 h. The catalysis using **16–18** with four carbodiimide substrates RN=C=NR (R = ^{i}Pr, Cy, ^{t}Bu, Dipp) was compared.[53] Of these, precatalyst **16** was found to be the most active with near quantitative yields for all substrates with longer reaction times for R = Dipp and ^{t}Bu (48 h). Precatalyst **17** showed dramatically reduced yields for bulkier substrates (R = Dipp, ^{t}Bu) and required further increased reaction times. The use of sterically demanding DippN=C=NDipp with precatalysts **15**, **17** or **19**, afforded the first reported example of a *C*-borylformamidine, identified in the ^{1}H NMR spectrum by a shift from 8.05 ppm of NC*H*N to 9.73 ppm of Ar-N*H*C (Scheme 9). Utilizing precatalyst **19**, the *C*-borylformamidine could be selectively synthesized in 60% yield. The rearrangement was explained by intermolecular hydrogen bonding formed by the *H*N=C moiety which increased the stability compared to the *N*-borylformamidine. Additionally, the electrophilicity of the boron of Bpin may allow for a shift from the C—N bond to more electron rich C=N bond.

Stoichiometric reactions between ^{i}PrN=C=N^{i}Pr and **15** form the expected aluminum formamidate **15a** (which has been structurally characterized), with a weak bonding interaction allowing for coordination of the carbodiimide allowing insertion into the Al—H bond, with a hydride shift to the carbon followed by sigma bond exchange of HBpin to complete the catalytic cycle (Scheme 10).[53] An analogous cycle can be proposed with precatalyst **16** but with an initiation step involving exchange of methyl groups with a hydride of HBpin.

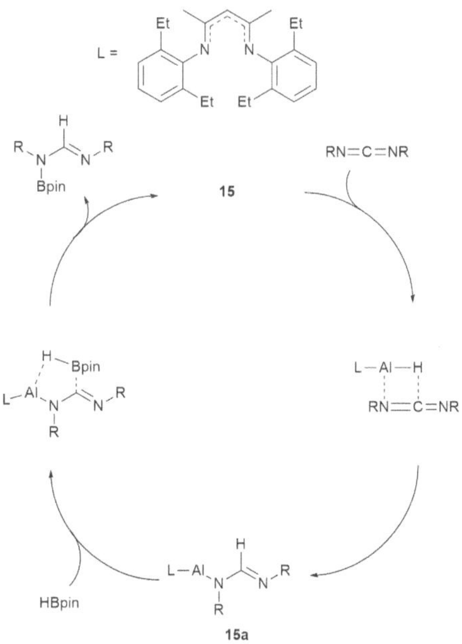

Scheme 10 Hydroboration of carbodiimides by precatalyst **15**.[53]

2.5 Catalyst-free

Frustrated Lewis pair (FLP) compounds reported by Stephan and co-workers using Piers' borane, HB(C$_6$F$_5$)$_2$ have been used for the hydroboration of carbodiimides (R = ^{i}Pr, ^{t}Bu) forming four-membered cyclic borylformamidines (**20**); this reaction is catalyst-free and proceeds at ambient temperature (Scheme 11, top).[55] In further extension of this work, Warren and co-workers found that when using a bulky isocyanate such as DippN=C=O, a six-membered ring formed by the insertion of two isocyanate molecules followed by oxophilic coordination of B(C$_6$F$_5$)$_2$ in compound **21**.[56] Use of an isocyanate substrate with a more sterically hindered aryl (2,6-Ph$_2$-4-^{t}BuC$_6$H$_2$), afforded a second species (**21a**), observed in equilibrium with **20** (Scheme 11, top). Fischer et al. present to the best of our knowledge the only example of the hydroboration of isothiocyanates.[57] Using HB(C$_6$F$_5$)$_2$, a four-membered ring is formed (R = Ph, adamantyl) with carbodiimides in compound **22** (Scheme 11, top). When R = Bn, **22** reveals a tetrameric structure in the solid state, **22a**, containing –S–C(H)=N(R)-HB(C$_6$F$_5$)$_2$–S–. Utilizing a different H—B source, H-BBN (BN=9-bicyclo[3.3.1]nonane), Koster et al. have shown at 100 °C CyN=C=NCy undergoes catalyst-free monohydroboration to yield **23a** (Scheme 11, bottom). At longer reaction times with another equivalent of H-BBN,

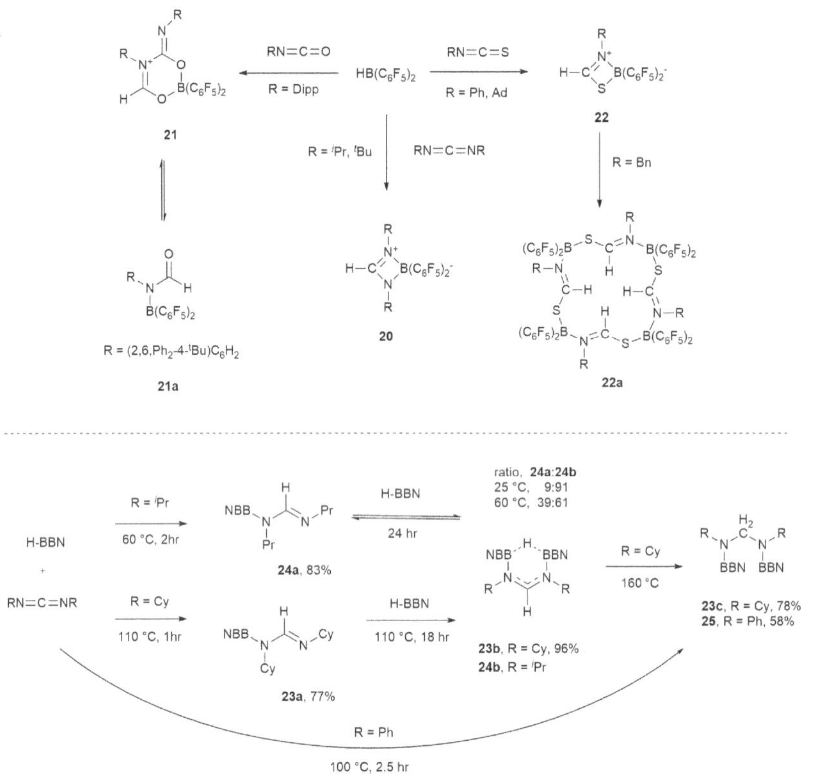

Scheme 11 Top: Piers' borane hydroboration of carbodiimides by Stephen et al.,[55] of isocyanates by Warren et al.[56] and of isothiocyanates by Fischer et al.[57] Bottom: H-BBN hydroboration of carbodiimides by Koster et al.[45] and Garcıa-Vivo et al.[58]

a cyclic double hydroboration intermediate **23b** is formed in excellent yields. Subsequent heating of **23b** at 160 °C afforded the monomeric diborylamidine. When PhN=C=NPh was used instead of CyN=C=NCy, less forcing conditions were required to form the free diborylamidine product **25** in moderate yields. Garcia-Vivo et al. have shown ^{i}PrN=C=N^{i}Pr shows similar reactivity to CyN=C=NCy also undergoing catalyst-free monohydroboration.

However, attempts at double hydroboration show that **24a** is in equilibrium with **24b** (Scheme 11, bottom).[58] A lower temperature of 25 °C favored intermediate **24b** in a 91:9 ratio, while heating at 60 °C lowered the ratio to 39:61.

3. Hydroamination

The hydroamination of isocyanates and isothiocyanates leads to the formation of urea and thiourea derivates. These are essential in a range of biological systems,[59,60] pharmaceuticals,[33,34,61,62] agrochemicals,[35,63–66] materials chemistry,[67,68] supramolecular chemistry[69,70] and synthetic chemistry.[35,36] Many synthetic methods for the synthesis of ureas involve the noble metals, for example in the oxidative carbonylation of amines,[71,72] reduction of nitro compounds with carbon monoxide[73] and cross-coupling of aryl chlorides with sodium cyanate,[74] or they utilize additional stoichiometric reagants.[75] The most widely used syntheses for thioureas are the condensation of primary and secondary amine with isothiocyanate,[76] thiophosgene[77] or its derivatives.[78] Additionally, the reaction between primary amines and carbon disulfide in the presence of mercury acetate, and the reaction of unsubstituted thioureas with primary alkyl amines are also used.[79,80] Hydroamination is a simple alternative route that avoids the use of excess of stoichiometric reagents and harsh conditions. It should be noted although some hydroamination reactions of isocyanates and isothiocyanates can occur without a catalyst, these generally require long reaction times and high temperatures and are limited to nucleophilic amines.[81]

3.1 Group 1 and 2 catalysis

In 2008, Hill and co-workers reported the hydroaminations of isocyanates using precatalysts **26–29** with bulky isocyanate substrates (Dipp, Ad) and HNPh$_2$ (Fig. 4).[82] The reaction using AdN=C=O proceeded at room temperature while the use of DippN=C=O required more forcing conditions (60 °C), with conversions to the urea products plateauing at 86% and 62%, respectively. The reaction is not first order in either amine or isocyanate but, importantly, at longer reaction times (or higher conversion) the rate of the reaction slows leading to pseudo first-order kinetics and product inhibition. The formation of intermolecularly hydrogen-bonded adducts between the product and substrates is plausible, with the urea product acting as a ligand for the catalytically active calcium species and hindering substrate coordination and activation at the metal. Using the strontium-based **28** provided an increase in the reaction rate compared to calcium-based **27**, revealing an effect of increased metal ionic radius. Use of the barium analogue **29** did

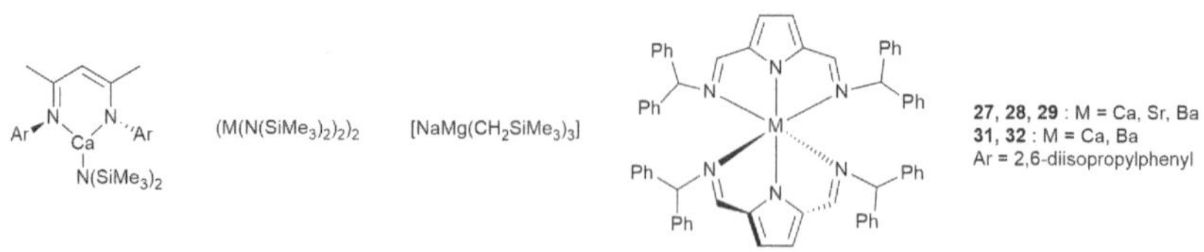

Fig. 4 Hydroamination precatalysts **26–29** by Hill and co-workers,[82] **30** by Hevia and co-workers[83] and **31–32** by Panda and co-workers.[84]

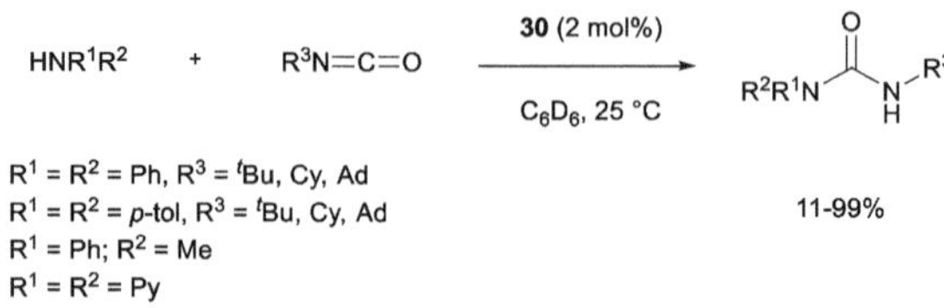

Scheme 12 Hydroamination of alkyl isocyanates by precatalyst **30**.[83]

not further increase the rate, but instead the precipitation of an insoluble species occurred, depleting the catalytically active species from solution. Catalyst loadings of 2 mol% of **30** catalyze the hydroamination of a variety of alkyl isocyanates and aryl secondary amines (Scheme 12). More sterically encumbered $HN(SiMe_3)_2$ was not tolerated, and $HNPy_2$ was also poorly tolerated, potentially due to the pyridyl donors forming a contact ion pair of sodium magnesiates reducing the coordination flexibility.[83] Treatment of the precatalyst **26** with three equivalents of $HNPh_2$ in hexane/THF solution afforded [(THF)NaMg(NPh$_2$)$_3$(THF)] **30a-THF** (Scheme 13, top). Subsequent reaction of **30a** with three equivalents of $^tBuN=C=O$ gave a metal tris(uredo) complex **30b-THF**. The hydroamination mechanism proceeds via protonation by three equivalents of $HNAr_2$ to yield **30a** followed by three isocyanate insertions to yield **30b** and completion of the catalytic cycle by protonation by $HNAr_2$ (Scheme 13, bottom). When the less electrophilic and bulkier p-tolN=C=O is used instead of PhN=C=O, the cyclotrimerization product, [p-tolN=C=O]$_3$ is formed.

Initially using precatalysts **31** and **32**, no hydroamination product was observed from the reaction between aryl isocyanates and primary aryl amines, with heating to 60 °C required to initiate the reaction. Further optimization found the use of solvents benzene, toluene and THF were inferior to performing the reaction without added solvent. The reaction between

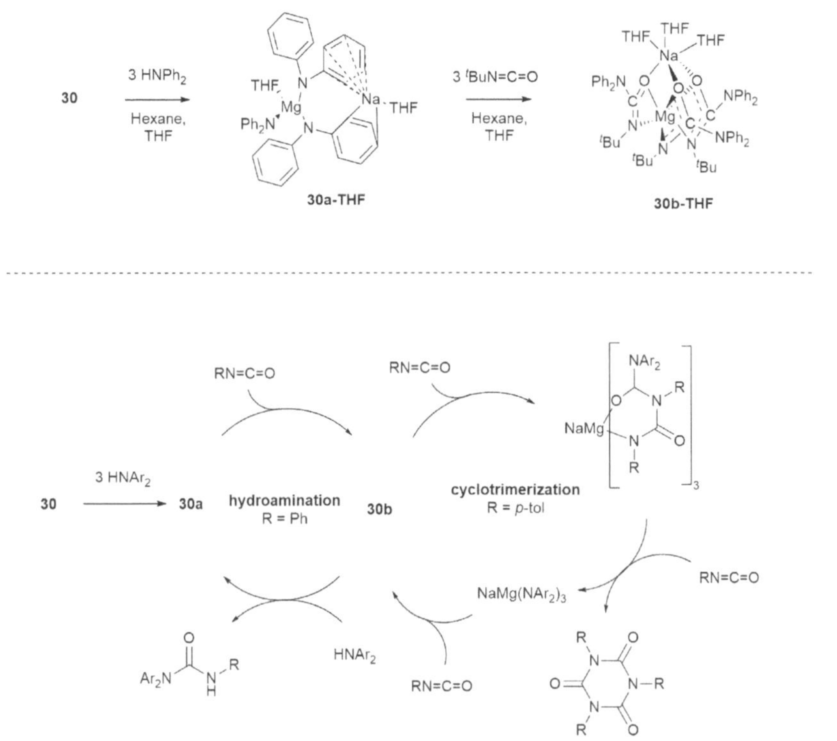

Scheme 13 Top: stoichiometric reactions, 3:1 reaction of HNPh$_2$ and **30** and reaction 3:1 reaction ^{t}BuN=C=O and **30a-THF**. Bottom: hydroamination mechanism for PhN=C=O and cyclotrimerization mechanism for p-tolN=C=O.[83]

p-tolN=C=O with H$_2$NPh and substituted secondary amines HNR$_2$ bearing electron-donating groups (R = 4-OMeC$_6$H$_4$, 2-MeC$_6$H$_4$, 2,6-Me$_2$C$_6$H$_3$, Bn), proceeded with excellent yields but those arylamines with electron-withdrawing groups, (R = 2-NO$_2$C$_6$H$_4$, 2-FC$_6$H$_4$, 2-BrC$_6$H$_4$) showed slightly lower conversions.[84] The same functional group scope of arylamines is also applicable to phenyl isothiocyanate. It is worth noting that in all cases only single hydroamination reactions occurred from the two possible N—H bonds; with no biuret product being observed.

3.2 Transition metal catalysis

In the presence of 5 mol% of imidazolin-2-iminato titanium-based precatalyst **33**, PhN=C=O undergoes hydroamination by a variety of amines, HNR^{1}R^2 (R^1=H; R^2 = ^{t}Bu, Et, Mes, Dipp, 2-FC$_6$H$_4$, R^1=R^2 = ^{i}Pr, and pyrrolidine)

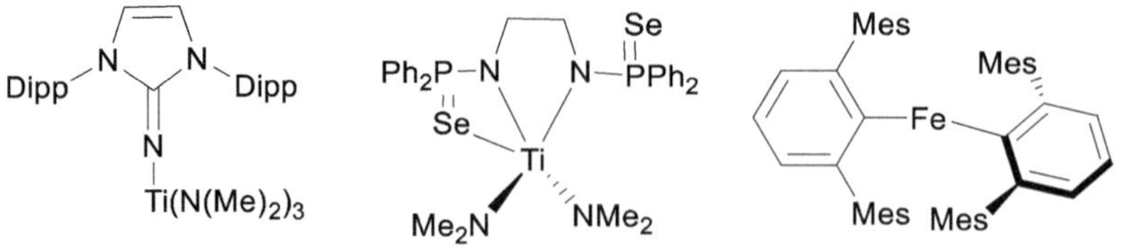

Fig. 5 Transition metal hydroamination precatalysts, titanium precatalysts **33** and **34**,[85,86] and iron complex **35**.[87]

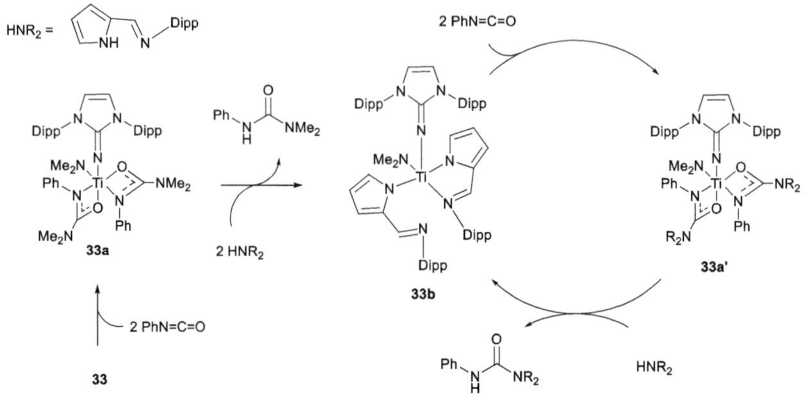

Scheme 14 Mechanism for the hydroamination of phenyl isocyanate based on stoichiometric reactions from **33**.[84]

with isocyanates, RN=C=O (R=Ph, *p*-tol and 4-ClC$_6$H$_4$) (Fig. 5).[85] When complex **33** was reacted with PhN=C=O in either a 1:1 or 1:2 stoichiometric ratio complex **33a** forms (Scheme 14). The C=O bond inserts into the Ti—N bond forming a κ^2-OC(NMe$_2$)NPh moiety, and using a large excess of PhN=C=O does not lead to the coordination of a third molecule of PhN=C=O, instead the cyclotrimerized product is formed in addition to **33a**. Attempts to obtain single crystals from the stoichiometric reaction of **33a** and H$_2$NDipp were unsuccessful; replacement of H$_2$NDipp with bulky and bidentate iminopyrrole ligand allowed for stabilization and isolation of the intermediate **33b**. The mechanism shown in Scheme 14 has been postulated, whereby two equivalents of PhN=C=O insert into the Ti—N bond to form **33a**, subsequent protonation by H$_2$NR generates the active catalysts **33b**. Insertion of two equivalents of isocyanate forms **33a′** analogous to **33a**, with protonation affording the urea product and reforming **33b′**.

Further work using **34** (Fig. 5) finds analogous tolerance to the primary amines utilized with **33**, and the heterocumulene scope expanded to utilize phenyl isothiocyanate to afford the hydroamination products in excellent yields.[86]

The use of low-coordinate iron precatalyst **35** (Fig. 5) in the reaction between HNPh$_2$ and PhN=C=O afforded a mixture of two products in a 88:12 ratio, the major was the urea product and the minor being the biuret, formed by the insertion of two equivalents of isocyanate into the N—H bond (Scheme 15, top).[87] Utilizing a 1:5 ratio increased selectivity for the biuret to 90–100%, but this was only successful for aryl isocyanates (R=Ph, C$_6$H$_3$-3,5-(OMe)$_2$ and C$_6$H$_4$-4-Br) as alkyl isocyanates ^{i}PrN=C=O and ^{t}BuN=C=O consistently formed the corresponding ureas.

For primary aliphatic isocyanate substrates there is a competing cyclotrimerization reaction of the isocyanate affording isocyanurates,[88] but the secondary and tertiary aliphatic isocyanates (R = ^{i}Pr and ^{t}Bu) are unaffected by this competing process. The use of solvent was also found to be key to reaction selectivity, with the use of THF resulting in the formation of urea products with high selectivity. The hydroamination of aniline proceeds without a catalyst, but the use of precatalyst **35** in this reaction afforded the double hydroamination product—a biuret in 68% yield (Scheme 15, bottom), whereby both N—H bonds react with the isocyanate; this is contrary reactivity to the titanium precatalysts **33** and **34**, where only one N—H

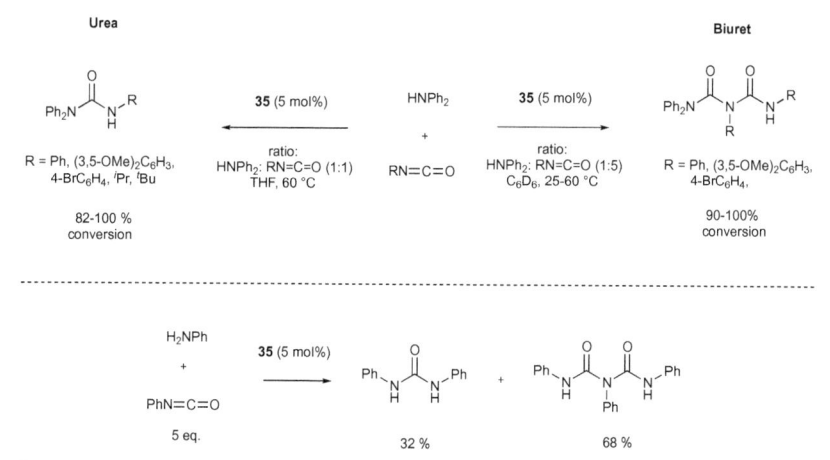

Scheme 15 Top: hydroamination of isocyanates by **35** affording the urea and biuret products. Bottom: double hydroamination of aniline by precatalyst **35**.[87]

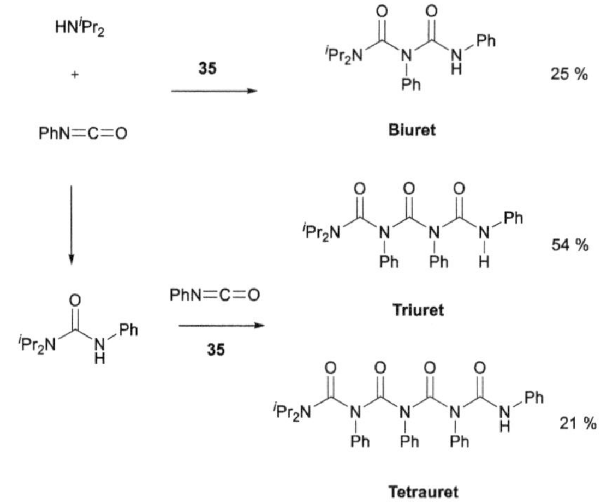

Scheme 16 Polyinsertion reactions of PhN=C=O and HN^{i}Pr$_2$ catalyzed using pre-catalyst 35.[87]

bond reacts. In the absence of catalyst, the more nucleophilic HN^{i}Pr$_2$ affords the respective urea products, using precatalyst **35** the biuret was obtained as the major product with small amounts of triuret product (Scheme 16). A PhN=C=O to NH^{i}Pr$_2$ ratio of 5:1 affords the tetrauret in 21% yield. Use of the urea as substrate leads to fast (<10 min) conversion to the same polyinsertion products in very similar ratios.

4. Hydroalkoxylation

The addition of alcohols to isocyanates has been known for well over a century, and is a rich and mature field perhaps best known for the formation of polyurethanes from diols and diisocyanates.[89–91] In general, isocyanates react rapidly with H$_2$O giving 1,3-disubstuted ureas and carbon dioxide and with alkyl alcohols and H$_2$O rapidly in the absence of a catalyst.[92–94] The same statement can be made with isothiocyanate and alcohols to a lesser extent.[92,95] For less nucleophilic aromatic alcohols, a wide variety of catalysts are available from bases such as pyridine, NaH, NaOH, Na$_2$CO$_3$ to simple metal halides such ZnCl$_2$ or AlCl$_3$, The most effective catalysts found have been Sn based salts such as Sn(^{n}Bu)$_3$(OC(O)Me) or Sn(^{n}Bu)$_3$H or SnCl$_2$.[93,96–99] Contrastingly, the less electrophilic carbodiimides in the absence of catalysts are fairly inert toward alcohols at room temperature. Under drastic conditions, PhN=C=NPh reacts additively with ethanol forming ethyl-N,N'-diphenylisourea.[100] Isoureas have additionally been prepared successfully using stoichiometric sodium ethoxide at room temperature.[101] Tertiary bases (for example, triethylamine, pyridine) are not suitable catalysts but certain copper salts such as CuCl$_2$ are good catalysts, particularly in the case of aliphatic carbodiimides.[102] In terms of other catalysts, copper sulfate exhibits catalytic activity, as do nickel and cobalt salts with reaction times varying from 4 h up to 28 days (e.g. for tertiary alcohols). Thus, the use of catalytic methodologies can be justified with carbodiimide substrates in order to achieve short reaction times, low reaction temperatures, high yields and if the functional group tolerance is high (particularly if the aryl or alkyl group contains other heteroatom substituents capable of further addition reactions).

4.1 Actinide catalysis

Eisen and co-workers have prepared U/Th(III) and U/Th(IV) precatalysts **36–38** for the hydroalkoxylation of carbodiimides.[103] Conceptually, the use of actinide complexes and alcohols is a challenging concept when considering the high bond strengths: Th–O = 208 kcal mol^{-1} and U–O = 181 kcal mol^{-1}.[106] Initial optimization reactions using several alkyl alcohols (R = Me, Et, ^{i}Pr,) and phenol with either ^{i}PrN=C=N^{i}Pr and p-tolN=C=Np-tol, did not show a conclusive trend between the activities of catalysts **37–39**, and the use of THF or methylcyclohexane solvents severely inhibited the catalytic activity. Generally, with alkyl alcohols (R = Me, Et, ^{i}Pr, ^{t}Bu) higher yields were obtained by utilizing more electrophilic p-tolN=C=Np-tol compared to less electrophilic ^{i}PrN=C=N^{i}Pr. The use of a proton trap (2,6-di-*tert*-butyl-4-methylpyridine) showed that Brønsted acids were not mediating the insertion of alcohols into carbodiimides, and kinetic studies demonstrated first-order behavior with respect to precatalyst and carbodiimide, but with inverse first-order behavior with respect to alcohols. Use of ethanol-d_6 and *tert*-butanol-d_{10} gave kinetic isotope effect (KIE) values of 0.98 and 0.99, respectively. A proposed mechanism for this reaction is shown in Scheme 17.

The reactions using precatalysts **39–44** (Fig. 6) with ^{i}PrN=C=N^{i}Pr and methanol substrates showed low yields over 24 h but a reduction in the reaction time to 1 h led to quantitative yields with **39–40** for ROH substrates (R = Me, Et, ^{i}Pr, ^{t}Bu, Ph), with longer reaction times required for ^{t}Bu-OH for near quantitative yields.[104,105] Bulkier organic groups on the alcohol required the use of longer reaction times to achieve good yields. This catalytic system also demonstrated high activity for diol and triol substrates (Scheme 18, top). In addition, more electron rich precatalysts **43** and

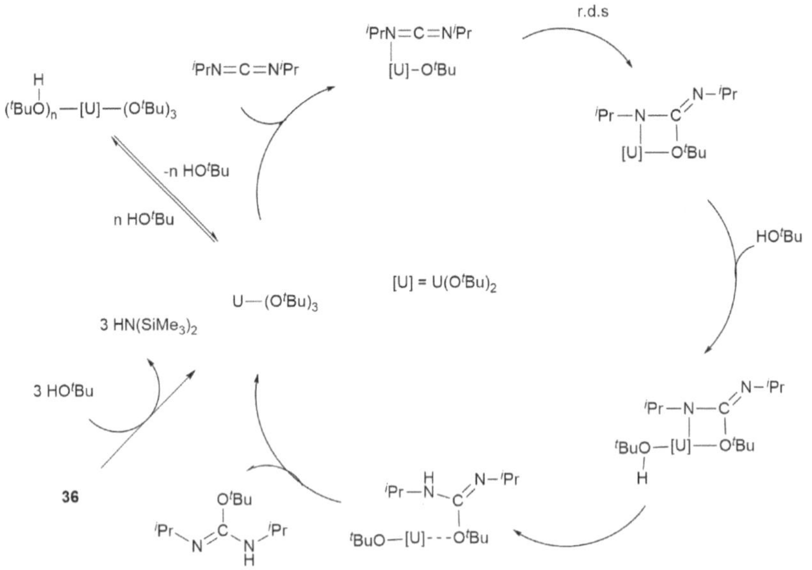

Scheme 17 Proposed mechanism for the hydroalkoxylation of carbodiimides by **36**.[103]

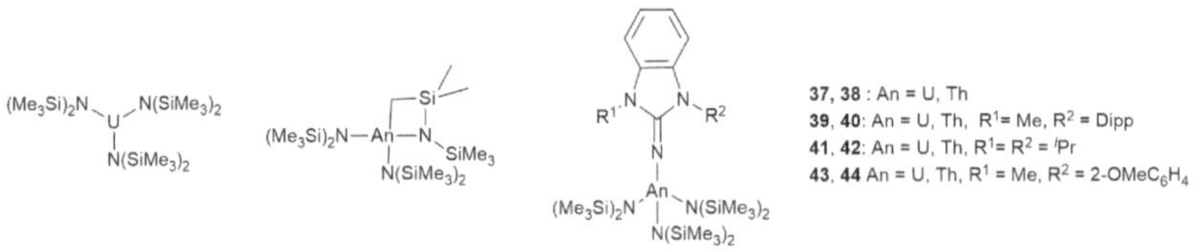

36 **37-38** **39-44**

Fig. 6 Precatalysts for the hydroalkoxylation of carbodiimides, **36–44**.[103–105]

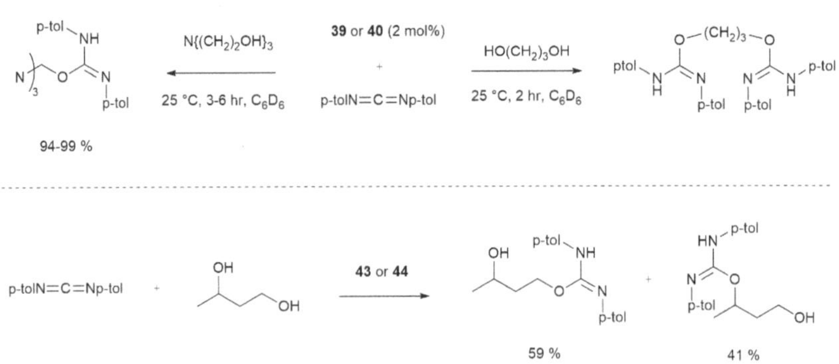

Scheme 18 Top: Hydroalkoxylation of diols and triols by precatalysts **39** or **40**,[105] bottom: selectivity of hydroxylation of 1,3-butanediol using **43** or **44**.[104]

Fig. 7 Hafnium based precatalysts **45–47** for the hydroalkoxylation of carbodiimides by Eisen and co-workers.[107]

44 tolerated N, S and O heteroatom-rich alcohols ROH (R= –CH$_2$-2-pyridyl, –CH$_2$-2-thienyl and –CH$_2$-2-furanyl) in quantitative yields.[104] Extension to functionalized carbodiimide substrates (containing Mes, 4-C$_6$H$_4$Cl and 4-C$_6$H$_4$OMe substituents) led to product formation in quantitative yields in less than an hour with methanol. The reaction of the non-symmetric diol, 1,3-butanediol, in a 1:1 ratio with ^{i}PrN=C=N^{i}Pr, afforded a 1:1.7 ratio of secondary to primary monoisourea products (Scheme 18, bottom), revealing a higher selectivity for the primary alcohol functionality.

4.2 Transition metal catalysis

Hafnium complexes **45–47**, similar to **39–44**, were proposed to have a more open coordination sphere, and of these **45** was found to be the most efficient precatalyst for the hydroalkoxylation of *p*-tolN=C=N*p*-tol using methanol or *tert*-butanol (Fig. 7).[107] A stronger steric effect was found for different carbodiimides resulting in lower reaction rates; methanol added to the *ortho*-substituted carbodiimides much slower than to isopropyl carbodiimide, whereas sterically hindered, non-symmetrical carbodiimide substrates experienced a larger inhibition of the catalytic activity, and two isomers were obtained.[107]

5. Hydrosilylation
5.1 Transition metal catalysis

In 1971, Ojima and co-workers reported the first example of the hydrosilylation of isocyanates and carbodiimides by PdCl$_2$, Pd—C and (Ph$_3$P)$_3$RhCl.[108] At temperatures ranging between 80 and 130 °C, the hydrosilylation of aromatic isocyanates with HSiEt$_3$ afforded *N*-triethylsilylformamides, but use of alkyl isocyanate substrates (R=Cy, ^{n}Bu) demonstrated opposite regioselectivity, affording the *C*-triethylsilylamide products (Scheme 19, top).

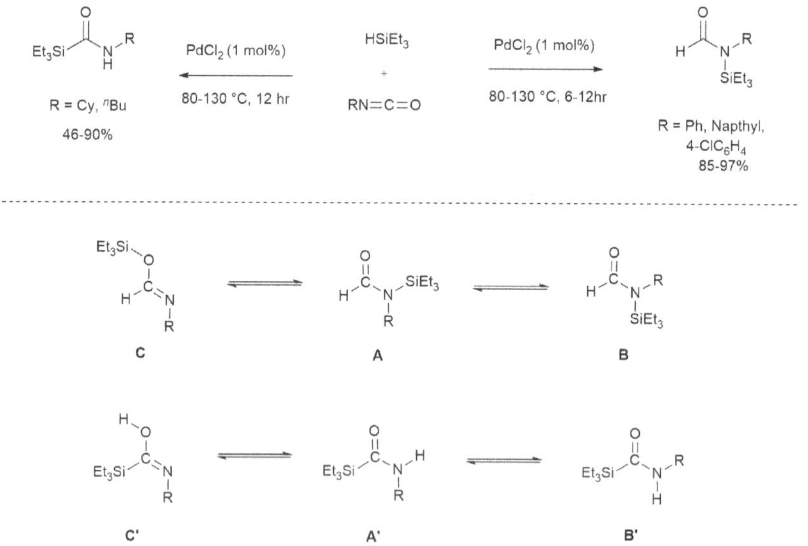

Scheme 19 Top: hydrosilylation of aryl isocyanates yielding *N*-triethylsilylamides and of alkyl isocyanates yielding *C*-triethylsilylformamides. Bottom: tautomeric forms between *N*-triethylsilylformamides and *C*-triethylsilylamides.[108]

As shown by ^{1}H NMR spectroscopy both *N*-silylformamides and *C*-silylformamides exist as a mixture of isomers (**A–C** and **A′–C′**). Process **A** to **B** is suggested to be caused by hindered rotation around the C—N bond and **A** to **C** is caused by amide-imidate tautomerism (Scheme 19, bottom). The *N*-silylformamides mainly exist as a mixture of **A** and **C**, with the *C*-silylformamides mainly existing as a mixture of **A′** and **B′**. When CyN=C=O is used as the substrate, the corresponding *C*-silylformamide shows only one set of N—H resonances in the ^{1}H NMR spectrum suggesting the presence of a **B′** isomer, which is potentially due to steric repulsion between the Cy and SiEt$_3$ groups forcing the equilibrium to shift from **A′** to **B′**. The hydrosilylation of carbodiimides proceeded at higher temperature in presence of 0.5 mol% of PdCl$_2$. In most cases, a high conversion is achieved at 140 °C in 15 h but use of higher temperature (200 °C) and longer reaction times (48 h) gave higher yields (Scheme 20). It is important to note that, even at temperatures of 200 °C, HSiEt$_3$ did not undergo addition to carbodiimides without a catalyst.

Homobimetallic rhodium complex **48** (Scheme 21) catalyzes the hydrosilylation of carbodiimides at elevated temperatures affording the *N*-silylformamidines, but the use of phenyl isocyanate afforded a product of type **C** (Scheme 19, bottom), a different isomer than the expected

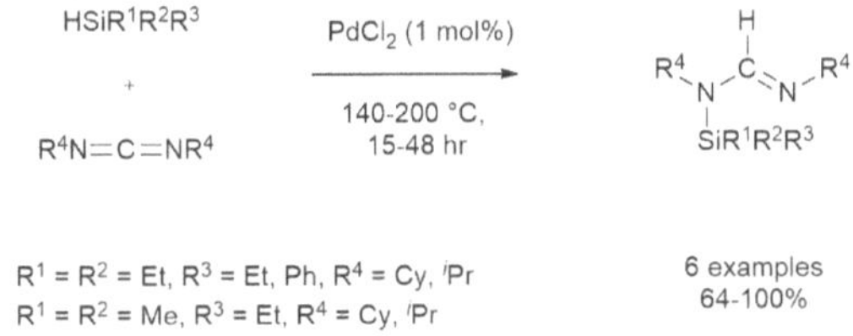

Scheme 20 Hydrosilylation of carbodiimides catalyzed by PdCl$_2$.[108]

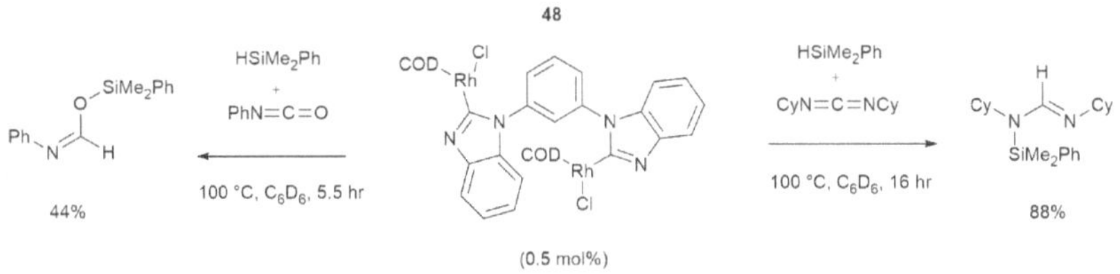

Scheme 21 Hydrosilylation of PhN=C=O and CyN=C=NCy by precatalyst **48**.[109]

N-silylformamide (type **A**).[109] This may be due to the increased steric bulk of the silane substituents leading to the disfavoring of isomers **A** and **B** due to steric clashes with the phenyl group.

6. P—H bond reactions

Within the scope of this discussion, the addition of a P—H bond is classified into three reactions: hydrophosphination involving the addition of a H–PR$_2$ moiety, hydrophosphinylation involving H–P(O)R$_2$, and hydrophosphonylation whereby addition of a H–P(O)(OR)$_2$ occurs (Fig. 8). The term "hydrophosphorylation" is sometimes used in-place of hydrophosphonylation but we feel this is a misnomer and is better used for the formation P–O–C containing molecules.

These reactions are divided in this such way as although seeming similar the products of the reactions have very different properties and catalytic challenges due to the degree of phosphorus atom functionalization. The hydrophosphination of carbodiimides, isocyanates and isothiocyanates is a straightforward route to phosphaguanidines, phosphaureas, and phosphathioureas, which have been classed as heteroatom-rich phosphorus analogues of guanidines, urea and thioureas. These compounds have found considerable utility as ligands in coordination compounds,[37,38,110–113] medicinal applications,[38] synthons for challenging organic transformations[38] and for the preparation of zinc phosphide films.[114] The less explored (thio)

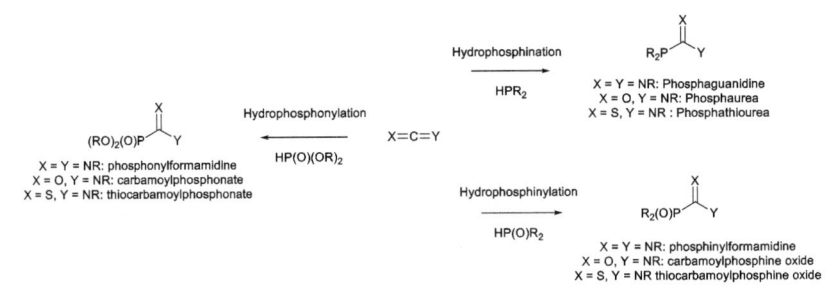

Fig. 8 Nomenclature of P—H unit addition reactions and commonly referred to names of products.

carbamoylphosphine oxides have received less attention likely owing at least in part to the challenges with their synthesis, despite this, however, these compounds have applications as ligands in coordination chemistry[39,40,115] and purification of lanthanide containing waste.[116,117] Carbamoylphosphonates are biologically active and have found significant uses in medicine as matrix metalloproteinase inhibitors[118–120] and also as polymers and monomers for lanthanide waste extraction.[121,122]

6.1 Hydrophosphination

6.1.1 Group 1 and 2 catalysis

The synthesis of neutral phosphaguanidines was first reported in 1980 by addition of p-tolN=C=Np-tol to HPPh$_2$ at elevated temperatures,[125,126] and, subsequently, silylated phosphaguanidines could be obtained by carbodiimide insertion into the P—Si bond of bis-silylated phosphines,[127] but it was not until 2006 that the catalytic synthesis of phosphaguanidines was reported.[123] In this work, readily available alkali metal compounds (**49–53**, Fig. 9) act as a precatalysts for the hydrophosphination of carbodiimides.

The activity increased as the metal size became larger (K > Na > Li). Utilizing 1 mol% of **51**, a wide scope of secondary aryl phosphines (2/3/4-Me-, 4-OMe-, 4-Cl-, 4-Br-, 3,5-Me$_2$- and 3,5-(OMe)$_2$-substituted aryl

Fig. 9 Group 1 and group 2 heterocumulene hydrophosphination catalysts. **49–53** by Zou et al.[123] and **54–58** by Procopiou et al.[124]

substituents in near quantitative yields) were shown to react with ^{i}PrN=C=N^{i}Pr. Additionally, a range of symmetric and asymmetric carbodiimides could be used: R^{1}N=C=NR^{2} where R^{1}, R^{2} = Cy, p-tol; ^{t}Bu, Et; Ph, Cy (Scheme 22). For the more challenging mixed alkyl/aryl phosphine HPPhEt and dialkyl phosphine HP^{i}Bu$_{2}$, a higher (3 mol%) loading of **51** was required for the hydrophosphination of ^{i}PrN=C=N^{i}Pr. Primary phosphines (H$_{2}$PPh and H$_{2}$PCy) were also well-tolerated and use of H$_{2}$PCy afforded the double-addition product (^{i}PrN=CNH^{i}Pr)$_{2}$(PCy) as a minor product of the reaction. Quantitative formation of the potassium phosphaguanidate complex, [Ph$_{2}$PC(NCy)$_{2}$K(OEt$_{2}$)]$_{2}$ (**51a**) occurred in the reaction between **51**, HPPh$_{2}$ and CyN=C=NCy. Subsequent reaction of **51a** with HPPh$_{2}$ afforded CyN=C=NCy and KPPh$_{2}$. The postulated mechanism is shown in Scheme 23.

Following this, homoleptic and heteroleptic group 2 amide complexes **54–58** (Fig. 9) have been investigated for the hydrophosphination of carbodiimides.[124] The activity of these catalysts toward the reaction between

Scheme 22 General scheme for the hydrophosphination of carbodiimides by pre-catalysts **51** and **57**.[123,124]

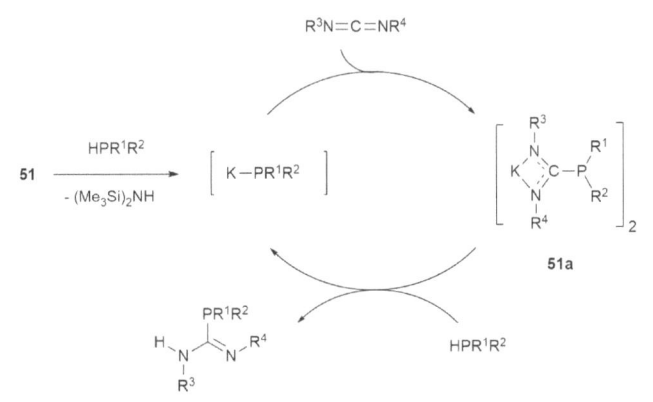

Scheme 23 Mechanism for the hydrophosphination of carbodiimides by **51**.[123]

HPPh$_2$ and CyN=C=NCy were determined to follow the order **57** > **58** > **55** > **56** > **54**. Using HPPh$_2$, hydrophosphination with alkyl carbodiimides R^{1}N=C=NR^2 (R^1=R^2= ^{i}Pr, Cy and R^1= ^{t}Bu, R^2=Et; 95–99% as determined by ^{1}H NMR spectroscopy) was successful, but the use of more sterically demanding ^{t}BuN=C=N^{t}Bu and the more electron rich phosphine HPCy$_2$ were unsuccessful. The more electron-donating HP(p-tol)$_2$ substrate could be utilized; however, this required the use of increased catalyst loadings.

It has been previously shown that complexes **55–58** react with HPPh$_2$ to produce group 2 complexes of the form [M(PPh$_2$)(L)$_n$] (M=Ca, Sr, Ba and L=THF, 18-crown-6).[128,129] These are poorly soluble in hydrocarbon solvents, making mechanistic investigations in solution challenging. It has been postulated that the reaction is likely to be initiated by HPPh$_2$ with **54** to yield **54a**, a transient intermediate, rapidly losing a THF molecule and being replaced by the carbodiimide (Scheme 24). Complex **54b** was isolated from the stoichiometric reaction of **54a** and ^{i}PrN=C=N^{i}Pr, and the addition of HPPh$_2$ did not result in the formation of **54a**. However, upon addition of carbodiimide, the phosphaguanidine product was formed. This was explained by an equilibrium between **54b** and HPPh$_2$ and **54d** and the product, with **54d** being in low concentration but readily reacting to form **54a**.

6.1.2 Lanthanide catalysis

Half-sandwich complexes **59–66** (Fig. 10) catalyze the addition of P—H bonds to carbodiimides, with the activity of the catalyst increasing with the size of the metal (La > Pr > Nd > Sm > Gd > Lu).[130] A wide range of diarylphosphines (HPAr$_2$ where Ar=2-MeC$_6$H$_4$, 3-MeC$_6$H$_4$, 4-MeC$_6$H$_4$, 4-MeOC$_6$H$_4$, 4-ClC$_6$H$_4$, 4-BrC$_6$H$_4$ and 3,5-Cl$_2$C$_6$H$_3$) could be reacted with ^{i}PrN=C=N^{i}Pr, and the reaction was not influenced by the position of the ring-substitution on the phosphine substrate. Similarly to the precatalyst **51** the reactions with PhEtPH and PhPH$_2$ required more forcing conditions for good conversion, and with the PhPH$_2$ substrate,[123] good selectively affording ^{i}PrN=C(PHPh)(NH^{i}Pr) with the E_{syn} isomer being obtained exclusively.[134,135] From the isolated intermediates and stoichiometric reactions a reaction mechanism was postulated (Scheme 25); an acid base reaction between **61** and HPR$_2$ followed by a nucleophilic addition to the carbodiimide to yield **61a**, the active catalyst and phosphaguanidine product is generated by protonation of HPR$_2$.

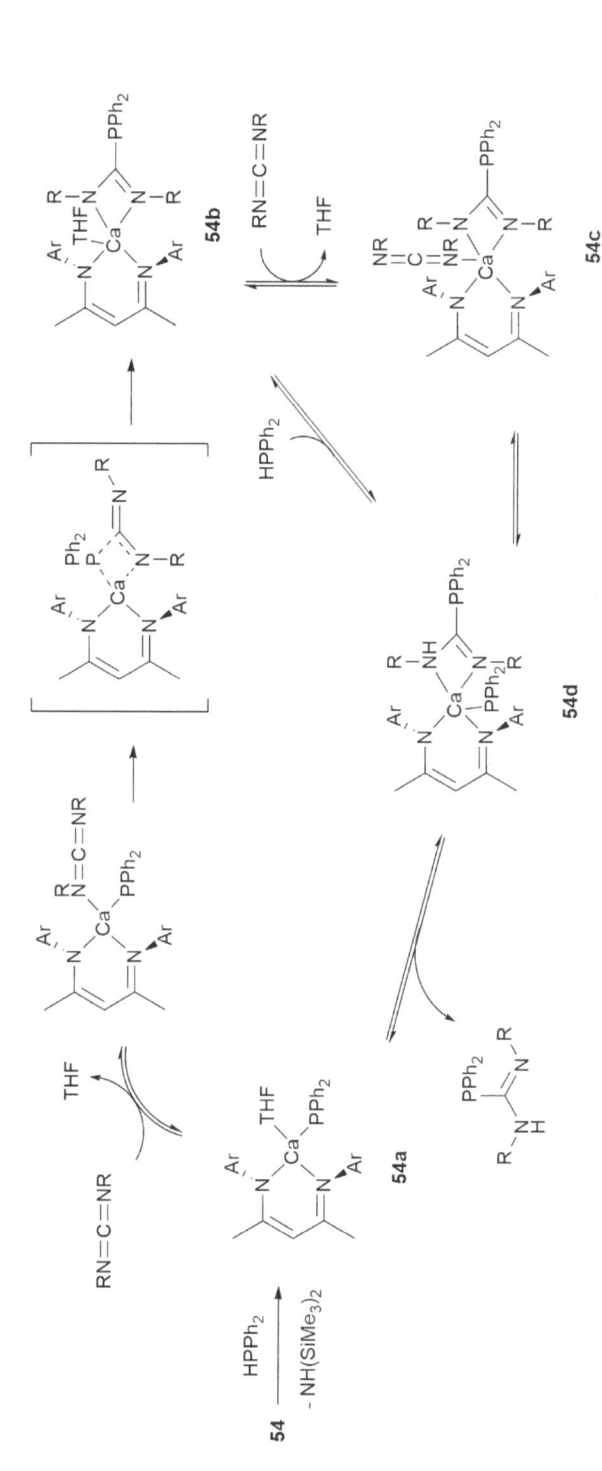

Scheme 24 Mechanism for the hydrophosphination of carbodiimides by **54**.[124]

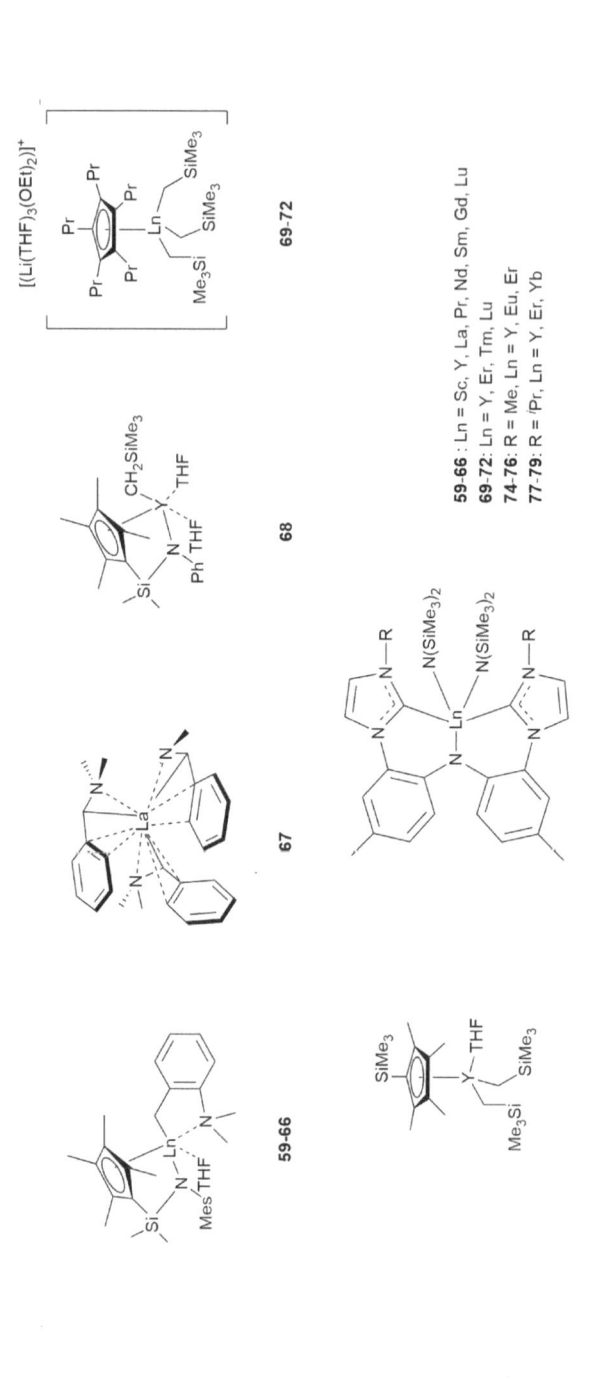

Fig. 10 Lanthanide hydrophosphination precatalysts **59–66** by Hou et al.,[130] **67** by Schimdt et al.,[131] **68–73** by Zhang et al.,[132] and **74–79** by Wang et al.[133]

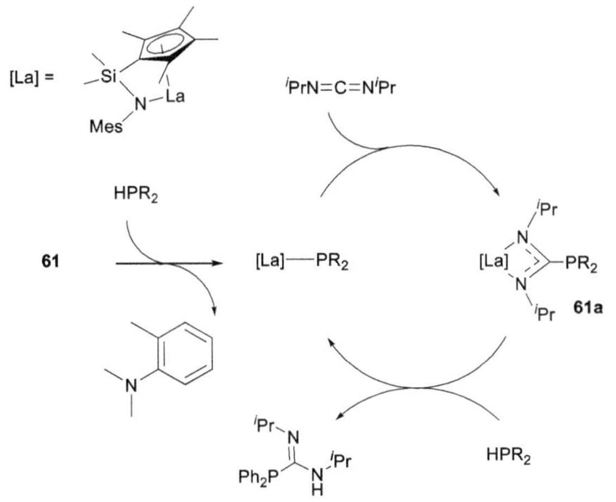

Scheme 25 Mechanism for the hydrophosphination of carbodiimides by **61**.[130]

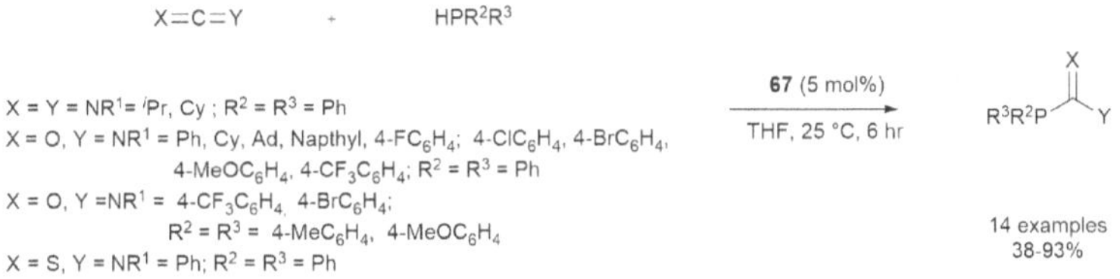

Scheme 26 Hydrophosphination of heterocumulenes by **67**.[131]

Improvements have been made on the lanthanide catalysts **59–66**; with the hydrophosphination reaction taking place at room temperature and additionally broadening the scope to isocyanate and isothiocyanate substrates.[131] The yttrium variant of **67** (Fig. 10) proved to be inferior for the hydrophosphination of carbodiimides as was the case for precatalysts **59–66**. A range of isocyanates, isothiocyanates and carbodiimides can be tolerated with near quantitative conversion for all substrates as determined by NMR spectroscopy, but the isolated yields were significantly lower (Scheme 26). Steric bulk was a limiting factor in the hydrophosphination the range with ^{t}BuN=C=O and AdN=C=O, which were poor-yielding reactions. The use of more acidic HP^{t}Bu$_2$ gave no product with ^{i}PrN=C=N^{i}Pr, even with heating to 80 °C for 36 h. Little effect of electronics on the phosphine was seen with HP(4-OMeC$_6$H$_4$)$_2$ and HP(4-MeC$_6$H$_4$)$_2$, which afforded products in similar yields to HPPh$_2$.

The order of addition of carbodiimide and phosphine is key to this catalysis. If the carbodiimide is added to the reaction mixture first, then three equivalents of carbodiimide insert into the La—C bond, ultimately reducing the maximum yield to 85% (at 5 mol% catalyst loading). Contrastingly, if HPPh$_2$ is added first, no such yield limitation is observed. The postulated mechanism involves the protonation of **67** by the phosphine substrate forming a species of the type L$_n$La(PPh$_3$)(THF)$_n$ (where L=neutral ligand). Subsequent insertion of the heterocumulene into the La—P bond followed by protonolysis yields the desired product and regenerates the active catalyst. This proposed mechanism is analogous to that for the hydrophosphination of carbodiimides precatalyst **61** (Scheme 25).[130] For the cyclopentadienyl-based catalysts **69–73** (Fig. 10), the anionic precatalyst **69** demonstrated greater activity than the neutral complexes.[132] The use of symmetrical substrates ^{i}PrN=C=N^{i}Pr and CyN=C=NCy afforded the phosphaguanidine products in excellent yields, and for unsymmetrical carbodiimides R^{1}N=C=NR^2 (R^1=Ph; R^2=cyclopentyl, Cy, cycloheptyl, cyclooctyl, CH$_2$Cy, or R^1 = p-tol; R^2 = Cy) lower yields of product were obtained, albeit with concomitant doubling of catalyst loading (to 6 mol%) and increase of the reaction time, due to the lower electrophilicity of the mixed aryl, alkyl carbodiimides. The use of functionalized aryl phosphines did not significantly affect yields. The use of very bulky DippN=C=NDipp and HPPh$_2$ was not tolerated, either due to increased steric bulk or lowered carbodiimide electrophilicity. When the primary phosphine PhPH$_2$ was reacted with one equivalent of ^{i}PrN=C=N^{i}Pr, both single- and double-addition products PhHP(^{i}PrN=C(NH^{i}Pr)) and PhP{^{i}PrN=C(NH^{i}Pr)}$_2$ could be detected by ^{31}P NMR spectroscopy, and increasing the stoichiometry of ^{i}PrN=C=N^{i}Pr to two equivalents, afforded PhP{^{i}PrN=C(NH^{i}Pr)}$_2$ exclusively.

The N-heterocyclic carbene complexes **74–79** (Fig. 10) catalyze the hydrophosphination reactions using HPPh$_2$ or HP(4-MeC$_6$H$_4$)$_2$ with CyN=C=NCy or ^{i}PrN=C=N^{i}Pr in near quantitative yields with 1 mol % catalyst loadings at ambient temperature, but, as with previous examples, the elusive ^{t}BuN=C=N^{t}Bu hydrophosphination product could not be obtained (Scheme 27).[124] In addition, excellent yields were obtained for a variety of functionalized aryl isothiocyanates (4-FC$_6$H$_4$, 4-BrC$_6$H$_4$, 4-MeC$_6$H$_4$, 4-MeOC$_6$H$_4$-substituted substrates in 94–99% yields); slightly lower yields were obtained for HP(4-MeC$_6$H$_4$)$_2$ compared to HPPh$_2$. Interestingly, the use of the alkyl isocyanate CyN=C=O afforded the hydrophosphination product, but the aromatic isocyanate 4-FC$_6$H$_4$N=C=O produced the isocyanurate, a product of isocyanate cyclotrimerization.

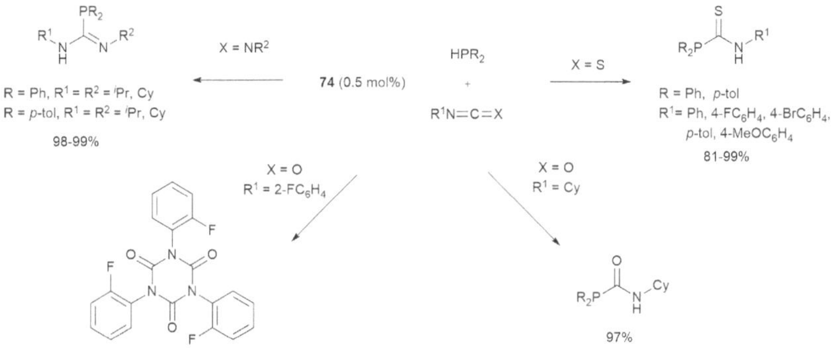

Scheme 27 Hydrophosphination of carbodiimides, isocyanates and isothiocyanates by **74**.[133]

6.1.3 Transition metal catalysis

The first transition metal catalyzed hydrophosphination of carbodiimides was reported by Waterman and co-workers in 2010 using Zr-based **80** at 5 mol% catalyst loadings at 120 °C affording the phosphaguanidine products.[136]

The hydrophosphination reaction using low-coordinate transition metal *m*-terphenyl precatalysts **35** and **81** with a 1:1 mixture of isocyanate: HPPh$_2$ affords a mixture of phosphinocarboxamides (Scheme 28, top) and phosphinodicarboxamides, resulting from the overall insertion of one of two isocyanate molecules into the P—H bond.[137] For aryl isocyanate substrates, the selectivity of the reaction can be controlled by the reaction conditions: use of THF solvent affording phosphinocarboxamides and addition of a weak acid (NEt$_3$·HCl) in benzene solution affording phosphinodicarboxamides. The reactions using secondary and tertiary isocyanates are selective for phosphinocarboxamide products, and primary alkyl isocyanates (^{n}HexN=C=O) afford a mixture of phosphinocarboxamides along with the isocyanurate produced through cyclotrimerization.[88] Attempts to form mixed phosphinodicarboxamides from a 1:1:1 mixture of PhN=C=O, CyN=C=O and HPPh$_2$ yields a mixture of four products from a proposed six, with the mixed aryl, alkyl phosphinodicarboxamide product being the least favored (Scheme 28, bottom).

A mechanism was proposed whereby the Fe center acts as a Lewis acid leading to coordination of the isocyanate to form the active catalyst, a subsequent nucleophilic attack occurs on intermediate **81b** and 1,3-proton shift with free isocyanate coordination generates the phosphinocarboxamide (Scheme 29).[137] If the substituents on the phosphinocarboxamides are small enough, the mechanism can repeat again proceeding to cycle 2 with a

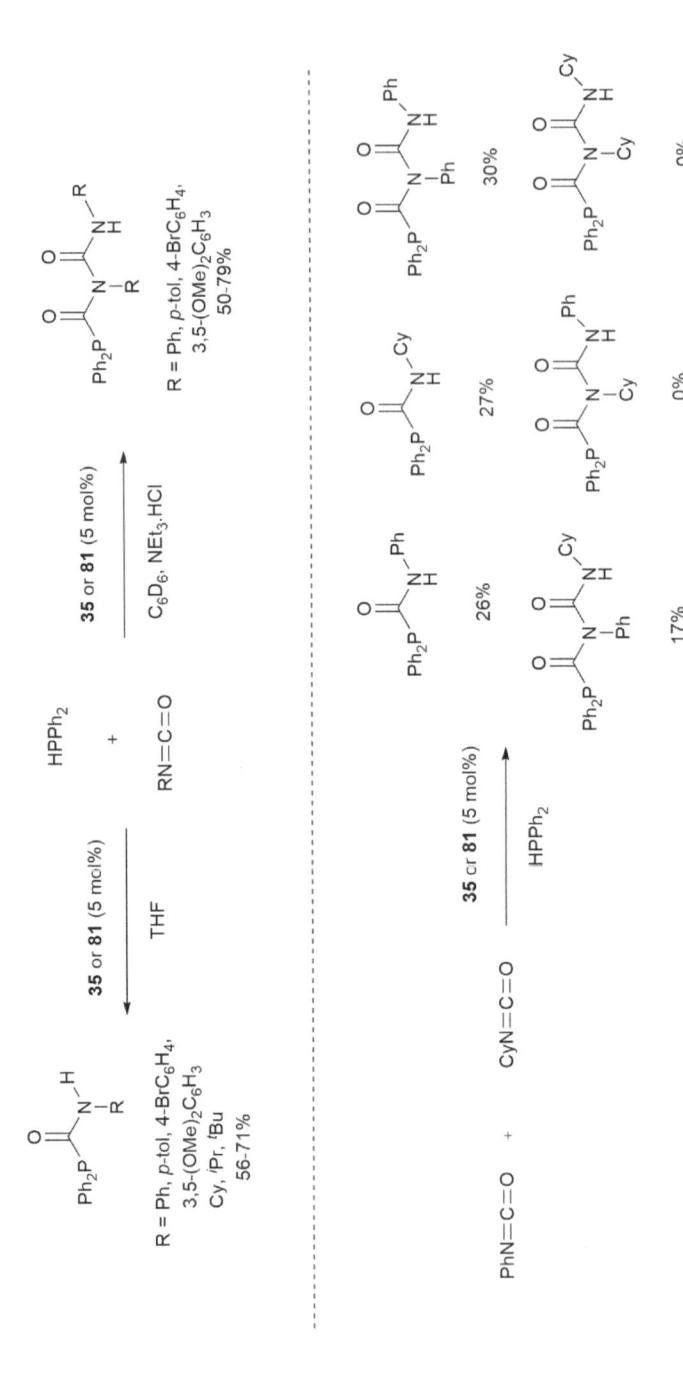

Scheme 28 Top: hydrophosphination selectivity by **35** and **81**. Bottom: hydrophosphination reaction of an isocyanate mixture.[137]

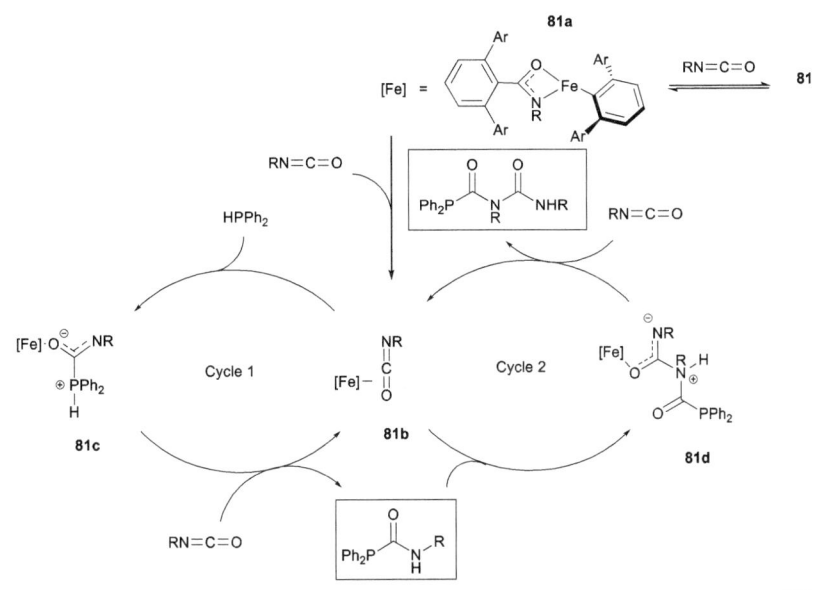

Scheme 29 Proposed mechanism for hydrophosphination of isocyanates by **81**.[137]

nucleophilic attack on **81b** by the phosphinocarboxamide to yield **81c**, proton transfer and isocyanate coordination regenerate **81b** and produce the phosphinodicarboxamide.

When using various neutral zirconium precatalysts **82–86** (Fig. 11) in the hydrophosphination of carbodiimides yields were very low (<11%), but addition of [Ph$_3$C][B(C$_6$F$_5$)$_4$] to generate the cationic precatalysts saw yields increase to 75–90%.[138] A similar argument in conversions was observed in the case of isocyanate substrates, and aryl isocyanates featuring electron-withdrawing or -donating groups were well-tolerated.

Ring expanded NHCs **87–90** (Fig. 11) tolerate a small number of isocyanates, precatalyst **90** was found to be the best performer, affording excellent yields for alkyl isocyanates (R= ^{i}Pr, ^{t}Bu, Cy); phenyl isocyanate underwent quantitative conversion, with 5% of the product being the phosphinodicarboxamide product.[139]

6.1.4 Catalyst-free

Recently, catalyst- and solvent-free hydrophosphination of isocyanates and phenyl isothiocyanate has been reported with HPPh$_2$ tolerating a range of aryl and primary alkyl isocyanates, affording the products in quantitative or near quantitative yields (Scheme 30).[140] Substrates featuring sterically demanding substituents (e.g. AdN=C=O) did not react. Using this methodology, dialkylphosphines such as HP^{i}Pr$_2$ and HP^{t}Bu$_2$ could also be

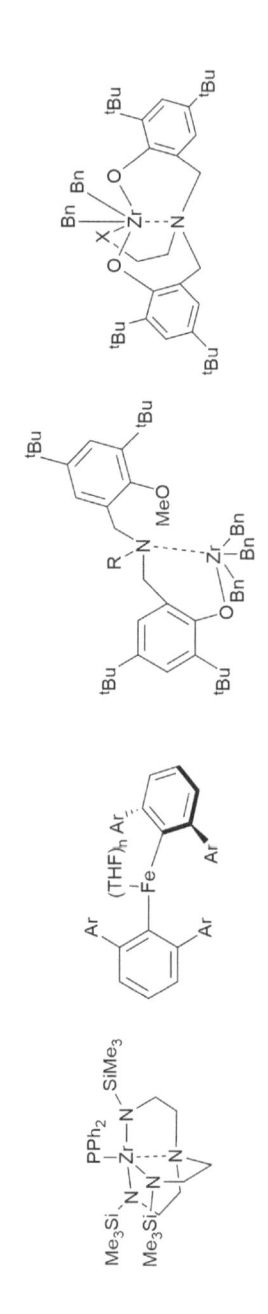

Fig. 11 Transition metal hydrophosphination precatalysts, **80** by Waterman and co-workers,[136] **35** and **81** by Kays and co-workers,[137] **82–86** by Shen and co-workers[138] and **87–90** by Liptrot and co-workers.[139]

35: n = 0, Ar = Mes
81: n = 1, Ar = Tmp
82: R = CH$_2$-3,5-(^{t}Bu)$_2$-6-OMeC$_6$H$_2$
83–84: X = OMe, NMe$_2$
85–86: X = N^{n}Bu, N(C$_4$H$_8$)N
87–88: Ar = Dipp (IPr), Mes (SIMes)
89–90: Ar = Dipp, Mes

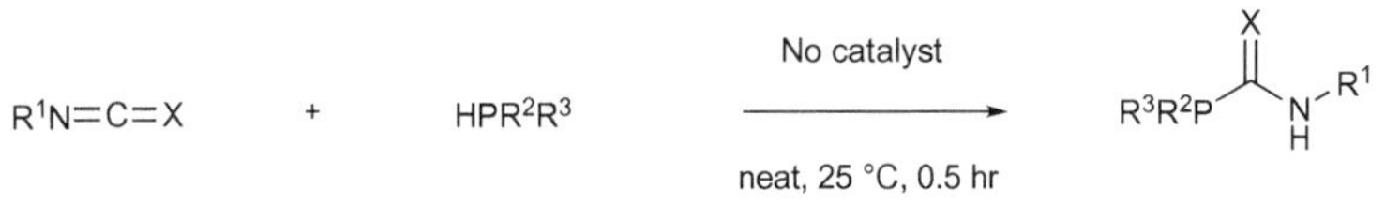

$R^1N=C=X$ + HPR^2R^3 →(No catalyst, neat, 25 °C, 0.5 hr) $R^3R^2P-C(=X)-N(H)-R^1$

17 examples
68-99%

X = O; R^1 = 4-MeOC$_6$H$_4$, 4-FC$_6$H$_4$, 4-ClC$_6$H$_4$, 4-BrC$_6$H$_4$,
4-CF$_3$C$_6$H$_4$, napthyl, (CH$_2$)$_2$Cl, (CH$_2$)$_3$SiMe$_2$Cl;
R^2 = R^3 = Ph
X = O; R^1 = Ph; R^2 = R^3 = 4-MeC$_6$H$_4$, 4-MeOC$_6$H$_4$,
4-FC$_6$H$_4$, ^{i}Pr, ^{t}Bu
X = O; R^1 = Ph; R^2 = Ph, R^3 = H (72 hr, 71%)
X = S, R^1 =Ph, R^2 =Ph R = Ph

Scheme 30 Catalyst-free neat hydrophosphination of isocyanates and phenyl isothiocyanate.[140]

employed affording the hydrophosphination products in high yields. The reaction with the primary phosphine H$_2$PPh took considerably longer than other phosphines (3 days vs 30 min), which was postulated to be a result of residual solvent in the commercial sample of H$_2$PPh used. Control experiments demonstrate that the key to the reaction is the solvent-free conditions (<10% solvent), similar to the catalyst-free hydrophosphination of alkenes and alkynes.[141]

6.2 Hydrophosphinylation

Hydrophosphinylation results in the direct synthesis of (thio)carbamoylphosphine oxides or phosphinylformamidine, while theoretically the oxidation of the P(II) analogue with oxidizing agent such as H$_2$O$_2$ could be proposed it has been shown that actually the expected P(V) products from use of ^{t}BuOOH and CyN=C(PPh$_2$)N(H)Cy is actually an [CyNC(H)NCy]$^+$[PPh$_2$O$_2$]$^-$ adduct joined by two NH···O hydrogen bonds forming an eight-membered cyclic compound.[37] Thus direct formation of phosphinylformamidine or (thio)carbamoylphosphine oxides has been pursued.

In 1963, Greenberg and co-workers reported the first synthesis of carbamoylphosphine oxides from LiPPh$_2$ with isocyanates followed by hydrolysis and oxidation with stoichiometric H$_2$O$_2$ in moderate yields after 12 h.[142] Subsequently in 1965 the first synthesis of thiocarbamoylphosphine oxides was reported by reaction PhN=C=S with tetradiarylphosphines by P—P bond cleavage.[143] The first catalytic synthesis of thiocarbamoylphosphine oxides was reported in 1969 showing at elevated temperature isothiocyanates would react diphenylphosphine oxide in the presence of catalytic NEt$_3$ catalyst in moderate to excellent yields.[144] Thus, in more recent times catalytic metal-mediated hydrophosphinylation has sought to avoid the use of such stoichiometric additives and develop catalysts which tolerate a wider range of heterocumulenes and functional groups.

6.2.1 Group 1 and 2 catalysis

Westerhausen and co-workers have utilized secondary phosphine oxides with calcium complex **91**, to undertake hydrophosphinylation catalysis (Fig. 12).[145] This complex is effective in this methodology as phosphine oxides readily form complexes with calcium, whereas phosphines do not represent effective Lewis bases in calcium chemistry.[148] Additionally, a previous attempt using a calcium-based catalyst, HPPh$_2$ and phenyl isocyanate led to cyclotrimerization[145] and a general dearth of Group 1 or group 2 catalysts for the hydrophosphination of (iso)thiocyanates. Complex **91** is a precatalyst for the hydrophosphinylation of a wide range of isocyanates (RN=C=O where R = ^{i}Pr, ^{t}Bu, Cy, Ph, 4-BrC$_6$H$_4$, 2,4,6-Me$_3$C$_6$H$_2$, 1-C$_{10}$H$_7$) and isothiocyanates (RN=C=S where R = ^{i}Pr, Cy, Ph, p-tol) in quantitative conversion as determined by ^{31}P NMR spectroscopy.[145] NMR spectroscopic investigation of the reaction between [(thf)$_4$Ca(PPh$_2$)$_2$] and HP(O)Ph$_2$ in THF showed an equilibrium ($K=0.013$) between these starting materials and the resulting [(thf)$_4$Ca(OPPh$_2$)] and HPPh$_2$ products, the large excess of HP(O)Ph$_2$ shifts the reaction equilibrium to [(thf)$_4$Ca(OPPh$_2$)]. Mechanistically the phosphinite anion adds to the carbon atom of RN=C=X under formation of a P–CCX bond, yielding the intermediate [Ph$_2$P(O)–C(X)–NR]$^-$ anion (Scheme 31). Protonolysis by diphenylphosphine oxide leads to the corresponding (thio)carbamoylphosphine oxides and reformation of the active catalyst.

Further work utilized the potassium catalyst **92** (Fig. 12) in reactions with bulky HP(O)Mes$_2$ and RN=C=O (R = ^{i}Pr, Cy), which required 42–48 h to reach full conversion, in contrast to ^{i}PrN=C=S reached quantitative conversion in 30 min (Scheme 32). There was no reaction between alkyl-substituted carbodiimides and HPMes$_2$, but the use of less bulky HPPh$_2$ resulted in quantitative product formation in 30 min, suggesting the steric bulk of the P-bound atom is key and the catalyst is compatible with carbodiimides.

Comparatively, the potassium-based phosphinite **92** is less reactive and promotes the addition of HP(O)Ar$_2$ to iso(thio)cyanates less efficiently than the corresponding calcium-based active catalyst [(thf)$_4$Ca(OPAr$_2$)$_2$].[146] Diverse reasons may account for this finding: the enhanced Lewis acidity of Ca^{2+} compared with K$^+$ ions results from a larger charge to radius ratio, additionally the calcium catalyst is mononuclear in THF solution while the potassium catalyst forms aggregates which may hinder substrate availability for catalytic centers. The authors proposed a mechanism based on (thio)phosphinite anion of the catalyst binding to heterocumulene, subsequent reaction of the carbon atom of the heterocumulene system via the phosphorus atom (Scheme 33).

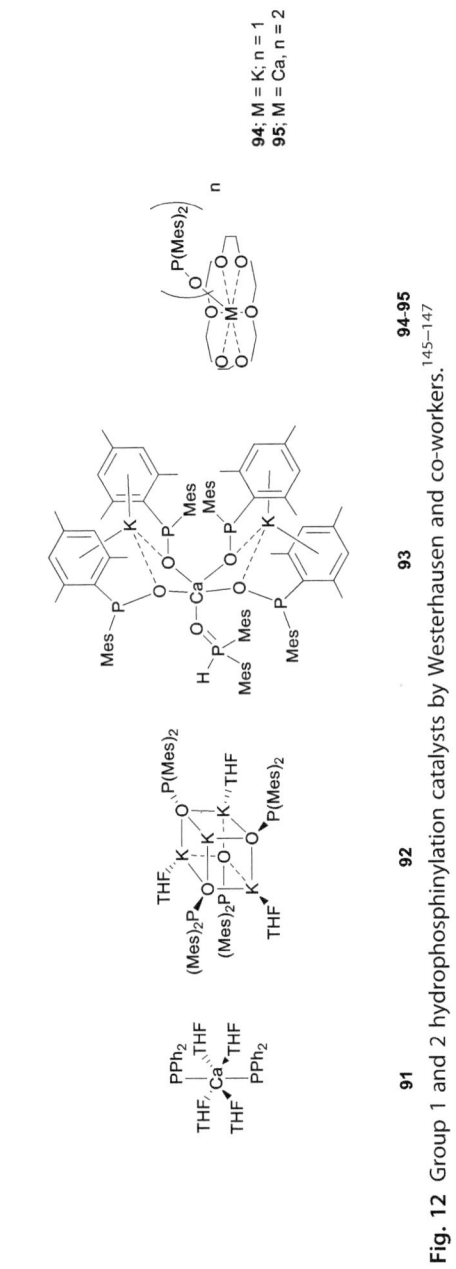

Fig. 12 Group 1 and 2 hydrophosphinylation catalysts by Westerhausen and co-workers.[145–147]

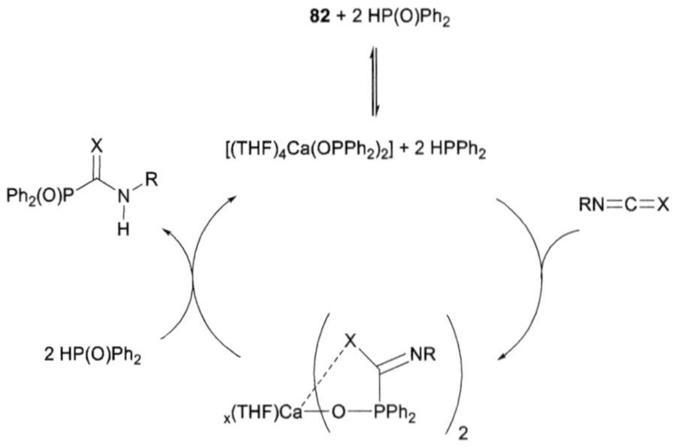

Scheme 31 Mechanism for the hydrophosphinylation of isocyanates and isothiocyanates by **91**.[145]

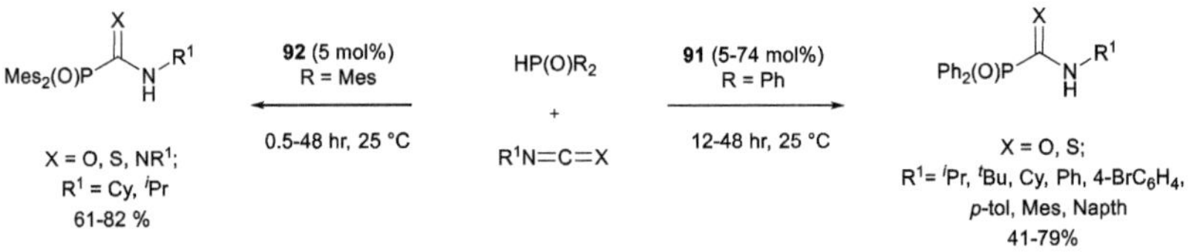

Scheme 32 Hydrophosphinylation of heterocumulenes by precatalysts **91** and **92**.[145,146]

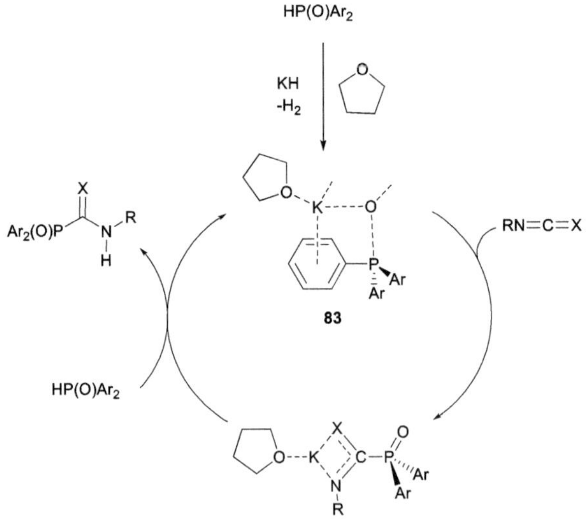

Scheme 33 Hydrophosphinylation of isocyanates and isothiocyanates by **92**.[146]

To increase the activity of the calcium and potassium complexes the formation of 18-crown-6 adducts has been used to prevent catalyst degradation (**94**, **95**, Fig. 12).[147] The potassium complex **94** is a more efficient catalyst than the calcium congeners in this metal-mediated hydrophosphinylation reaction, due to the significantly larger Lewis acidity of the Ca^{2+} ion, hampering dissociation of these complexes. In addition, the divalent cations are bound to two bulky anions, increasing the shielding of the catalytic site. The latter argument also explains the decreasing reactivity of the catalyst (regardless of the metals) after the addition of 18-crown-6 because the crown ethers shield the metal ions much more effectively than monodentate THF.

6.3 Hydrophosphonylation

The synthesis of carbamoylphosphonates was reported by the base-catalyzed reaction of aryl isocyanate (4-BrC$_6$H$_4$, 4-ClC$_6$H$_4$, 4-NO$_2$C$_6$H$_4$) and with dialkylphosphite by either NaCN, Na$_2$CO$_3$ or NEt$_3$.[149,150] Only solid carbamoylphosphonates derived from aryl isocyanates could isolated with liquid carbamoylphosphonates decomposing to starting materials upon distillation. Additionally, via an Arbuzov reaction of carbamoyl chlorides and trialkyl phosphites or by amidation of ethyl diethoxyphosphinylformate.[151,152] Carbamoylphosphonates obtained by these highly reactive reagents suffer from toxicity and low selectivity that limit general use and scope. Zahedi and co-workers have reported that 5 mol% CaCl$_2$ can act as catalyst for the hydrophosphonylation of a range of alkyl and aryl isocyanates and isothiocyanates in moderate yields.[153] Purification of the products required the use of column chromatography.

7. Hydrothiolation

The chemistry of hydrothiolation is analogous to that of hydroalkoxylation but to a slightly less reactive extent.[154] The reaction of isocyanates and isothiocyanates can proceed without a catalyst, for example, in 2008 Soleiman-Beigi and co-workers have developed a catalyst and solvent-free approach to the synthesis of more challenging S-alkyl thiocarbamates in high yields at temperatures in the range 25–70 °C, where both aryl and primary alkyl thiols can be utilized with a range of isocyanates, however for purification the reaction column chromatography is required.[155] Additionally, tertiary amines, AlCl$_3$ or Zn have been used as catalysts in hydrothiolation reactions.[156–158] There are a few reported examples with CuCl$_2$ or base-catalyzed addition of S—H bonds to carbodiimides at elevated temperatures which afford the products in moderate to high yields.[159–162] Although most reactions can proceed either in the absence of catalyst or in high yields with a simple catalyst, the use of more developed catalysts can be justified if reaction temperatures and reaction times are lowered and if more substituted heteroatom-rich thiols can be used.

7.1 Lanthanide catalysis

Yao et al. have sought to develop a simple and efficient catalytic to produce thiocarbamates with good functional tolerance.[163] The hydrothiolation reaction between PhN=C=O and HSBn using Ln[N(SiMe$_3$)$_2$]$_2$ precatalysts found activity in the order La (**96**) > Sm = Y > Yb. The optimal solvent was non-polar toluene with more polar DMSO and MeCN significantly lowering yields. A range of aromatic isocyanates and functionalized benzylthiols were used in the reaction, but using sterically hindered HSCPh$_3$ required longer reaction times, and afforded lowered yields. DFT calculations indicate two main reaction pathways to product formation (Scheme 34): (i) catalyst activation by protonation of HSBn to yield the active catalyst La–SBn, isocyanate coordination via the oxygen and insertion of SBn into the isocyanate, followed by protonation from another equivalent of HSBn

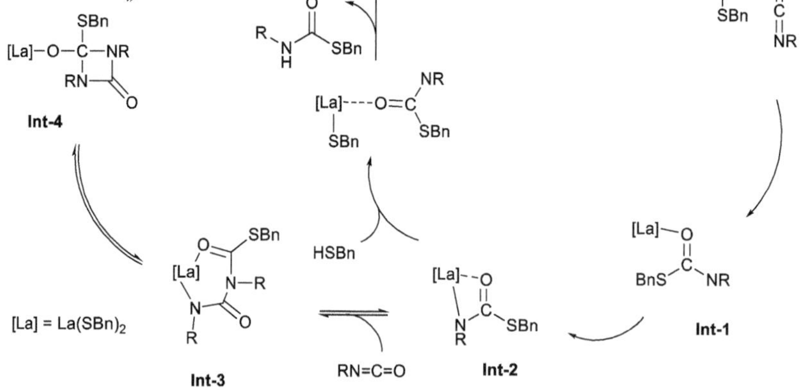

Scheme 34 Mechanism for hydrothiolation of isocyanates by precatalyst **96**.[163]

to complete the cycle and leave the product; (ii) an equivalent of isocyanate inserts into **Int-2** via the La—N bond to produce **Int-3**, intramolecular nucleophilic attack yields **Int-4** with protonolysis by HSBn yields heterocyclic compound, **P**. Compound **P** could be isolated by utilizing a 2:1 ratio of isocyanate to the thiol substrate. Subsequent stoichiometric reaction of compound **P** and HSBn with 1 mol% of **96** yields the desired S-benzyl thiocarbamate, with the formation of **P** being reversible.

8. Multi E–H hydrofunctionalization
8.1 Actinide catalysis

In 2015 Eisen and co-workers reported the first example of multi-element hydrofunctionalization utilizing protic E–H bond containing nucleophiles from N—H, P—H and S—H compounds.[164] Precatalyst **97** (Fig. 13) displays an unusual tolerance and reactivity toward these heteroatoms containing substrates and shows moderate to excellent yields across for heterocumulene substrates; ^{i}PrN=C=N^{i}Pr(54–99%), o-tolN=C=No-tol (95–97%), PhN=C=O (76–83%) and PhN=C=S (78–87%). Stoichiometric reactions of **97** with phenyl isocyanate, phenyl isothiocyanate and diisopropylcarbodiimide showed in all cases insertion of heterocumulene into the Th–N(SiMe$_3$)$_2$ bond, subsequent addition of HPPh$_2$, H$_2$NPh, or HSBn nucleophiles showed addition of two equivalents of the E–H moiety. Based on these finding the mechanism involves (a) insertion of two equivalents of heterocumulene into the Th–N(SiMe$_3$)$_2$ bond, (b) protonolysis by E–H to yield the active catalyst and releasing some heterocumulene substituted with N(SiMe$_3$)$_2$, and (c) repetition of steps (a) and (b) regenerate the active catalyst and the desired product. Kinetic experiments with DPPh$_2$ and heterocumulenes indicate a k_H/k_D ratio of 1.98 suggesting that the

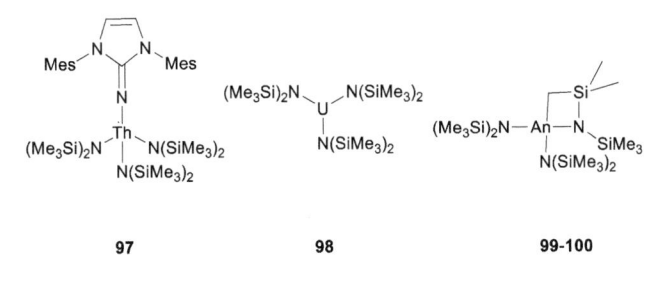

97   98   99-100

99, 100: An = U, Th

Fig. 13 Multi E–H hydrofunctionalization catalysts by Eisen and co-workers.[164,165]

protonolysis is rate determining. More recently, Eisen and co-workers have shown that simple actinide amido complexes can be used to catalyze the hydrofunctionalization of heterocumulenes. Thorium precatalyst **100** was found to be much more efficient than uranium analogues **98–99** at temperatures of 75 °C.[165] Use of ^{i}PrN=C=N^{i}Pr or p-tolN=C=Np-tol achieves quantitative yields of product when using H$_2$NAr and BnSH, but the use of alkyl amines substrates significantly reduces yields and requires longer reaction times. However, the hydrophosphination and hydrothiolation of PhN=C=O and PhN=C=S only proceeded with uranium precatalysts **98–99**; this disparity in reactivity can be justified by the higher bond strength of Th–O/S bonds compared to U–O/S bonds.

8.2 Transition metal catalysis

The dinuclear titanium(IV) complex **101** (Fig. 13), shows remarkable reactivity and selectivity in the room temperature of addition of the E–H bond (E=NR$_2$, OR, SR, P(O)R$_2$), toward a wide variety of heterocumulenes (Scheme 35).[166]

The reaction of a substrate such as 2-mercaptopyridine, which consists of amine and thiol groups as different nucleophiles in its two tautomeric forms, with ^{i}PrN=C=N^{i}Pr yields N,N'-diisopropyl-2-thioxopyridine-1 (2H)-carboximidamide in high yields, indicating that the reactivity of the NH group is greater compared to that of the SH group toward the

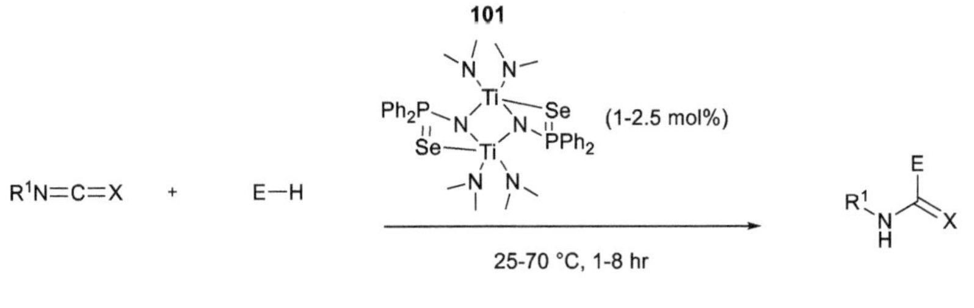

E = OR^3; X = NR^2; R^1 = R^2 = ^{i}Pr; R^3 = Ph, 2-MeC$_6$H$_4$, 2-ClC$_6$H$_4$, Napth,Bn, Et
E = OR^3; X = NR^2; R^1 = R^2 = Cy; R^3 = Ph, 4-IC$_6$H$_4$, 2-IC$_6$H$_4$, 2-ClC$_6$H$_4$, Bn, Et, ^{n}Bu
E = OR^3; X = S; R^1 = Ph; R^3 = ^{n}Bu
E = SR^3; X = NR^2; R^1 = R^2 = ^{i}Pr, Cy; R^3 = Ph, 4-ClC$_6$H$_4$, 2-C$_5$H$_4$N, 2-(6-CF$_3$)C$_5$H$_3$N
E = SR^3; X = S; R^1= Ph; R^3 = Ph, 4-ClC$_6$H$_4$
E = NR^{3}R^4; X = S; R^1 = Ph; R^3 = R^4 = ^{i}Pr; R^3 = H, R^4 = o-tol, 2-FC$_6$H$_4$
E= P(O)Ph$_2$; X = NR^2; R^1 = R^2 = Cy, ^{i}Pr
E= P(O)Ph$_2$; X = NR^2; R^1 = ^{t}Bu, R^2 = Et
E= P(O)Ph$_2$; X = O; R^1 = p-tol, 4-OMeC$_6$H$_4$

32 examples
72-99%

Scheme 35 Multi E–H hydrofunctionalization of heterocumulenes by **101**.[166]

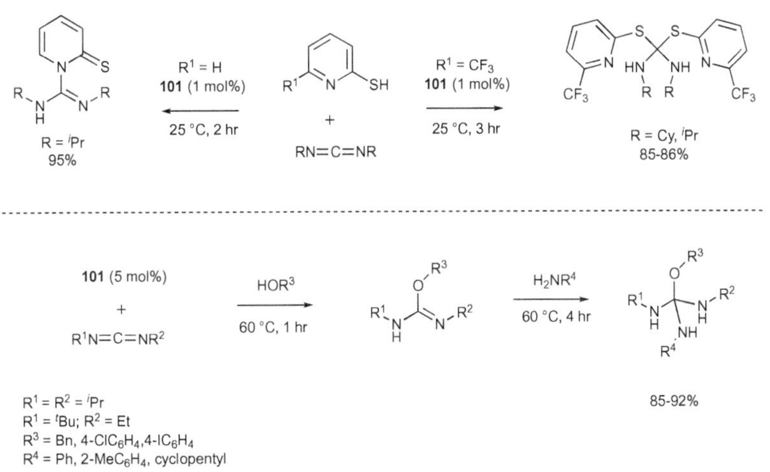

Scheme 36 Top: (left) mono hydroamination of carbodiimides and (right) double hydrothiolation of carbodiimides and bottom: sequential hydroalkoxylation-hydroamiantion of carbodiimides.[166]

carbodiimides, despite the fact the NH form is the favored tautomer (Scheme 36, top left).[166] However, if the 2-mercaptopyridine is substituted with a strongly electron-withdrawing group such as CF$_3$, the carbodiimide undergoes a double hydrofunctionalization and additionally the thiol functionality undergoes the insertion reaction, rather than the amine functionality (Scheme 36, top right).[166]

This catalytic system is the first example of a transition metal-mediated hydrophosphinylation reaction. Hydrophosphinylation was tested for three unsymmetrical carbodiimides at 70 °C.[166] Both E and Z isomers formed, and the use of the bulkier carbodiimide EtN=C=N^{t}Bu successfully yielded only the ^{t}Bu=N=C{P(O)Ph$_2$}NH(Et) isomer. Additionally, both PhN=C=O and p-tolN=C=O could be utilized in the reaction. This investigation demonstrated the first example of unsaturated bonds within a heterocumulene undergoing a double reaction by two different E–H sources, in the sequential hydroalkoxylation-hydroamination of carbodiimides to afford N,N',N''-trialkylmethanetriamines (Scheme 36, bottom). Kinetic experiments suggest a mechanism of rapid protolysis by the E–H nucleophile of two of the Ti–NMe$_2$ groups, migratory insertion, followed by protolytic cleavage by E–H to complete the catalytic cycle. It should be noted that for the hydrophosphinylation reaction tautomerization will have to occur first to generate a P(OH) nucleophile, as reflected by the much more sluggish reaction times reported compared to other E–H nucleophiles.

9. Concluding remarks

The hydrofunctionalization of heterocumulenes has become a well explored area, with a clear distinction for the dominance of carbodiimide substrates, followed by isocyanates, then by isothiocyanates. Of these, the hydrofunctionalization of one multiple bond is the predominant reaction, showing a near complete tolerance for most alkyl or aryl heterocumulenes with most E–H substrates. Only two selected examples of double hydroboration of carbodiimides and isocyanates and double hydrothiolation of carbodiimides have been presented, with no examples for isothiocyanates.

In most cases, the C/N bond insertion selectivity is predetermined by the inherent resonance electrophilicity of the isocyanate or isothiocyanate, with the "E" being selective for the carbon atom and "H" being selective for nitrogen. However, in the case of hydrosilylation with a bulky $SiMe_2Ph$, the silyl group is selective for the oxygen atom of the isocyanate and "H" selective for the carbon. Less bulky $SiEt_3$ shows C-silyl selectivity for alkyl isocyanates and N-silyl selectivity for aryl isocyanates. Additionally, the hydroboration of the very bulky DippN=C=NDipp substrate shows the unexpected formation of a C-borylformamidine rather than the much more common N-borylformamidine.

Notable successes have come from the use of iron precatalysts **31** and **73** yielding novel compounds that involve coupling of two (or more) isocyanates with concomitant hydrophosphination or hydroamination reactions.

Some catalyst-free approaches, particularly in the case of hydrophosphination, have shown a remarkable reactivity and scope for the isocyanate substrates, especially compared to the conditions in the catalytic reactions. The key for this reaction was that it was performed without added solvent. Plausibly, isothiocyanates could also be utilized as phenyl isothiocyanate showed excellent reactivity, however it is unlikely this methodology can be applied to the much less electrophilic carbodiimides. Multihydrofunctionalization precatalyst **101** is the first example of the sequential hydroalkoxylation-hydroamination of carbodiimides, highlighting potential future areas that could synthesize more complex heteroatom-rich products. These particular successes described in these remarks venture beyond the expected well defined reactions of hydrofunctionalization, and highlight the future potential of this area via polyinsertion reactions, mixed double hydrofunctionalization reactions and reversed C/N bond hydrofunctionalization selectivity.

Acknowledgments

We thank the Engineering and Physical Sciences Research Council (EPSRC) through grant number EP/R004064/1 (for D.L.K.) and for the support of the Centre for Doctoral Training in Sustainable Chemistry through grant number EP/L015633/1 (PhD studentship to C.D.H.), and the University of Nottingham.

References

1. Failla S, Finocchiaro P, Consiglio GA. Syntheses, characterization, stereochemistry and complexing properties of acyclic and macrocyclic compounds possessing α-amino- or α-hydroxyphosphonate units: a review article. *Heteroat Chem.* 2000;11(7):493–504. 10.1002/1098-1071(2000)11:7<493::AID-HC7>3.0.CO;2-A.
2. Zhang WX, Xu L, Xi Z. Recent development of synthetic preparation methods for guanidines via transition metal catalysis. *Chem Commun.* 2015;51(2):254–265. https://doi.org/10.1039/C4CC05291A.
3. Ertl P, Altmann E, Mckenna JM. The most common functional groups in bioactive molecules and how their popularity has evolved over time. *J Med Chem.* 2020;63 (15):8408–8418. https://doi.org/10.1021/acs.jmedchem.0c00754.
4. Eto M. Functions of phosphorus moiety in agrochemical molecules. *Biosci Biotechnol Biochem.* 1997;61(1):1–11. https://doi.org/10.1271/bbb.61.1.
5. Bourget-Merle L, Lappert MF, Severn JR. The chemistry of β-diketiminatometal complexes. *Chem Rev.* 2002;102(9):3031–3065. https://doi.org/10.1021/cr010424r.
6. Schlummer B, Scholz U. Palladium-catalyzed C-N and C-O coupling—a practical guide from an industrial vantage point. *Adv Synth Catal.* 2004;346(13–15):1599–1626. https://doi.org/10.1002/adsc.200404216.
7. Liu X, Hamon JR. Recent developments in penta-, hexa- and heptadentate Schiff base ligands and their metal complexes. *Coord Chem Rev.* 2019;389:94–118. https://doi.org/10.1016/j.ccr.2019.03.010.
8. Taylor JE, Bull SD, Williams JMJ. Amidines, isothioureas, and guanidines as nucleophilic catalysts. *Chem Soc Rev.* 2012;41(6):2109–2121. https://doi.org/10.1039/c2cs15288f.
9. Selig P. Guanidine organocatalysis. *Synthesis.* 2013;45(6):703–718. https://doi.org/10.1055/s-0032-1318154.
10. Beddoe RH, Andrews KG, Magné V, et al. Redox-neutral organocatalytic Mitsunobu reactions. *Science.* 2019;365(6456):910–914. https://doi.org/10.1126/science.aax3353.
11. Rodriguez-Ruiz V, Carlino R, Bezzenine-Lafollée S, et al. Recent developments in alkene hydro-functionalisation promoted by homogeneous catalysts based on earth abundant elements: formation of C-N, C-O and C-P bond. *Dalton Trans.* 2015;44 (27):12029–12059. https://doi.org/10.1039/c5dt00280j.
12. Bezzenine-Lafollée S, Gil R, Prim D, Hannedouch J. First-row late transition metals for catalytic alkene hydrofunctionalisation: recent advances in C-N, C-O and C-P bond formation. *Molecules.* 2017;22(11):1901–1929. https://doi.org/10.3390/molecules22111901.
13. Obligacion JV, Chirik PJ. Earth-abundant transition metal catalysts for alkene hydrosilylation and hydroboration. *Nat Rev Chem.* 2018;2(5):15–34. https://doi.org/10.1038/s41570-018-0001-2.
14. Shenvi RA, Matos JLM, Green SA. Hydrofunctionalization of alkenes by hydrogenatom transfer. *Org React.* 2019;100:383–470. https://doi.org/10.1002/0471264180.or100.07.

15. Zhang Y, Xu X, Zhu S. Nickel-catalysed selective migratory hydrothiolation of alkenes and alkynes with thiols. *Nat Commun.* 2019;10(1):1–9. https://doi.org/10.1038/s41467-019-09783-w.
16. Trost BM, Ball ZT. Alkyne hydrosilylation catalyzed by a cationic ruthenium complex: efficient and general trans addition. *J Am Chem Soc.* 2005;127(50):17644–17655. https://doi.org/10.1021/ja0528580.
17. Kamitani M, Itazaki M, Tamiya C, Nakazawa H. Regioselective double hydrophosphination of terminal arylacetylenes catalyzed by an iron complex. *J Am Chem Soc.* 2012;134(29):11932–11935. https://doi.org/10.1021/ja304818c.
18. Shi SL, Buchwald SL. Copper-catalysed selective hydroamination reactions of alkynes. *Nat Chem.* 2015;7(1):38–44. https://doi.org/10.1038/nchem.2131.
19. Yamamoto K, Mohara Y, Mutoh Y, Saito S. Ruthenium-catalyzed (Z)-selective hydroboration of terminal alkynes with naphthalene-1,8-diaminatoborane. *J Am Chem Soc.* 2019;141(43):17042–17047. https://doi.org/10.1021/jacs.9b06910.
20. Cheng Z, Guo J, Lu Z. Recent advances in metal-catalysed asymmetric sequential double hydrofunctionalization of alkynes. *Chem Commun.* 2020;56(15):2229–2239. https://doi.org/10.1039/d0cc00068j.
21. Zhao CQ, Han LB, Tanaka M. Palladium-catalyzed hydrophosphorylation of allenes leading to regio- and stereoselective formation of allylphosphonates. *Organometallics.* 2000;19(21):4196–4198. https://doi.org/10.1021/om000513e.
22. Krause N, Winter C. Gold-catalyzed nucleophilic cyclization of functionalized allenes: a powerful access to carbo-and heterocycles. *Chem Rev.* 2011;111(3):1994–2009. https://doi.org/10.1021/cr1004088.
23. Jung MS, Kim WS, Shin YH, Jin HJ, Kim YS, Kang EJ. Chemoselective activities of Fe(III) catalysts in the hydrofunctionalization of allenes. *Org Lett.* 2012;14(24):6262–6265. https://doi.org/10.1021/ol303021d.
24. Koschker P, Breit B. Branching out: rhodium-catalyzed allylation with alkynes and allenes. *Acc Chem Res.* 2016;49(8):1524–1536. https://doi.org/10.1021/acs.accounts.6b00252.
25. Wang C, Teo WJ, Ge S. Access to stereodefined (Z)-allylsilanes and (Z)-allylic alcohols via cobalt-catalyzed regioselective hydrosilylation of allenes. *Nat Commun.* 2017;8(1):1–9. https://doi.org/10.1038/s41467-017-02382-7.
26. Li G, Huo X, Jiang X, Zhang W. Asymmetric synthesis of allylic compounds: via hydrofunctionalisation and difunctionalisation of dienes, allenes, and alkynes. *Chem Soc Rev.* 2020;49(7):2060–2118. https://doi.org/10.1039/c9cs00400a.
27. Bagherzadeh S, Mankad NP. Extremely efficient hydroboration of ketones and aldehydes by copper carbene catalysis. *Chem Commun.* 2016;52(19):3844–3846. https://doi.org/10.1039/c5cc09162d.
28. Sahoo RK, Mahato M, Jana A, Nembenna S. Zinc hydride-catalyzed hydrofuntionalization of ketones. *J Org Chem.* 2020;85(17):11200–11210. https://doi.org/10.1021/acs.joc.0c01285.
29. Wang J, Xu F, Cai T, Shen Q. Addition of amines to nitriles catalyzed by ytterbium amides: an efficient one-step synthesis of monosubstituted N-arylamidines. *Org Lett.* 2008;10(3):445–448. https://doi.org/10.1021/ol702739c.
30. Ito M, Itazaki M, Nakazawa H. Selective double hydrosilylation of nitriles catalyzed by an iron complex containing indium trihalide. *ChemCatChem.* 2016;8(21):3323–3325. https://doi.org/10.1002/cctc.201600940.
31. Ibrahim AD, Entsminger SW, Fout AR. Insights into a chemoselective cobalt catalyst for the hydroboration of alkenes and nitriles. *ACS Catal.* 2017;7(5):3730–3734. https://doi.org/10.1021/acscatal.7b00362.
32. Banerjee I, Harinath A, Panda TK. Alkali metal catalysed double hydrophosphorylation of nitriles and alkynes. *Eur J Inorg Chem.* 2019;2019(16):2224–2230. https://doi.org/10.1002/ejic.201900164.

33. Ballinger E, Mosior J, Hartman T, et al. Opposing reactions in coenzyme A metabolism sensitize *Mycobacterium tuberculosis* to enzyme inhibition. *Science*. 2019;363(6426):eaau8959. https://doi.org/10.1126/science.aau8959.
34. Kim IH, Tsai HJ, Nishi K, Kasagami T, Morisseau C, Hammock BD. 1,3-Disubstituted ureas functionalized with ether groups are potent inhibitors of the soluble epoxide hydrolase with improved pharmacokinetic properties. *J Med Chem*. 2007;50(21):5217–5226. https://doi.org/10.1021/jm070705c.
35. McCall JM, TenBrink RE, Ursprung JJ. A new approach to triaminopyrimidine N-oxides. *J Org Chem*. 1975;40(22):3304–3306. https://doi.org/10.1021/jo00910a040.
36. Froidevaux V, Negrell C, Caillol S, Pascault JP, Boutevin B. Biobased amines: from synthesis to polymers; present and future. *Chem Rev*. 2016;116(22):14181–14224. https://doi.org/10.1021/acs.chemrev.6b00486.
37. Mansfield NE, Coles MP, Hitchcock PB. Examining the stability of phospha(III)guanidines: formation of a formamidinium: phosphinate ion-pair and an N-protonated phospha(III)guanidinium chloride. *Polyhedron*. 2012;37(1):9–13. https://doi.org/10.1016/j.poly.2012.01.021.
38. Geeson MB, Jupp AR, Mcgrady JE, Goicoechea JM. On the coordination chemistry of phosphinecarboxamide: assessing ligand basicity. *Chem Commun*. 2014;50:12281–12284. https://doi.org/10.1039/c4cc06094f.
39. Sharova EV, Artyushin OI, Nelyubina YV, Lyssenko KA, Passechnik MP, Odinets IL. Complexation of N-alkyl (aryl)- and N,N-dialkylcarbamoylmethylphosphin e oxides with the f-elements. *Russ Chem Bull*. 2008;57(9):1890–1896. https://doi.org/10.1007/s11172-008-0255-9.
40. Boehme C, Wipff G, Msm L, Cnrs UMR, De Chimie I, Pascal B. Carbamoylphosphine oxide complexes of trivalent lanthanide cations: role of counterions, ligand binding mode, and protonation investigated by quantum mechanical calculations. *Inorg Chem*. 2002;41(4):494–499. https://doi.org/10.1021/ic010658t.
41. Alonso-Moreno C, Antiñolo A, Carrillo-Hermosilla F, Otero A. Guanidines: from classical approaches to efficient catalytic syntheses. *Chem Soc Rev*. 2014;43(10):3406–3425. https://doi.org/10.1039/c4cs00013g.
42. Xu L, Zhang WX, Xi Z. Mechanistic considerations of the catalytic guanylation reaction of amines with carbodiimides for guanidine synthesis. *Organometallics*. 2015;34(10):1787–1801. https://doi.org/10.1021/acs.organomet.5b00251.
43. Burgess K, Ohlmeyer MJ. Transition-metal-promoted hydroborations of alkenes, emerging methodology for organic transformations. *Chem Rev*. 1991;91(6):1179–1191. https://doi.org/10.1021/cr00006a003.
44. Geier S, Vogels CM, Westcott SA. Current developments in the catalyzed hydroboration reaction. In: Coca A, ed. *Boron Reagents in Synthesis*. Oxford University Press; 2016. ACS Symposium Series; https://doi.org/10.1021/bk-2016-1236.ch006.
45. Boese R, Koster R, Yalpanib M. Produkte der Carbodiimid-Hydroborierung mit (9H -9-BBN)2. *Z Naturforsch B*. 1994;49(11):1453–1458. https://doi.org/10.1515/znb-1994-1101.
46. Mukherjee D, Shirase S, Spaniol TP, Mashima K, Okuda J. Magnesium hydridotriphenylborate [Mg(thf)6[HBPh3]2: a versatile hydroboration catalyst. *Chem Commun*. 2016;52(89):13155–13158. https://doi.org/10.1039/c6cc06805g.
47. Weetman C, Hill MS, Mahon MF. Magnesium catalysis for the hydroboration of carbodiimides. *Chem Eur J*. 2016;22(21):7158–7162. https://doi.org/10.1002/chem.201600681.
48. Rauch M, Ruccolo S, Parkin G. Synthesis, structure, and reactivity of a terminal magnesium hydride compound with a carbatrane motif, [TismPriBenz]MgH: a multifunctional catalyst for hydrosilylation and hydroboration. *J Am Chem Soc*. 2017;139(38):13264–13267. https://doi.org/10.1021/jacs.7b06719.

49. Anker MD, Arrowsmith M, Arrowsmith RL, Hill MS, Mahon MF. Alkaline-earth derivatives of the reactive [HB(C6F5)3]- anion. *Inorg Chem.* 2017;56(10):5976–5983. https://doi.org/10.1021/acs.inorgchem.7b00678.
50. Yang Y, Anker MD, Fang J, et al. Hydrodeoxygenation of isocyanates: Snapshots of a magnesium-mediated CO bond cleavage. *Chem Sci.* 2017;8(5):3529–3537. https://doi.org/10.1039/c7sc00117g.
51. Abeliovich H. An empirical extremum principle for the hill coefficient in ligand-protein interactions showing negative cooperativity. *Biophys J.* 2005;89(1):76–79. https://doi.org/10.1529/biophysj.105.060194.
52. Liu H, Kulbitski K, Tamm M, Eisen MS. Organoactinide-catalyzed monohydroboration of carbodiimides. *Chem Eur J.* 2018;24(22):5738–5742. https://doi.org/10.1002/chem.201705987.
53. Shen Q, Ma X, Li W, et al. Organoaluminum compounds as catalysts for monohydroboration of carbodiimides. *Chem Eur J.* 2019;25(51):11918–11923. https://doi.org/10.1002/chem.201902000.
54. Ding Y, Ma X, Liu Y, Liu W, Yang Z, Roesky HW. Alkylaluminum complexes as precatalysts in hydroboration of nitriles and carbodiimides. *Organometallics.* 2019;38(15):3092–3097. https://doi.org/10.1021/acs.organomet.9b00421.
55. Dureen MA, Stephan DW, Ms C. Reactions of boron amidinates with CO2 and CO and other small molecules. *J Am Chem Soc.* 2010;132(38):13559–13568. https://doi.org/10.1021/ja1064153.
56. Mcquilken AC, Dao QM, Cardenas AJP, Bertke JA, Grimme S, Warren TH. A frustrated and confused Lewis pair. *Angew Chem Int Ed.* 2016;55(46):14335–14339. https://doi.org/10.1002/anie.201608968.
57. Fischer M, Schmidtmann M. B(C6F5)3- and HB(C6F5)2-mediated transformations of isothiocyanates. *Chem Commun.* 2020;56:6205–6208. https://doi.org/10.1039/D0CC02626C.
58. Ramos A, Antiñolo A, Carrillo-Hermosilla F, Fernández-Galán R, Rodríguez-Diéguez A, García-Vivó D. Carbodiimides as catalysts for the reduction of CO2 with boranes. *Chem Commun.* 2018;54(37):4700–4703. https://doi.org/10.1039/c8cc02139b.
59. Liu W, Tang Y, Guo Y, et al. Synthesis, characterization and bioactivity determination of ferrocenyl urea derivatives. *Appl Organomet Chem.* 2012;26(4):189–193. https://doi.org/10.1002/aoc.2837.
60. Shakeel A. Thiourea derivatives in drug design and medicinal chemistry: a short review. *J Drug Des Med Chem.* 2016;2(1):10. https://doi.org/10.11648/j.jddmc.20160201.12.
61. Jia Q, Cai T, Huang M, et al. Isoform-selective substrates of nitric oxide synthase. *J Med Chem.* 2003;46(12):2271–2274. https://doi.org/10.1021/jm0340703.
62. McMorris TC, Chimmani R, Alisala K, Staake MD, Banda G, Kelner MJ. Structure-activity studies of urea, carbamate, and sulfonamide derivatives of acylfulvene. *J Med Chem.* 2010;53(3):1109–1116. https://doi.org/10.1021/jm901384s.
63. Yonova PA, Stoilkova GM. Synthesis and biological activity of urea and thiourea derivatives from 2-aminoheterocyclic compounds. *J Plant Growth Regul.* 2004;23(4):280–291. https://doi.org/10.1007/s00344-003-0054-3.
64. Rodríguez-Fernández E, Manzano JL, Benito JJ, Hermosa R, Monte E, Criado JJ. Thiourea, triazole and thiadiazine compounds and their metal complexes as antifungal agents. *J Inorg Biochem.* 2005;99(8):1558–1572. https://doi.org/10.1016/j.jinorgbio.2005.05.004.
65. Wu J, Shi Q, Chen Z, He M, Jin L, Hu D. Synthesis and bioactivity of pyrazole acyl thiourea derivatives. *Molecules.* 2012;17(5):5139–5150. https://doi.org/10.3390/molecules17055139.
66. Guan A, Liu C, Yang X, Dekeyser M. Application of the intermediate derivatization approach in agrochemical discovery. *Chem Rev.* 2014;114(14):7079–7107. https://doi.org/10.1021/cr4005605.

67. Kurien M, Claramma NM, Kuriakose AP. A new secondary accelerator for the sulfur vulcanization of natural rubber latex and its effect on the rheological properties. *J Appl Polym Sci*. 2004;93(6):2781–2789. https://doi.org/10.1002/app.20770.
68. Lin B, Waymouth RM. Urea anions: SIMPLE, fast, and selective catalysts for ring-opening polymerizations. *J Am Chem Soc*. 2017;139(4):1645–1652. https://doi.org/10.1021/jacs.6b11864.
69. Dawn S, Salpage SR, Koscher BA, et al. Applications of a bis-urea phenylethynylene self-assembled nanoreactor for [2 + 2] photodimerizations. *J Phys Chem A*. 2014;118 (45):10563–10574. https://doi.org/10.1021/jp505304n.
70. Gumus I, Solmaz U, Binzet G, Keskin E, Arslan B, Arslan H. Supramolecular self-assembly of new thiourea derivatives directed by intermolecular hydrogen bonds and weak interactions: crystal structures and Hirshfeld surface analysis. *Res Chem Intermed*. 2019;45(2):169–198. https://doi.org/10.1007/s11164-018-3596-5.
71. Zhao J, Li Z, Yan S, et al. Pd/C catalyzed carbonylation of azides in the presence of amines. *Org Lett*. 2016;18(8):1736–1739. https://doi.org/10.1021/acs.orglett.6b00381.
72. Krishnakumar V, Chatterjee B, Gunanathan C. Ruthenium-catalyzed urea synthesis by N-H activation of amines. *Inorg Chem*. 2017;56(12):7278–7284. https://doi.org/10.1021/acs.inorgchem.7b00962.
73. Paul F, Fischer J, Ochsenbein P, Osborn JA. The palladium-catalyzed carbonylation of nitrobenzene into phenyl isocyanate: the structural characterization of a metallacylic intermediate. *Organometallics*. 1998;17(11):2199–2206. https://doi.org/10.1021/om9708485.
74. Vinogradova EV, Fors BP, Buchwald SL. Palladium-catalyzed cross-coupling of aryl chlorides and triflates with sodium cyanate: a practical synthesis of unsymmetrical ureas. *J Am Chem Soc*. 2012;134(27):11132–11135. https://doi.org/10.1021/ja305212v.
75. Yagodkin A, Löschcke K, Weisell J, Azhayev A. Straightforward carbamoylation of nucleophilic compounds employing organic azides, phosphines, and aqueous trialkylammonium hydrogen carbonate. *Tetrahedron*. 2010;66(12):2210–2221. https://doi.org/10.1016/j.tet.2010.01.017.
76. Katritzky AR, Ledoux S, Witek RM, Nair SK. 1-(Alkyl/Arylthiocarbamoyl)benzotriazoles as stable isothiocyanate equivalents: synthesis of di- and trisubstituted thioureas. *J Org Chem*. 2004;69(9):2976–2982. https://doi.org/10.1021/jo035680d.
77. Sharma S. Thiophosgene in organic synthesis. *Synthesis*. 1978;11:803–820. https://doi.org/10.1055/s-1978-24896.
78. Staab HA. New methods of preparative organic chemistry IV. Syntheses using heterocyclic amides (azolides). *Angew Chem Int Ed*. 1967;1(7):351–367. https://doi.org/10.1002/anie.196203511.
79. Bernstein J, Yale HL, Losee K, Holsing M, Martins J, Lott WA. The chemotherapy of experimental tuberculosis. III. The synthesis of thiosemicarbazones and related compounds. *J Am Chem Soc*. 1951;73(3):906–912. https://doi.org/10.1021/ja01147a007.
80. Erickson JG. Reactions of long-chain amines. VI. Preparation of thioureas. *J Org Chem*. 1956;21(4):483–484. https://doi.org/10.1021/jo01110a611.
81. Farkas A, Mills GA. Catalytic effects in isocyanate reactions. *Adv Catal*. 1962;13 (C):393–446. https://doi.org/10.1016/S0360-0564(08)60290-4.
82. Barrett AGM, Boorman TC, Crimmin MR, Hill MS, Kociok-Köhn G, Procopiou PA. Heavier group 2 element-catalysed hydroamination of isocyanates. *Chem Commun*. 2008;4356(41):5206–5208. https://doi.org/10.1039/b809649j.
83. Hernán-Gómez A, Bradley TD, Kennedy AR, Livingstone Z, Robertson SD, Hevia E. Developing catalytic applications of cooperative bimetallics: competitive hydroamination/trimerization reactions of isocyanates catalysed by sodium magnesiates. *Chem Commun*. 2013;49(77):8659–8661. https://doi.org/10.1039/c3cc45167d.

84. Bano K, Anga S, Jain A, Nayek HP, Panda TK. Hydroamination of isocyanates and isothiocyanates by alkaline earth metal initiators supported by a bulky iminopyrrolyl ligand. *New J Chem.* 2020;44(22):9419–9428. https://doi.org/10.1039/d0nj01509a.
85. Naktode K, Das S, Bhattacharjee J, Nayek HP, Panda TK. Imidazolin-2-iminato ligand-supported titanium complexes as catalysts for the synthesis of urea derivatives. *Inorg Chem.* 2016;55(3):1142–1153. https://doi.org/10.1021/acs.inorgchem.5b02302.
86. Bhattacharjee J, Das S, Kottalanka RK, Panda TK. Hydroamination of carbodiimides, isocyanates, and isothiocyanates by a bis(phosphinoselenoic amide) supported titanium(IV) complex. *Dalton Trans.* 2016;45(44):17824–17832. https://doi.org/10.1039/c6dt03063g.
87. South AJ, Geer AM, Taylor LJ, et al. Iron(II)-catalyzed hydroamination of isocyanates. *Organometallics.* 2019;38(21):4115–4120. https://doi.org/10.1021/acs.organomet.9b00393.
88. Sharpe HR, Geer AM, Williams HEL, et al. Cyclotrimerisation of isocyanates catalysed by low-coordinate Mn(ii) and Fe(ii) m-terphenyl complexes. *Chem Commun.* 2017;53(5):937–940. https://doi.org/10.1039/C6CC07243G.
89. Six C, Richter F. Isocyanates, organic. *Ullmann's Encycl Ind Chem.* 2003;20:63–82. https://doi.org/10.1002/14356007.a14.
90. Ozaki S. Recent advances in isocyanate chemistry. *Chem Rev.* 1972;72(5):457–496. https://doi.org/10.1021/cr50013a002.
91. Arnold RG, Nelson JA, Verbanc JJ. Recent advances in isocyanate chemistry. *Chem Rev.* 1957;57(1):47–76. https://doi.org/10.1021/cr50013a002.
92. Mohr E. Mitkheilnngen aus dem chemischen Institut der. 50. Beitrag zur Kenntnis der L ossenschen Umlagerung. *J Park Chem.* 1904;71(1):133–149. https://doi.org/10.1002/prac.19050710108.
93. Baker JW, Gaunt J. The mechanism of the reaction of aryl isocyanates with alcohols and amines, part V. Kinetic investigations of the reaction between phenyl isocyanate and methyl and ethyl alcohols in benzene solution. *J Chem Soc.* 1949;(0):27–31. https://doi.org/10.1039/JR9490000027.
94. Baker JW, Gaunt J. The mechanism of the reaction, etc. Part III. The mechanism of the reaction of aryl isocyanates with alcohols and amines. Part III. The "spontaneous" reaction of phenyl isocyanate with various alcohols. Further evidence relating to the anomalous effect. *J Chem Soc.* 1949;(0):19–24. https://doi.org/10.1039/JR9490000019.
95. Li Z, Mayer RJ, Ofial AR, Mayr H. From carbodiimides to carbon dioxide: quantification of the electrophilic reactivities of heteroallenes. *J Am Chem Soc.* 2020;142(18):8383–8402. https://doi.org/10.1021/jacs.0c01960.
96. Wesifeld LB. The estimation of catalytic parameters of metal acetylacetonates in isocyanate polymerization reactions. *J Appl Polym Sci.* 1961;5(16):424–427. https://doi.org/10.1002/app.1961.070051607.
97. Leuckart R. Ueber einige Reaktionen der aromatischen cyanate. *Ber Dtsch Chem Ges.* 1885;18(1):873–877. https://doi.org/10.1002/cber.188501801182.
98. Michael A, Cobb PH. Phenylisocyanat als Reagens zur Feststellung der Constitution merotroper Verbindungen. *Justus Liebigs Ann Chem.* 1908;363(1):64–93. https://doi.org/10.1002/jlac.19083630105.
99. French HE, Wirtel AF. Alpha-naphthylisocyanate as a reagent for phenols and aliphatic amines. *J Am Chem Soc.* 1926;48(6):1736–1739. https://doi.org/10.1021/ja01417a041.
100. Dabritz E. Syntheses and reactions of O,N,N'-trisubstituted isoureas. *Angew Chem Int Ed.* 1996;5(5):470–477. https://doi.org/10.1002/anie.196604701.
101. Dains FB. On the isourea ethers and other derivatives of ureas. *J Am Chem Soc.* 1899;21(2):136–192. https://doi.org/10.1021/ja02052a004.
102. Schimdt E, Carl W. Zur Kenntnis Aliphatischer Carbodiimide, XII. *Justus Liebigs Ann Chem.* 1961;639(1):24–31. https://doi.org/10.1002/jlac.19616390104.

103. Batrice RJ, Kefalidis CE, Maron L, Eisen MS. Actinide-catalyzed intermolecular addition of alcohols to carbodiimides. *J Am Chem Soc*. 2016;138(7):2114–2117. https://doi.org/10.1021/jacs.5b12731.
104. Liu H, Fridman N, Tamm M, Eisen MS. Catalytic addition of alcohols to carbodiimides mediated by benzimidazolin-2-iminato actinide complexes. *Organometallics*. 2017;36(23):4600–4610. https://doi.org/10.1021/acs.organomet.7b00432.
105. Liu H, Khononov M, Fridman N, Tamm M, Eisen MS. Catalytic addition of alcohols into carbodiimides promoted by organoactinide complexes. *Inorg Chem*. 2017;56(6):3153–3157. https://doi.org/10.1021/acs.inorgchem.7b00357.
106. Haynes WM. *CRC Handbook of Chemistry and Physics*. 96th ed. CRC Press; 2015.
107. Khononov M, Liu H, Fridman N, Tamm M, Eisen MS. Benzimidazolin-2-iminato hafnium complexes: synthesis, characterization, and catalytic addition of alcohols to carbodiimides. *Organometallics*. 2020;39:3021–3033. https://doi.org/10.1021/acs.organomet.0c00384.
108. Ojima I, Ichi IS. Reduction of isocyanates and carbodiimides via hydrosilylation. *J Organomet Chem*. 1977;140(1):97–111. https://doi.org/10.1016/S0022-328X(00)84400-9.
109. Huckaba AJ, Hollis TK, Reilly SW. Homobimetallic Rh NHC complexes as highly versatile catalysts for hydrosilylation in the presence of air and moisture. *Organometallics*. 2013;32:6248–6256.
110. Mansfield NE, Coles MP, Hitchcock PB. Lithium and aluminium complexes supported by chelating phosphaguanidinates. *Dalton Trans*. 2005;17:2833–2841. https://doi.org/10.1039/b506332a.
111. Mansfield NE, Coles MP, Hitchcock PB. The effect of P-cyclohexyl groups on the coordination chemistry of phosphaguanidinates. *J Chem Soc Dalton Trans*. 2006;44(17):2052–2054. https://doi.org/10.1039/b601235c.
112. Grundy J, Mansfield NE, Coles MP, Hitchcock PB. Phosphaguanidines as scaffolds for multimetallic complexes containing metal-functionalized phosphines. *Inorg Chem*. 2008;47(7):2258–2260. https://doi.org/10.1021/ic7024457.
113. Jin G, Jones C, Junk PC, Lippert KA, Rose RP, Stasch A. Synthesis and characterisation of bulky guanidines and phosphaguanidines: precursors for low oxidation state metallacycles. *New J Chem*. 2009;33(1):64–75. https://doi.org/10.1039/b809120j.
114. Beddoe SVF, Cosham SD, Kulak AN, Jupp AR, Goicoechea JM, Hyett G. Phosphinecarboxamide as an unexpected phosphorus precursor in the chemical vapour deposition of zinc phosphide thin film. *Dalton Trans*. 2018;47:9221–9225. https://doi.org/10.1039/c8dt00544c.
115. Ouizem S, Rosario-Amorin D, Dickie DA, et al. Synthesis and f-element ligation properties of NCMPO-decorated pyridine N-oxide platforms. *Dalton Trans*. 2014;43:8368–8386. https://doi.org/10.1039/c3dt53611d.
116. Horwitz EP, Kalina DC, Diamond H, Vandegrift GF, Schulz W. The truex process—a process for the extraction of the transuranic elements from nitric acid wastes utilizing modified purex solvent. *Solvent Extr Ion Exch*. 1985;3(1–2):75–109. https://doi.org/10.1080/07366298508918504.
117. Ozawa M, Koma Y, Nomura K, Tanaka Y. Separation of actinides and fission products in high-level liquid wastes by the improved TRUEX process. *J Alloy Compd*. 1998;271–273:538–543. https://doi.org/10.1016/S0925-8388(98)00147-9.
118. Breuer E, Salomon CJ, Katz Y, Chen W, Lu S, Ro G. Carbamoylphosphonates, a new class of in vivo active matrix metalloproteinase inhibitors. 1. Alkyl- and cycloalkylcarbamoylphosphonic. *J Med Chem*. 2004;47:2826–2832. https://doi.org/10.1021/jm030386z.
119. Reich R, Hoffman A, Veerendhar A, et al. Carbamoylphosphonates control tumor cell proliferation and dissemination by simultaneously inhibiting carbonic anhydrase IX and

matrix metalloproteinase-2. Toward nontoxic chemotherapy targeting tumor microenvironment. *J Med Chem.* 2012;55(17):7875–7882. https://doi.org/10.1021/jm300981b.
120. Reich R, Hoffman A, Suresh RR, et al. Carbamoylphosphonates inhibit autotaxin and metastasis formation in vivo. *J Enzyme Inhib Med Chem.* 2015;30(5):767–772. https://doi.org/10.3109/14756366.2014.968146.
121. Icnatious F, Sein A, Cabasso I, Smid J. Novel carbamoyl phosphonate monomers and polymers from unsaturated lsocyanates. *J Polym Sci A Polym Chem.* 1993;31(1):239–247. https://doi.org/10.1002/pola.1993.080310128.
122. Turgis R, Leydier A, Arrachart G, et al. Carbamoylalkylphosphonates for dramatic enhancement of uranium extraction from phosphates ores. *Solvent Extr Ion Exch.* 2014;32(14):685–702. https://doi.org/10.1080/07366299.2014.951279.
123. Zhang WX, Nishiura M, Hou Z. Alkali-metal-catalyzed addition of primary and secondary phosphines to carbodiimides. A general and efficient route to substituted phosphaguanidines. *Chem Commun.* 2006;36:3812–3814. https://doi.org/10.1039/b609198a.
124. Crimmin MR, Barrett AGM, Hill MS, Hitchcock PB, Procopiou PA. Heavier group 2 element catalyzed hydrophosphination of carbodiimides. *Organometallics.* 2008;27(4):497–499. https://doi.org/10.1021/om7011198.
125. Thewissen D, Ambrosius HPMM. The reactions of hetero-allenes RN=C=X with phosphine derivatives containing a P-H bond. *Recl Trav Chim Pays-Bas.* 1980;99(11):344–350. https://doi.org/10.1002/recl.19800991105.
126. Ambrosius HPMM, Van Der Linden AHIM, Steggerda JJ. Molybdenum- and tungsten-cyclopentadienyl carbonyl complexes with hetero-allyl derivatives as ligand. *J Organomet Chem.* 1981;204(2):211–220. https://doi.org/10.1016/S0022-328X(00)84587-8.
127. Issleib K, Schmidt H, Meyer H. Phospha-guanidine I. Eine neue verbindungsklassemit zweifach-koordiniertem dreibindigem phosphor. *J Organomet Chem.* 1980;192(1):33–39. https://doi.org/10.1016/S0022-328X(00)93328-X.
128. Fischer R, Langer J, Go H, Walther D, Westerhausen M. Syntheses and structures of alkaline earth metal bis (diphenylamides). *Inorg Chem.* 2007;46(12):5118–5124. https://doi.org/10.1021/ic700459d.
129. Gärtner M, Görls H, Westerhausen M. Arylphosphanide complexes of the heavy alkaline earth metals calcium, strontium and barium of the formula (thf)nM[P(R) Aryl]2. *Z Anorg Allg Chem.* 2007;633(11 – 12):2025–2031. https://doi.org/10.1002/zaac.200700256.
130. Zhang WX, Nishiura M, Mashiko T, Hou Z. Half-sandwich o-N, N-dimethylaminobenzyl complexes over the full size range of group 3 and lanthanide metals. synthesis, structural characterization, and catalysis of phosphine P-H bond addition to carbodiimides. *Chem Eur J.* 2008;14(7):2167–2179. https://doi.org/10.1002/chem.200701300.
131. Behrle AC, Schmidt JAR. Insertion reactions and catalytic hydrophosphination of heterocumulenes using α-metalated N, N-dimethylbenzylamine rare-earth-metal complexes. *Organometallics.* 2013;32(5):1141–1149. https://doi.org/10.1021/om300807k.
132. Ma W, Xu L, Zhang WX, Xi Z. Half-sandwich rare-earth metal tris(alkyl) ate complexes catalyzed phosphaguanylation reaction of phosphines with carbodiimides: an efficient synthesis of phosphaguanidines. *New J Chem.* 2015;39(10):7649–7655. https://doi.org/10.1039/c5nj01136a.
133. Gu X, Zhang L, Zhu X, et al. Synthesis of Bis(NHC)-based CNC-pincer rare-earth-metal amido complexes and their application for the hydrophosphination of heterocumulenes. *Organometallics.* 2015;34(18):4553–4559. https://doi.org/10.1021/acs.organomet.5b00628.

134. Boeré R, Klassen K, Wolmershäuser G. Synthesis of some very bulky N, N′-disubstituted amidines and initial studies of their coordination chemistry. *J Chem Soc Dalton Trans*. 1998;(24):4147–4154. https://doi.org/10.1039/A805548C.
135. Mansfield NE, Grundy J, Coles MP, Avent AG, Hitchcock PB. A conformational study of phospha(III)- and phospha(V)-guanidine compounds. *J Am Chem Soc*. 2006;128(42): 13879–13893. https://doi.org/10.1021/ja064212t.
136. Roering AJ, Leshinski SE, Chan SM, et al. Insertion reactions and catalytic hydrophosphination by triamidoamine-supported zirconium complexes. *Organometallics*. 2010;29(11):2557–2565. https://doi.org/10.1021/om100216f.
137. Sharpe HR, Geer AM, Lewis W, Blake AJ, Kays DL. Iron(II)-catalyzed hydrophosphination of isocyanates. *Angew Chem Int Ed*. 2017;56(17):4845–4848. https://doi.org/10.1002/anie.201701051.
138. Zhang Y, Qu L, Wang Y, Yuan D, Yao Y, Shen Q. Neutral and cationic zirconium complexes bearing multidentate aminophenolato ligands for hydrophosphination reactions of alkenes and heterocumulenes. *Inorg Chem*. 2018;57:139–149. https://doi.org/10.1021/acs.inorgchem.7b02248.
139. Horsley Downie TM, Hall JW, Collier Finn TP, Liptrot DJ, et al. The first ring expanded NHC-copper(I) phosphides as catalysts in highly selective hydrophosphination of isocyanates. *Chem Commun*. 2020;56:13359–13362. https://doi.org/10.1039/d0cc05694d.
140. Itazaki M, Matsutani T, Nochida T, Moriuchi T, Nakazawa H. Convenient synthesis of phosphinecarboxamide and phosphinecarbothioamide by hydrophosphination of isocyanates and isothiocyanates. *Chem Commun*. 2020;56(3):443–445. https://doi.org/10.1039/c9cc08329d.
141. Moglie Y, González-soria MJ, Martín-garcía I, Radivoy G, Alonso F. Catalyst- and solvent-free hydrophosphination and multicomponent hydrothiophosphination of alkenes and alkynes. *Green Chem*. 2016;18(18):4896–4907. https://doi.org/10.1039/c6gc00903d.
142. Aguiar AM, Giacin J, Greenberg HJ. Reaction of lithium diphenylphosphide and carbonyl compounds. *J Org Chem*. 1963;28(12):3545–3547. https://doi.org/10.1021/jo01047a503.
143. Issleib K, Harzfeld G. Thiocarbamoylphosphine. *Angew Chem Int Ed*. 1965;77(20): 917–918. https://doi.org/10.1002/ange.19650772021.
144. Iwao O, Kin-ya A, Naoki I. Reactions of diarylphosphines, their oxides and sulfides with isothiocyanates and thiocyanic acid. *Bull Chem Soc Jpn*. 1969;42(10):2975–2981. https://doi.org/10.1246/bcsj.42.2975.
145. Härling S, Greiser J, Al-Shboul TMA, Görls H, Krieck S, Westerhausen M. Calcium-mediated hydrophosphorylation of organic isocyanates with diphenylphosphane oxide. *Aust J Chem*. 2013;66(10):1264–1273. https://doi.org/10.1071/CH13259.
146. Härling SM, Görls H, Krieck S, Westerhausen M. Potassium-mediated hydrophosphorylation of heterocumulenes with diarylphosphane oxide and sulfide. *Inorg Chem*. 2016;55(20):10741–10750. https://doi.org/10.1021/acs.inorgchem.6b01973.
147. Härling SM, Krieck S, Görls H, Westerhausen M. Influence of 18-crown-6 ether coordination on the catalytic activity of potassium and calcium diarylphosphinites in hydrophosphorylation reactions. *Inorg Chem*. 2017;56(15):9255–9263. https://doi.org/10.1021/acs.inorgchem.7b01314.
148. Westerhausen M, Krieck S, Langer J, Al-Shboul TMA, Görls H. Phosphanides of calcium and their oxidation products. *Coord Chem Rev*. 2013;257(5–6):1049–1066. https://doi.org/10.1016/j.ccr.2012.06.018.
149. Fox RB, Venezky DL. The base-catalyzed reaction of dialkyl phosphonates with isocyanates. *J Am Chem Soc*. 1956;78(8):1661–1663. https://doi.org/10.1021/ja01589a046.
150. Tashma Z. (N-alkylthiocarbamoyl)phosphonic acid esters. 1. Preparation and spectral properties. *J Org Chem*. 1982;47(15):3012–3015. https://doi.org/10.1021/jo00136a044.

151. Nylen P. 203. Paul Nylen: Beitrag sur Kenntnis der organischen phosphorverbindungen. *Ber Dtsch Chem Ges*. 1924;57(6):1023–1038. https://doi.org/10.1002/cber.9240570625.
152. Reetz T, Chadwick DH, Hardy EE, Kaufman S. Carbamoylphosphonates. *J Am Chem Soc*. 1955;77(14):3813–3816. https://doi.org/10.1021/ja01619a040.
153. Kaboudin B, Zahedi H. A novel and convenient method for synthesis of carbamoyl and thiocarbamoyl phosphonates. *Heteroat Chem*. 2009;20(4):250–253. https://doi.org/10.1002/hc.20538.
154. Petersen S. Niedermolekulare Umsetzungsprodukte aliphatischer diisocyanate. *Justus Liebigs Ann Chem*. 1949;562(3):205–229. https://doi.org/10.1002/jlac.19495620306.
155. Movassagh B, Soleiman-Beigi M. Synthesis of thiocarbamates from thiols and isocyanates under catalyst- and solvent-free conditions. *Monatsh Chem*. 2008;139(2):spi6;137–140. https://doi.org/10.1007/s00706-007-0762-7.
156. Smith JF, Friedrich EC. Urethans of 2-mercaptoethanol. *J Am Chem Soc*. 1959;81(1):161–163. https://doi.org/10.1021/ja01510a037.
157. Dyer E, Glenn JF. The kinetics of the reactions of phenyl isocyanate with certain thiols. *J Am Chem Soc*. 1957;79(2):366–369. https://doi.org/10.1021/ja01559a034.
158. Iwakura Y, Okada H. The kinetics of the tertiary-amine-catalyzed reaction of organic isocyanates with thiols. *Can J Chem*. 1960;38(12):2418–2424. https://doi.org/10.1139/v60-328.
159. Schlack P, Keil G. Neue carbodiimid-synthesen. *Justus Liebigs Ann Chem*. 1890;661(1):164–172. https://doi.org/10.1002/jlac.19636610111.
160. Kurzer F, Davies PR, Langer SS. Diisophorone and related compounds. 20. Idiisophoranes incorporating the 1,3-thiazine ring system: 8,lla-methanocycloocta[d,e][3,*l*]benzothiazines. *J Org Chem*. 1987;52(22):4966–4973. https://doi.org/10.1021/jo00231a024.
161. Wang T, Xu XM, Ke XX, Liu XY, Luo J, Yi BX. Iminophosphorane-mediated efficient synthesis of new fluorine-containing triazolo[4,5-d]pyrimidin-7-ones. *Phosphorus Sulfur Silicon Relat Elem*. 2012;187(2):155–164. https://doi.org/10.1080/10426507.2011.590169.
162. Zhu YP, Mampuys P, Sergeyev S, Ballet S, Maes BUW. Amine activation: N-arylamino acid amide synthesis from isothioureas and amino acids. *Adv Synth Catal*. 2017;359(14):2481–2498. https://doi.org/10.1002/adsc.201700134.
163. Lu C, Hu L, Zhao B, Yao Y. Addition of thiols to isocyanates catalyzed by simple rare-earth-metal amides: synthesis of S-alkyl thiocarbamates and dithiocarbamates. *Organometallics*. 2019;38(9):2167–2173. https://doi.org/10.1021/acs.organomet.9b00147.
164. Karmel ISR, Tamm M, Eisen MS. Actinide-mediated catalytic addition of E-H bonds (E=N, P, S) to carbodiimides, isocyanates, and isothiocyanates. *Angew Chem Int Ed*. 2015;54(42):12422–12425. https://doi.org/10.1002/anie.201502041.
165. Batrice RJ, Eisen MS. Catalytic insertion of E-H bonds (E = C, N, P, S) into heterocumulenes by amido-actinide complexes. *Chem Sci*. 2016;7(2):939–944. https://doi.org/10.1039/c5sc02746b.
166. Bhattacharjee J, Harinath A, Banerjee I, Nayek HP, Panda TK. Highly active dinuclear titanium(IV) complexes for the catalytic formation of a carbon–heteroatom bond. *Inorg Chem*. 2018;57:12610–12623. https://doi.org/10.1021/acs.inorgchem.8b01766.

CHAPTER TWO

Polymerization of terpenes and terpenoids using metal catalysts

Miguel Palenzuela, David Sánchez-Roa, Jesús Damián, Valentina Sessini, and Marta E.G. Mosquera*

Department of Organic and Inorganic Chemistry, Institute of Chemical Research "Andrés M. del Río" (IQAR), Universidad de Alcalá, Campus Universitario, Alcalá de Henares, Madrid, Spain
*Corresponding author: e-mail address: martaeg.mosquera@uah.es

Contents

1. Introduction: Terpenes and terpenoids, origin and classification — 55
2. Polymerization of terpenes — 57
   2.1 Polymerization of acyclic terpenes — 58
   2.2 Polymerization of cyclic terpenes — 68
3. Polymerization of terpenoids — 75
4. Copolymerization of terpenoids — 81
   4.1 Polyesters — 81
   4.2 Polycarbonates — 83
5. Concluding remarks — 85
Acknowledgments — 86
References — 86

1. Introduction: Terpenes and terpenoids, origin and classification

During this century, there has been a significant focus on the use of renewable feedstock as raw materials to prepare chemicals.[1–8] Within renewables, those coming from plants are of particular interest, where terpenoids stand out since they are feedstock of natural origin coming from cheap non-food crops.[9–11]

Terpenes and their oxygenated derivatives, terpenoids, are synthesized mainly by plants and are found in flowers, fruits, trees and spices.[12] They are one of the largest family (>50,000 members) of natural products synthesized as secondary metabolites. They can be extracted from the volatile fraction of some conifers, called turpentines, which is composed of mainly α-pinene (45%–97%) and β-pinene (0.5%–28%). As well, other terpenes

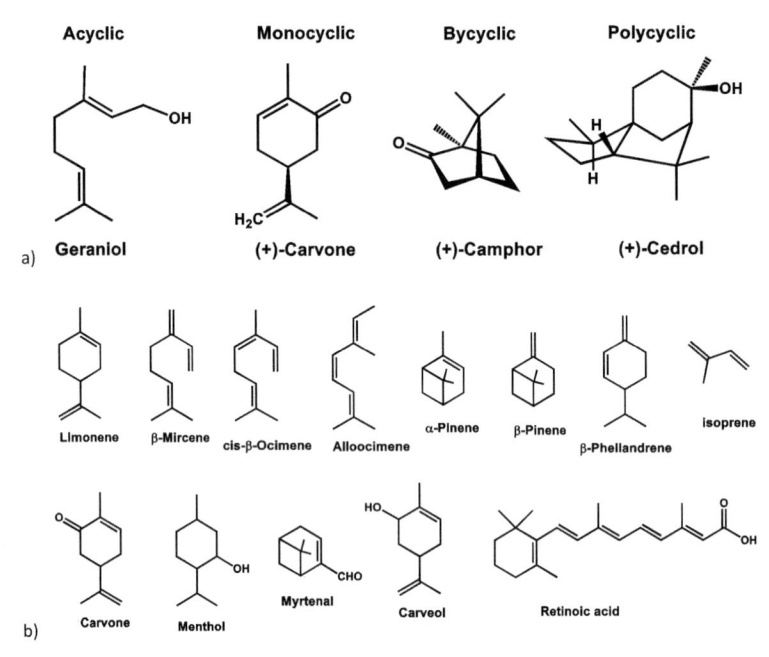

Fig. 1 (A) Different classes of terpenes. (B) Common terpenes and terpenoids.

can be obtained from essential oils from some plants like lemon and orange trees (S- and R-limonene), rose (geraniol), mint (menthol), lemon grass (citronellal), among others. Furthermore, terpenes can be produced in biorefineries from biomass coming from different types of waste such as agricultural residues. Due to the increasing demand of terpenes and terpenoids, arising from their potential uses, interest in producing them in biorefineries has significantly increased over recent years.[13]

As can be seen in Fig. 1, a remarkable structurally different terpenes can be found, including acyclic, monocyclic, bicyclic and polycyclic structures. In all cases, they present functional groups which render them as interesting precursors for the synthesis of new higher-value chemicals.

Terpenes play an important biological role, for example, as a defense mechanism in plants since they can act as repellents to some insects. As well terpenes can be involved in the stimulation of pollination processes.[14,15] Besides, terpenes show many applications in areas such as cosmetics, pharmacy, food or insecticides, amongst other applications.[16]

Even though the use of these compounds as essential oils has been well known since ancient times, the study of terpene chemistry was only really started at the end of the XIXth century by Otto Wallach. In 1884, Wallach

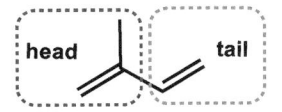

Fig. 2 Isoprene structure and the head and tail denomination.

recognized the structure of the isoprene as the repetitive unit in these species and proposed that all terpenes and terpenoids are derived from the condensation "head-tail" of isoprene molecules (Fig. 2), although in some cases heavy terpenes do not follow this rule.[17]

Considering this description, terpene derivatives can be classified depending on the number of isoprene units that they contain. In this way, they can be defined as hemiterpenes (C_5), monoterpenes (C_{10}), sesquiterpenes (C_{15}), diterpenes (C_{20}), sesterterpenes (C_{25}), triterpenes (C_{30}) and rubbers (C_n), depending on the isoprene molecule content.[11] Terpenoids are terpene derivatives that contain oxygenated functional groups such as carbonyl, epoxide, ester, ether or hydroxides. The high structural diversity found in terpenes and terpenoids arises from the different number of carbon atoms, isomers, and stereochemistry of their stereogenic centres.[16]

2. Polymerization of terpenes

The most popularly studied terpenes are monoterpenes ($C_{10}H_{16}$), which are mainly produced from turpentines.[18–23] From a chemical point of view, terpenes are hydrocarbon units with double bonds, hence there are suitable natural substitutes of oil-derived monomers such as ethylene, propylene and styrene. However, terpene polymerization is an area that has been little explored, which is surprising since some terpenes like isoprene have been used for decades, and even centuries, i.e., natural rubber (*cis*-1,4-polyisoprene) is a natural terpene-based polymer. Only a handful of terpenes have been studied for polymerization, where maybe the reason for being overlooked in this area is the great structural diversity and complexity of this family, which at the same time is their biggest strength.[24] Although the number of reports on terpene polymerization is not very abundant, recently the development of processes to polymerize them is attracting great interest and the number of articles or patents has significantly increased over the last few years. The polyterpenes described to date are mainly elastomers while from terpenoids there is a wider range of possibilities both in properties and applications, particularly when copolymerized with other monomers.

The terpenes that have been studied in polymerization can be classified in two main groups: cyclic and acyclic.[25–28] Examples of linear terpenes used as monomers include myrcene, ocimene and alloocimene, while the main cyclic terpenes studied in polymerization are β-pinene, α-pinene, limonene and phellandrene (Fig. 1). Different mechanisms have been used for terpene polymerization, such as radical, ionic, coordination-insertion and coordinative chain-growth. In this chapter, an overview of the most commonly studied terpenes in polymerization processes catalyzed by metal compounds, excluding radical-based transformations, will be provided.

2.1 Polymerization of acyclic terpenes

Some acyclic terpenes are similar to well-known petrochemical-based monomers, as such myrcene, ocimene, and alloocimene present clear similarities to the olefins (Fig. 3).[25,29,30]

2.1.1 Myrcene polymerization

Myrcene (7-methyl-3-methylene-octa-1,6-diene, $C_{10}H_{16}$), is a natural conjugated diene, so it is a perfect candidate to be studied as a monomer. Surprisingly, after some initial studies in the 1960s,[31] this monomer has been quite forgotten about until recently.[32] There are two possible isomers of myrcene, α and β, but the natural occurring one is the β-isomer.

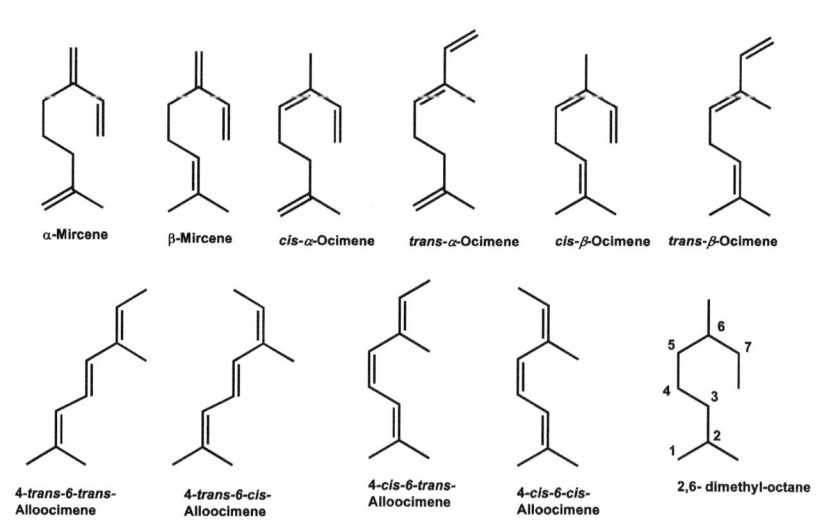

Fig. 3 Acyclic terpenes studied in polymerization reactions. *Reproduce from Behr A, Johnen L. Myrcene as a natural base chemical in sustainable chemistry: a critical review. ChemSusChem 2009;2:1072–1095, with permission from Wiley-VCH.*

β-Myrcene is a readily available terpene that can be obtained from various essential oils, such as ylang-ylang, hops, bay, thyme or cannabis, but in fact it is more economical to produce it from the pyrolysis of β-pinene,[9,33] since β-pinene is more abundant and can be easily isolate from turpentine oils. The isomer ocimene is produced from the thermal cracking of α-pinene and can isomerize at high temperature to give alloocimene. These three terpenes are isomers, in that their structures vary in the position of the double bonds as shown in Fig. 3, however, most of the studies have been focused on myrcene.

Myrcene is structurally very similar to isoprene, it could be considered as an isoprene with a bulky substituent. Hence, it is not surprising that it can be polymerized in a similar way. Like isoprene, the polymerization can take place through nearly all types of polymerization mechanisms: anionic, free-radical, coordination and cationic. The mechanism of polymerization is going to influence heavily the microstructure, which for myrcene will depend on the regioselectivity of the addition since 1,4 (*cis* or *trans*), 3,4 or 1,2 can take place (Fig. 4). Polymer properties can be tuned by controlling their microstructure, and as in the case of isoprene a small difference (only 1%–2%) provokes important discrepancies in the final properties.[34] The obtained poly-β-myrcenes display features comparable to those of polyisoprene in terms of T_g and mechanical properties and they are mainly elastomers.

Myrcene can polymerize naturally, and in fact, natural polymyrcene has been used since old times as a type of gum (gum mastic) that has found applications in adhesives, coatings and even medicine. In the 1990s, the structure of this natural polymer was elucidated and found to be a *cis*-1,4-polymer (ca. 75% *cis*) with broad molecular-weight dispersity ($Đ_M$) and molecular weights of up to 100 kDa.[35]

Myrcene radical polymerization has been studied since the 1940s,[36] the first studies on anionic and cationic polymerization were also performed as early as that, but they were considered industrially relevant and so mostly reported in patents.[37] In the 1960s, Marvel described an anionic polymerization using *n*BuLi in a non-polar solvent, the polymer produced presented mainly a 3,4-structure, as detected using IR spectroscopy, in similarity to the microstructure obtained by the radical mechanism.[31]

Later studies identified the influence of the solvent on the final structure of the polymers. As such, Newmark and Majumdar synthesized polymyrcene with *sec*-BuLi, in cyclohexane, to give a polymer containing around 85% of 1,4-*cis* units, along with ca. 15% of 3,4-defects, in this case

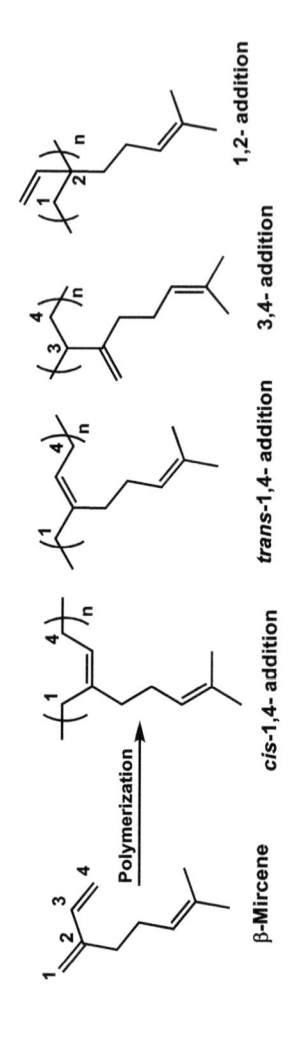

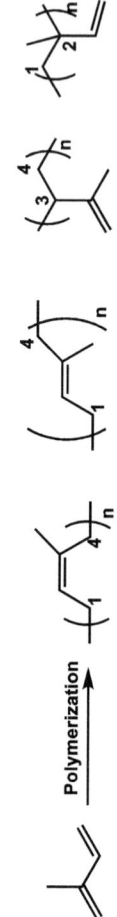

Fig. 4 Possible microstructures depending on the polymerization regioselectivity. Top: poly-β-myrcene. Bottom: polyisoprene.

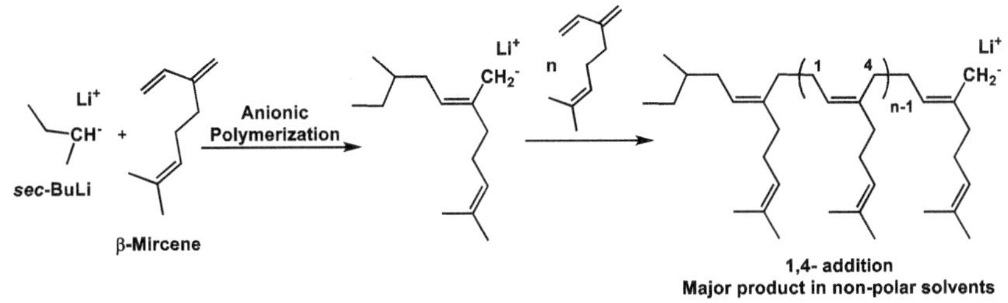

Scheme 1 Myrcene anionic polymerization. *Adapted from Ávila-Ortega A, Aguilar-Vega M, Loría Bastarrachea MI, Carrera-Figueiras C, Campos-Covarrubias M. Anionic synthesis of amine ω-terminated β-myrcene polymers. J Polym Res 2015;22:226, with permission from Springer Nature.*

analysis of the microstructure was performed in a more reliable way using NMR spectroscopy (Scheme 1).[38] When the polymerization was performed in a polar solvent such as THF a higher presence of 3,4-units (39%–44%) was observed. Further studies also had the same outcome, a high content of 1,4-units was obtained when polymerized in non-polar solvents.[39–41] This behavior has also been detected during the synthesis of polyisoprene and polybutadiene. Recent kinetic studies have evidenced that the incorporation mode is 4,1-*trans* and 4,1-*cis*, as the carbanionic chain end attacks preferentially the 3,4 double bond since it is more reactive.[39] The temperature or the initiator concentration can also influence the 1,4 and 3,4 ratio, with an increase of either of these two factors leading to a decrease in the amount of the 1,4-units.[30] In the myrcene polymerization, the formation of 1,2-units was observed to a very minor extent, which can be attributed to the steric hindrance of the double bond. These double bonds of the alkyl side chains remain inactive through the polymerization. The poly-β-myrcenes obtained by anionic polymerization with alkyl lithium compounds show molecular weights in the range 5000–30,000 Da and good to moderate molecular-weight dispersity ($Đ_M = M_w/M_n$ in the range 1.1–1.6).[42]

Copolymers with styrene or isoprene have also been prepared *via* anionic copolymerization, mainly to give block-like copolymers. These materials are very interesting as precursors for bio-based thermoplastic elastomers or multiblock architectures.[43,44]

Cationic polymerization can also afford poly-β-myrcene, as reported by Marvel in 1960. In their studies, $BF_3 \cdot OEt_2$ was used as Lewis acid, and the polymerization took place at low temperature (from −78 °C to 0 °C).[31] The structure obtained differs from the one generated by radical polymerization. Although the structure is not clearly described, the analysis performed showed that the products only had one double bond per monomer unit, which would imply that a cyclopolymerization *via* the formation of a cyclohexene ring had taken place. Alloocimene can also be polymerized using BF$_3$·OEt$_2$ as catalyst. Other metal halides such as AlCl$_3$ or SnCl$_4$, have been reported as active catalysts for the cationic polymerization.[45]

In addition, Ziegler-type catalyst have proven to be effective for the polymerization of this monomer, in particular Al(^{i}Bu)$_3$ combined with TiCl$_4$ or VCl$_4$, resulting in predominantly 1,4-structures. The IR spectra are nearly identical to the ones obtained in free radical polymerization.[31]

Ionic catalysis in aqueous media has also been recently described using a new type of catalyst approach termed "Lewis Acid Surfactant combined Complexes" (LASC).[46] Here, an Yb catalyst bearing dodecylbenzenesulfonate ligands initiates the cationic polymerization of isoprene, styrene and myrcene (Scheme 2). The polymerization takes place *via* a cationic mechanism,

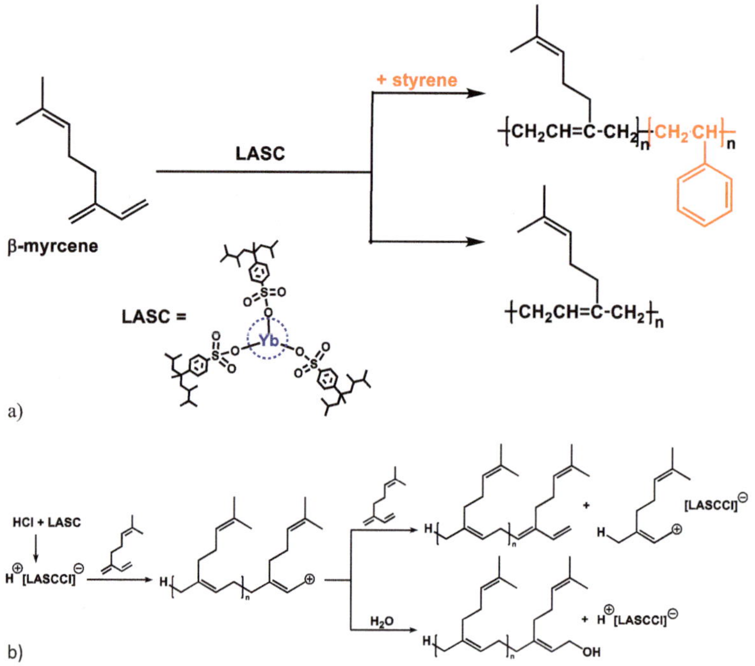

Scheme 2 (A) β-Myrcene polymerization with a LASC catalyst in aqueous media. (B) Reactions inside of the β-myrcene monomer droplets in the LASC-mediated polymerization. *Adapted from Hulnik MI, Vasilenko IV, Radchenko AV, Peruch F, Ganachaud F, Kostjuk SV. Aqueous cationic homo- and co-polymerizations of β-myrcene and styrene: a green route toward terpene-based rubbery polymers. Polym Chem 2018;9:5690–5700, with permission from The Royal Society of Chemistry.*

inside the monomer droplets present in the media (Scheme 2B). The poly-β-myrcene obtained shows a predominant 1,4-microstructure, with almost equal amount of *cis* (ca. 43%) and *trans* (ca. 50%) units and high molecular weights ($M_n > 100$, kDa) although the dispersity was not good (M_w/M_n range 2.3–4.1). The copolymerization of myrcene and styrene was also achieved but resulted in random copolymers.

Another strategy for the polymerization that leads to a good control of the stereoregularity is a coordination polymerization using metal complexes as catalysts. The first example was reported by Zinck,[32] where the coordination polymerization of myrcene using a Nd borohydride-based, [Nd(BH$_4$)$_3$(THF)$_3$], was achieved to give poly-β-myrcene with high selectivity (Scheme 3A). This catalyst had previously shown to be active in the polymerization of isoprene, as in that case, stoichiometric amounts of Mg^{n}BuEt are needed in the reaction media to attain myrcene polymerization. A very good stereoselectivity was achieved and mainly *cis*-1,4 polymyrcene (up to 90%) was isolated. Even though the data seems to indicate a *trans*-polymerization, which is thermodynamically preferred, the isolated polymer is *cis*-regular. The analysis of the mechanism indicates that the *anti-syn* isomerization is directed toward the *cis*-isomer, which could be caused by the coordination of the side double bond in the growing polymer chain (Scheme 3B).

Interestingly, when an excess of Mg^{n}BuEt was present, a narrowing of the molecular-weight dispersity (Đ$_M$) values together with a decrease in the molecular weight is observed which may indicate an effective Nd/Mg transfer process. As well, the selectivity changes and a higher number of 3,4-defects appear. These results would agree with the propagation taking place not only on the Nd but also on the Mg, which would be acting as a transfer agent in a kind of chain shuttling polymerization. This type of polymerization has been named *Coordinative Chain Transfer Polymerization* (CCTP),[25] and implies the transfer of the polymeric chain from the transition metal catalysts to a *Chain Transfer Agent* (CTA), usually a main group

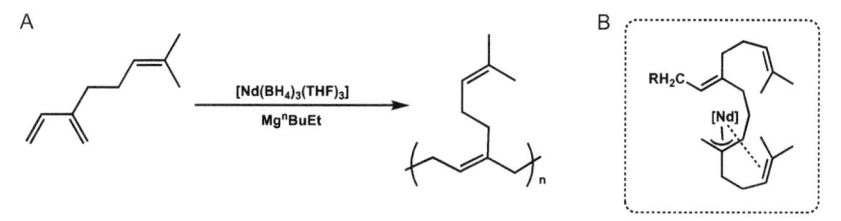

Scheme 3 (A) Polymerization of β-myrcene with [Nd(BH$_4$)$_3$(THF)$_3$]/Mg^{n}BuEt in toluene (B) Proposed active species bearing an additional coordination.[32]

metal alkyl. The chain transfer during the process has to be reversible and faster than the propagation and other chain-end processes. The advantage of the CCTP is that allows the growth of several polymer chains per catalyst molecule which improves the overall efficiency of the process.

Further studies were made with a metallocene catalyst generated *in situ*, [Nd(BH$_4$)$_2$(Cp*)(THF)$_2$] (Cp* = 1,2,3,4,5-pentamethylcyclopentadienyl), in this case an even better *cis*-stereoregularity was observed (up to 98.5%). The same authors also explored the use of the borohydride pre-catalyst activated by a boron derivative and using Al^{i}Bu$_3$ as alkylating agent. In this case although the polymer formed had higher molecular weight, the dispersity and the stereoregularity was negatively impacted and some cross-linking was apparent.[32,47–49]

Using this CCTP strategy, it is also possible to prepare myrcene copolymers and terpolymers with styrene and isoprene. The polymerization has been achieved with [LaCp*(BH$_4$)$_2$(THF)$_2$] combined with magnesium dialkyl and aluminum dialkyl compounds (Scheme 4). The polymers obtained showed a high degree of stereoregularity. When one equivalent of magnesium dialkyl was used, the formation of highly stereoregular poly(1,4-*trans*-myrcene-*co*-styrene), poly(1,4-*trans*-myrcene-co-1,4-transisoprene) copolymers and poly(1,4-*trans*-myrcene-co-1,4-transisoprene-co-styrene) terpolymers was attained. The addition of Al(^{i}Bu)$_3$ influenced the percentage of comonomers in the copolymer or terpolymer as well as the microstructure, so by controlling the CTA added it was possible to tune the microstructure.[47]

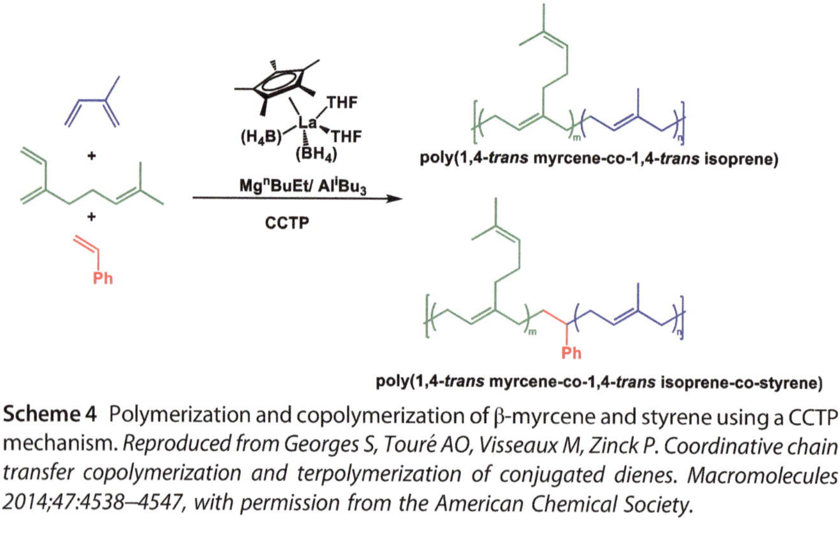

Scheme 4 Polymerization and copolymerization of β-myrcene and styrene using a CCTP mechanism. *Reproduced from Georges S, Touré AO, Visseaux M, Zinck P. Coordinative chain transfer copolymerization and terpolymerization of conjugated dienes. Macromolecules 2014;47:4538–4547, with permission from the American Chemical Society.*

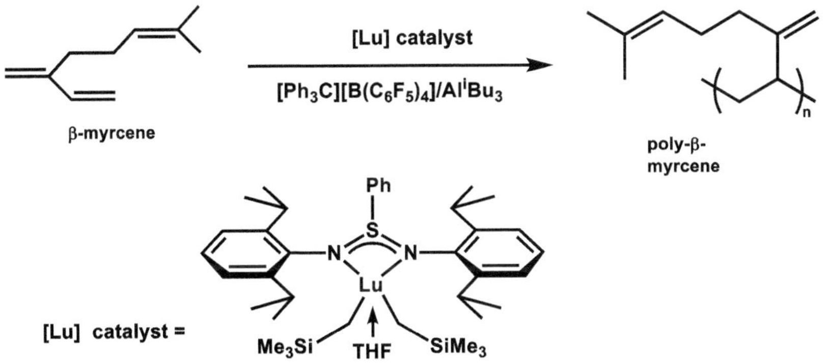

Scheme 5 Isoselective polymerization of myrcene with a lutetium dialkyl complex. Adapted from Liu B, Li L, Sun G, Liu D, Li S, Cui D. Isoselective 3,4-(co)polymerization of bio-renewable myrcene using NSN-ligated rare-earth metal precursor: an approach to a new elastomer. Chem Commun. 2015;51:1039-1041, with permission from The Royal Society of Chemistry.

More recently, Lu and Cu have reported a very effective system with a lutetium dialkyl complex bearing a bidentate β-diimidosulfonate ligand that exhibited perfect 3,4-regio- and isospecific stereoselectivity that lead to the formation of an isotactic 3,4-polymyrcene,[50a] this catalyst had also been shown to exhibit similar high selectivity for isoprene polymerization.[50b] The pre-catalyst is activated with [Ph$_3$C][B(C$_6$F$_5$)$_4$] and AlBu$_3$ to *in situ* generate a cationic β-diimidosulfonate lutetium alkyl complex (Scheme 5). The polymyrcene obtained present perfect 3,4-regio- (>99%) and isospecific stereoselectivity, and the control was further improved when the polymerization was performed at low temperatures.

The alkyl aluminum species used influence the stereoregularity of the compounds obtained, and when moving from Al^{i}Bu$_3$ to AlEt$_3$ and AlMe$_3$, a reduction of the 3,4-content was observed, likely due to the formation of different active species. As well, a marked drop in molecular weight is observed which can be attributed to the strong chain transfer characteristics of AlEt$_3$ and AlMe$_3$.[51,52] Interestingly, a clear effect of the aluminum *co*-initiator on the stereocontrol of the process has also been observed for the polymerization of isoprene with an yttrium imino complex, [Y(NCNdipp)(*o*-CH$_2$C$_6$H$_4$NMe$_2$)$_2$] (NCNdipp = PhC(NC$_6$H$_4^i$Pr$_2$-2,6)$_2$). After activation with [Ph$_3$C][B(C$_6$F$_5$)$_4$], the addition of AlMe$_3$ provoked the switching from being 3,4-isospecific to 1,4-*cis* selective.[53] Hence in this type of polymerization the choice of the CTA ratio or the nature of the alkyl groups has a strong effect in the resultant polymer. This lutetium catalyst is also active for the copolymerization of myrcene and isoprene, in this case isoprene–myrcene di-block and random copolymers are generated.

Further, Carpentier has reported a very active lanthanidocene catalyst for styrene polymerization that can produce copolymers with myrcene which are highly syndiotactic on the styrene part of the chain.[54] This catalyst is also able to homopolymerize myrcene in combination with Mg^nBu_2 as scavenger, although this results in poor activity and control of the microstructure (Scheme 6).

Very active iron(II) iminopyridine complexes which can polymerize isoprene as well as myrcene have also been reported. In this case, the regioselectivity towards the 1,4-addition is high, besides, the formation of *cis* or *trans* can be regulated by the modification of the ligand substituents (Scheme 7). As such, the imine moiety in the ligand controls the stereo and regioselectivity of the monomer insertion, when supermesityl substituents are present the polymer formed shows a majority of *cis*-1,4 units.

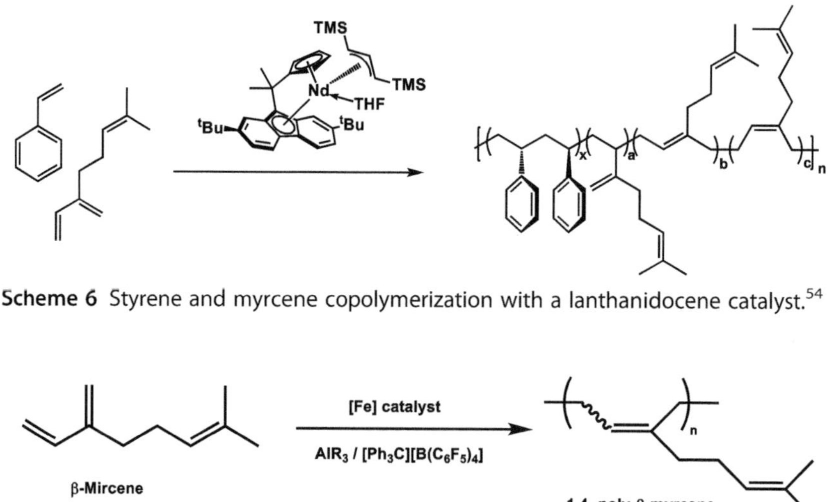

Scheme 6 Styrene and myrcene copolymerization with a lanthanidocene catalyst.[54]

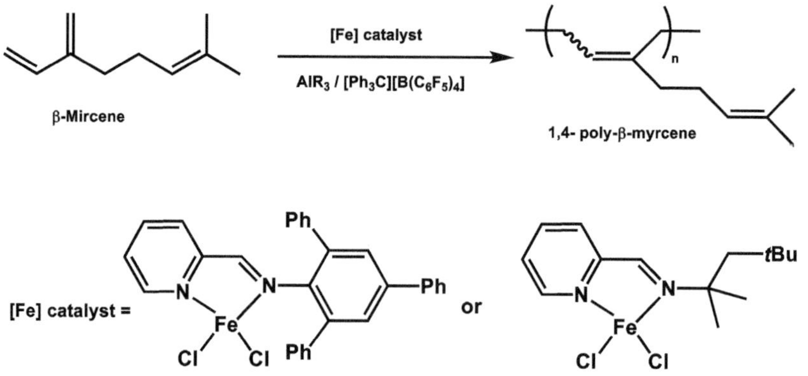

Scheme 7 β-Myrcene polymerization to give 1,4-poly-β-myrcene using as catalysts an iron(II) iminopyridine complexes.[49]

Although, the origin of the selectivity could not be determined due to the lack of isolated intermediates, and the diene coordination (σ-*cis* or σ-*trans*), migratory insertion into an η^2- or η^4-coordinated diene, and σ-π-σ rearrangements of the iron–allyl complexes could be significant. For the compounds to be active, the presence of an alkylating agent (trisisobutylaluminum or triethylaluminum) and a dealkylating reagent (Ph$_3$CB(C$_6$F$_5$)$_4$, trityl BArF) in an aprotic reaction media are needed. The alkyl aluminum reagents may replace the chloride ligands while the trityl BArF abstracts one of the alkyl groups to generate the active iron catalyst. The resultant polymers present high molecular weights, >200,000 Da and controlled Đ$_M$ and can behave as elastomers.[49]

Another example where the stereoselectivity was influenced by the ligands has been described by Capacchione (Fig. 5).[55] In this work, titanium catalysts that polymerize myrcene when activated by MAO were reported. From the two types of compounds studied, higher molecular weights are achieved (>100,000 Da) with the titanocene catalysts. The stereoselectivity observed depends on the ancillary ligand and for the titanocene compound the polymer generated showed a prevalent 1,4-*cis* microstructure (92%) while with the [OSSO]-type catalyst the polymers obtained were predominately 1,4-*trans*. Interestingly, the catalyst with the bulkier [OSSO] ligand gives a higher quantity of 3,4-units, which could be due to the tendency of the more sterically demanding metal center favoring the 3,4-insertion of the monomer. As well, the [OSSO]-type catalyst is able to promote the copolymerization with styrene to give multiblock copolymers, with very good control of the structure of the styrene segments which are isotactic.

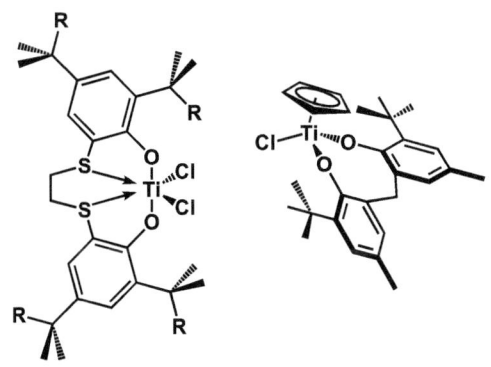

Fig. 5 Titanium catalysts for myrcene polymerization.[55]

2.2 Polymerization of cyclic terpenes

Pinene and limonene are the most popular cyclic terpenes studied for polymerization (Fig. 1). In this case, the anionic polymerization is not appropriated as there are not two conjugated double bonds. Since these cyclic monomers present a tertiary carbon that can give a stable carbocations, a cationic mechanism has been more frequently studied, as well as the radical one. However, polymers with low molecular weights, frequently <5000 Da, and with diverse stereochemistry are most often generated, in processes difficult to control.[1] These cyclic molecules present stereocenters, so the stereoregularity in the polymerization process can significantly affect the final properties of the polymers, hence, to find catalysts able to provide a good control over the polymerization is greatly desired.

2.2.1 Pinene polymerization

The polymerization of pinene leads to polymers that are considered terpene resins and have found uses as thermoplastics, pressure sensitive adhesives, coatings or tackifier additives. Although they are more expensive than petrol based resins, their unique properties justify their higher cost.[56,57] Even though α-pinene is more abundant, β-pinene has been the most studied due to the higher reactivity of the exocyclic double bond.[58] Both monomers can be polymerized *via* a cationic mechanism, usually with Lewis acids, such as metal chlorides, as initiators.[58–60] The activity is clearly related to the strength of the Lewis acid used, and the effectiveness decreases in the order: $AlBr_3 > AlCl_3 > ZrCl_4 > AlCl_3 \cdot Et_3O > BF_3 > BF_3 \cdot Et_2O > SnCl_4 > SbCl_3 > BiCl_3 > ZnCl_2$.[58] In this process, the presence of adventitious water has been considered necessary for catalysis to take place (Scheme 8). In a first step, the reaction between the Lewis acid and H_2O occurs and the $H[MX_nOH]$ species are generated. This species transfers the proton to the monomer to form a cyclic carbocation that isomerizes to give the *p*-menthenile cation. The formation of this less encumbered carbenium ion by ring opening isomerization is a key part of the β-pinene cationic polymerization as this species is responsible for the polymerization propagation.[12,61] The poly(β-pinene) formed has a structure of an alternating copolymer of isobutylene and cyclohexene.

For α-pinene, the isomerization of the carbocation initially generated can give two distinct carbenium ions able to be involved in the propagation of the chain; hence, two different polymers can be obtained.[59] Similar Lewis acid catalysts have been explored for α-pinene as for the β-isomer, and for a given initiator the yields and the molecular weights are poorer for α-pinene

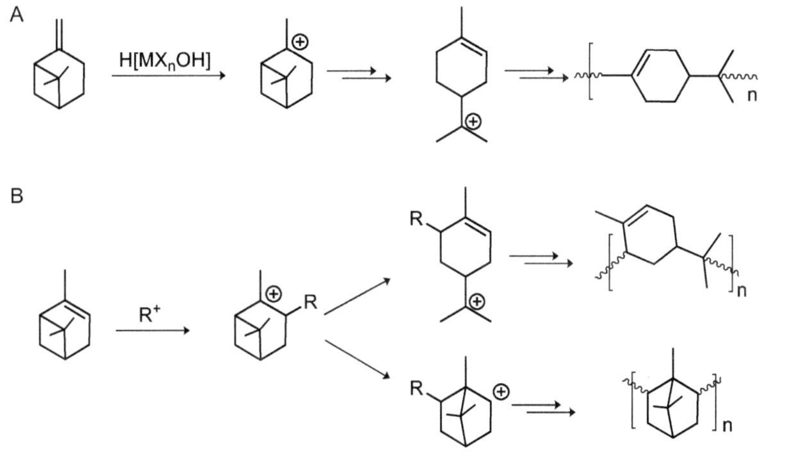

Scheme 8 Mechanism for the cationic polymerization of (A) β-pinene, (B) α-pinene.

than β-pinene. In order to achieve better values for the molecular weights, the polymerization has to be carried out at low temperatures (0–40 °C) and high concentrations of Lewis acid (1–4 mol%) are also required. An improvement in control of the processes had been attained using aluminum chlorides together with different additives such as SbCl$_3$, SnCl$_2$ and other group 14 chlorides. The presence of those agents slows down the process and avoids the formation of oligomers, hence leading to higher molecular weights than the AlCl$_3$·H$_2$O system alone.[59] The reason for this improvement may arise from the formation of a bulkier counter-anion than [AlCl$_3$OH]$^-$, then the generation of the secondary cyclic carbenium ion will be less favored, and one type of chain growth preferred.

There are significantly more studies performed for β-pinene polymerization to gain better controlled processes and higher molecular weights. One of the main reasons for the formation of low molecular weight polymers is the presence of chain transfer reactions, since those may lead to the deactivation of the propagating polymer chain-end, and at the same time the generation of species able to reinitiate polymerization in a cationic mechanism. To achieve a good control over the molecular weight, the chain transfer is therefore an undesirable side reaction. An approach to limit these side reactions is to develop living polymerization processes where chain transfer reactions are not detected on a laboratory timescale.[62]

The first report of living cationic isomerization polymerization of β-pinene was described by Deng[63] using a Lewis acidic titanium compound, [TiCl$_3$(O^{i}Pr)]. The presence of a HCl-2-chloroethyl vinyl ether adduct

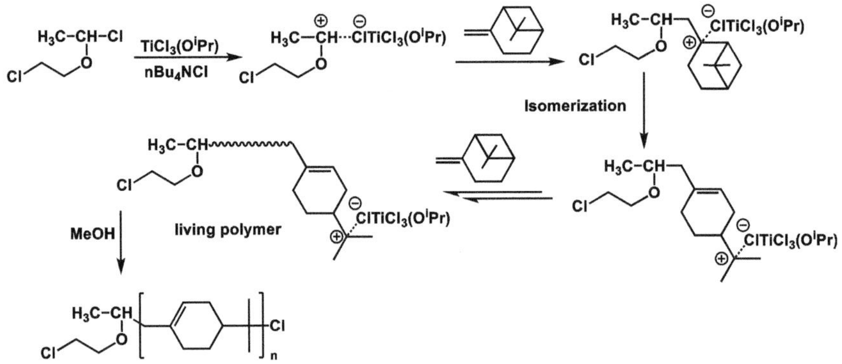

Scheme 9 Living cationic polymerization of β-pinene. *Adapted from Lu J, Kamigaito M, Sawamoto M, Higashimura T, Deng Y. Living cationic isomerization polymerization of β-Pinene. 1. Initiation with HCl – 2-chloroethyl vinyl ether adduct/TiCl₃(O'Pr) in conjunction with ⁿBu₄NCl. Macromolecules 1997;30:22–26, with permission from the American Chemical Society.*

[CH$_3$CH(OCH$_2$-CH$_2$Cl)Cl] and *tetra*-n-butylammonium chloride (ⁿBu$_4$NCl) was necessary, with the reaction being performed in CH$_2$Cl$_2$ at low temperature (Scheme 9). This titanium derivative is also active for living cationic polymerization of styrene, which suggests some similarity for the reactivity of pinene and styrene.

In this process, only one living polymer chain grows in the presence of [TiCl$_3$(O^{i}Pr)] and Bu$_4$NCl. As shown in Scheme 9, the C-Cl bond in the HCl-vinyl ether adduct is activated by the Lewis acid [TiCl$_3$(O^{i}Pr)] and a vinyl ether cation is generated. This cation adds to the *exo*-methylene double bond in β-pinene to give a sterically hindered cation. The isomerization of this strained cation takes place through the ring opening of the cyclobutane fragment and a stable tertiary cation, structurally similar to the one derived from isobutene, is formed. This cation can now be involved in the propagation of the chain growth. The poly(β-pinene) produced had controlled molecular weights (up to 4000 Da), narrow molecular-weight dispersities (~1.3) and controlled chain-end groups.

Later, Kamigaito accomplished the production of polymers with higher molecular weights (up to 40,000 Da) in a living cationic polymerization processes using aluminum chloroalkyl derivatives, AlEtCl$_2$ or AlEt$_{1.5}$Cl$_{1.5}$, at low temperature (Scheme 10).[64] In this case, the judicious choice of the Lewis acid and also the solvent lead to a better control of the chain growth. The same authors also published a controlled living cationic polymerization using an initiation system that consisted of a chloro alkyl (RCl), a Lewis acid

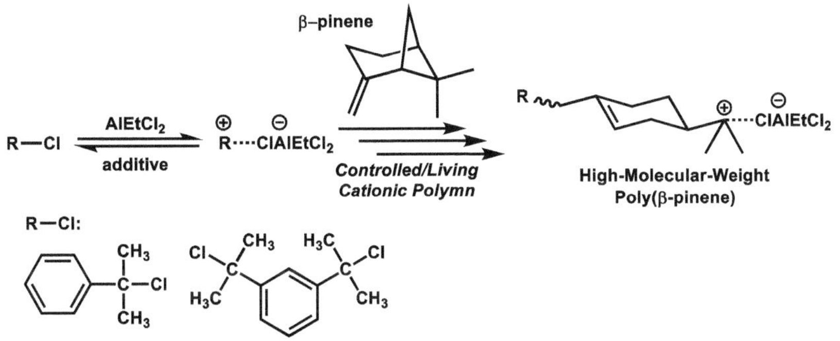

Scheme 10 Controlled living cationic polymerization of β-pinene with aluminum alkyl catalysts. *Adapted from Satoh K, Nakahara A, Mukunoki K, Sugiyama H, Saito H, Kamigaito M. Sustainable cycloolefin polymer from pine tree oil for optoelectronics material: living cationic polymerization of β-pinene and catalytic hydrogenation of high-molecular-weight hydrogenated poly(β-pinene). Polym Chem 2014;5:3222–3230, with permission from The Royal Society of Chemistry.*

and a base, and combined it with an incremental monomer addition technique. The RCl/EtAlCl$_2$/Et$_2$O catalytic system led to higher molecular weights, of up to 100 kDa. The effect of the presence of weak Lewis bases, such as Et$_2$O was to retard the reaction and hence to achieve a more controlled polymerization.[65]

The need of using additives to modulate the reactivity of the Lewis acids comes from the fact that strong Lewis acids are needed to achieve the polymerization of these monomers since they are not very active. As such, to accomplish the cationic polymerization of monomers with low reactivity, highly active Lewis acid catalysts must be used, however, to perform a controlled polymerization an appropriate dormant–active equilibrium should be induced.[62,66] To achieve this, an initiating system where the catalytic activity can be finely modulated is needed. Such a system may consist of a Lewis acid and a weak Lewis base, in fact, base-assisted initiating systems that combines a metal halide and an ester or an ether have proven to be efficient for the living cationic polymerization of a series of vinyl ethers and styrene derivatives.[67]

For β-pinene, the use of ethers as additives to access higher molecular weights had been reported by Kostjuk.[20] This report observed that the modification of the aluminum ligands environment gives a more controlled polymerization and polymers with good physical and optical properties, and relatively high molecular weights are attained. In the combined action of AlCl$_3$ with a weak base such as an ether Ph$_2$O, Bu$_2$O or EtOAc the chain

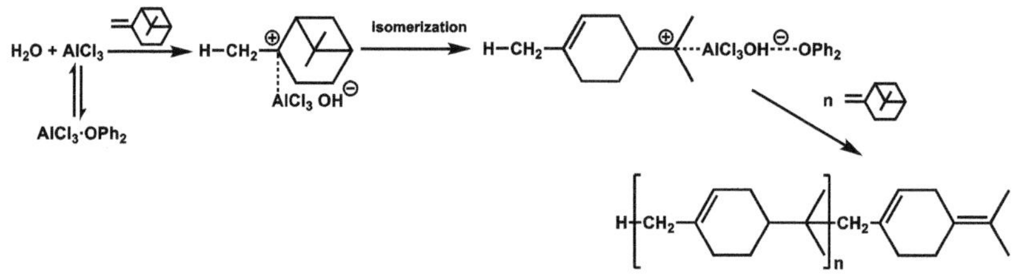

Scheme 11 Cationic polymerization of β-pinene using ethers as additives. *Adapted from Kukhta NA, Vasilenko IV, Kostjuk SV. Room temperature cationic polymerization of β-pinene using modified AlCl3 catalyst: toward sustainable plastics from renewable biomass resources. Green Chem 2011;13:2362–2364, with permission from The Royal Society of Chemistry.*

grows closer to the metallic center and poly(β-pinene) with higher molecular weights are produced (ca. 14,000 Da) (Scheme 11). The *in situ* generation of a weakly nucleophilic counter-anion due to the interactions of [AlCl$_3$OH]$^-$ with the ether leads to the elimination of chain transfer reactions. The nature of the base used for the catalysts adduct AlCl$_3$·ED, ED = Ph$_2$O, Bu$_2$O or EtOAc, heavily influences the molecular weight values. If the base is too strong for the system, such as Bu$_2$O, polymers with lower molecular weights are obtained due to the β-H abstraction in the monomer by the free bases, as detected by NMR spectroscopy. When the base is weaker, e.g. Ph$_2$O, higher molecular weights are obtained because of the inertness of those bases toward the β-H abstraction. In this system, the molecular weight obtained did not depend on the temperature and lower catalytic loading was needed.

In addition, alkylbenzenes can be used as additives since they have very weak basicity and can form π-complexes with metal halides.[68–71] In 2015, Aoshima brought this idea to fruition and the combination of different alkylbenzenes with metal halides, such as GaCl$_3$, TiCl$_4$, and SnCl$_4$, as the Lewis acid catalyst in conjunction with 2-chloro-2,4,4-trimethylpentane (TMPCl), proved to be an active system for the controlled cationic polymerization of β-pinene.[72] From the three Lewis acids studied, the GaCl$_3$ gave the best activity and control over the polymerization, which took place at low temperature (−78 °C). In relation to the alkylbenzenes toluene, mesitylene, pentamethylbenzene (C$_6$HMe$_5$), and hexamethylbenzene (C$_6$Me$_6$) were tested, and the hexamethylbenzene, which has the highest basicity, gave the most controlled polymerization and produced polymers with narrow molecular-weight dispersity. Further, NMR analysis confirmed that the polymerization took place without side reactions to give a polymer with a main chain structure coming from the isomerization mechanism.

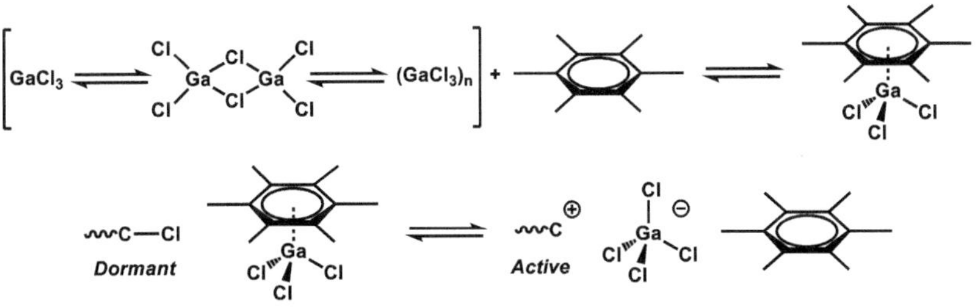

Scheme 12 Cationic polymerization of β-pinene using GaCl₃ and alkylbenzenes as additives. *Reproduced from Karasawa Y, Kimura M, Kanazawa A, Kanaoka S, Aoshima S. New initiating systems for cationic polymerization of plant-derived monomers: GaCl₃/alkylbenzene-induced controlled cationic polymerization of β-pinene. Polym J 2015;47:152–157, with permission from Springer Nature.*

In this controlled polymerization, the interaction between the GaCl₃ and the hexamethylbenzene to form a π-complex is probably responsible for the modification of the gallium Lewis acidity. Indeed, the GaCl₃·C₆Me₆ complex can be involved in a dormant-active equilibrium that leads to successful control over the reaction. However, the possibility that the alkylbenzene interacts with the propagating carbocation to eliminate the side reactions, like β-proton elimination, cannot be fully ruled out (Scheme 12).

Ziegler-type catalysts have also been explored for the polymerization of pinene, where catalytic systems combining ⁱBu₃Al or ⁱBu₂AlCl and TiCl₄ or VOCl₃, had been explored and the formation of low molecular weight polymers was attained, although the mechanism is unknown.[73] More recently, in 2007, relatively high molecular weight β-pinene polymers (~10,500 Da), were produced at 40°C using a salen nickel catalyst in combination with MAO (Fig. 6).[57] On their own, only the Ni compounds or the MAO did not result in polymerization, which suggest that the mechanism operates by a Ziegler-Natta type mechanism. However, the analysis of the obtained polymer revealed the same structure as when the cationic mechanism is used. Likely the initiation is promoted by the nickel catalyst cation formed by the reaction of the complex and MAO, and the propagation takes places *via* a conventional carbocationic polymerization. Further indication of the carbocationic polymerization was evidenced when quenching the polymerization with methanol at low monomer conversion which led to the incorporation of a methoxy end group on the polymer chain.[57]

Copolymers of both α- and β-pinene and styrene have also been produced *via* the cationic copolymerization with Lewis acids, such as titanium or aluminum compounds, in non-polar solvents at low temperatures.

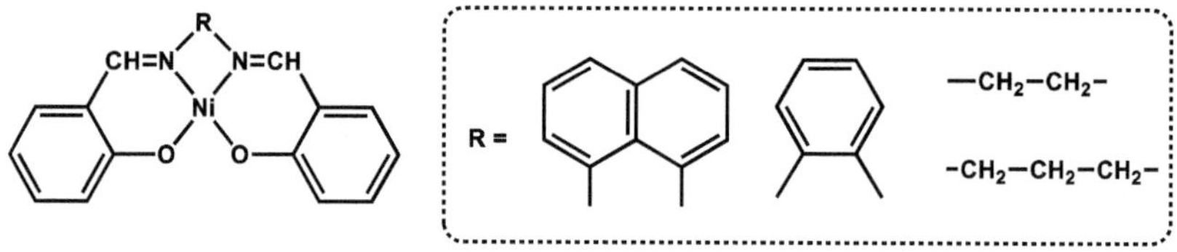

Fig. 6 Salen nickel catalysts for β-pinene polymerization.[57]

Various well-controlled copolymer architectures such as block, random, grafted and end-functionalized polymers have been described.[30,74,75]

2.2.2 Limonene polymerization

Limonene is one of the most important terpenes that can be found in various citrus fruits. In fact, it is the main component of the citrus oil and the estimated worldwide production is over 70,000 tons per year and demand is constantly increasing.[6,76] Limonene has found uses as a flavoring agent and also possesses anti-cancer, anti-inflammatory and immunomodulatory activities.[6,77,78] Although limonene has been considered an emerging platform chemical,[6,79] there are very few reports on its polymerization, and mostly *via* radical or cationic mechanisms.[80] Since it contains not very accessible and unconjugated C=C double bonds, the polymerization of limonene to give high molecular weight polymers is challenging. After initial studies more than 60 years ago, most of the works published since had been in the form of patents.

The cationic polymerization was studied in the 1950s using $TiCl_4$ or $BF_3 \cdot OEt_2$ as Lewis acid catalysts, where solid material was obtained but the structure of these polymers was not fully determined.[58] More recently, limonene has been polymerized using $AlCl_3$ *via* a cationic mechanism in toluene at 40–45 °C, to give low molecular weight polymers (~1000–1200 Da), its use as an antioxidant additive for food contact polymers such as High Impact Polystyrene (HIPS) has been explored.[81]

In 1965, Marvel studied limonene polymerization, using a Ziegler-type catalyst: $AlCl^iBu_2$ or Al^iBu_3 and $TiCl_4$, TiI_4 or $VOCl_3$.[73] The production of polylimonene was achieved although in a very slow rate and the polymers obtained displayed low molecular weights. The proposed mechanism is shown in Scheme 13. The polymerization was considered to take place *via* a cationic mechanism through the addition of the cationic propagating species to the exocyclic C=C double bond. The formation of more than one type of ring due to isomerization processes to give camphane or pinane-type units was also suggested based on the low quantity of unsaturated

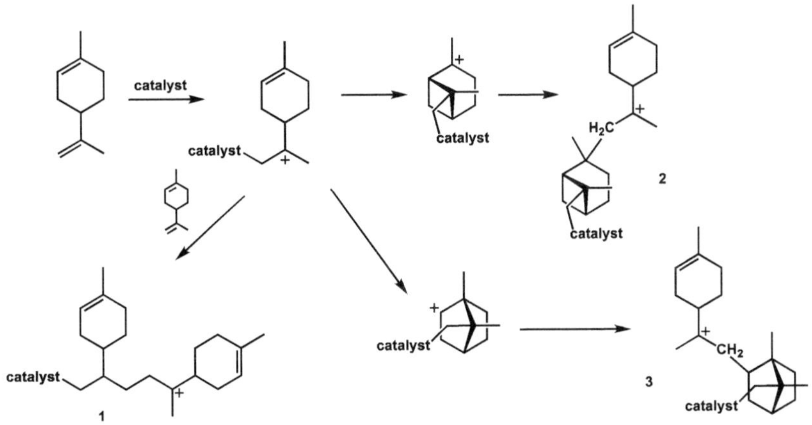

Polymerization of terpenes and terpenoids 75

Scheme 13 Limonene polymerization *via* a cationic mechanism.[73]

bonds in the final structure. In this mechanism, the bicyclic type of structures (2 and 3) are more favored than the monocyclic structures 1 (Scheme 13).

The copolymerization with various comonomers such as maleic anhydride, methyl methacrylate, styrene or acrylonitrile has also been reported.[82–84]

3. Polymerization of terpenoids

As mentioned, terpenoids are terpene derivatives with oxygenated functionalities. In terms of the natural occurring structures, the copolymerization of the aldehydes (1R)-(−)-myrtenal, (S)-(−)-perillaldehyde and β-cyclocitral with isobutyl vinyl ether *via* a cationic mechanism using the EtSO$_3$H/GaCl$_3$ as catalysts in the presence of a Lewis base (1,4-dioxane) has been reported (Scheme 14).[85] The copolymerization took place *via* 1,2-carbonyl addition of the aldehydes, displaying good control and alternating copolymers with molecular weights of up to 20,000 Da and narrow molecular-weight dispersities (1.2–1.4) were attained. The best controlled process were achieved for the myrtenal monomer which indicates that the side groups can have an effect on the control of the process, with a bulkier group preferred for controlled copolymerization.

As well, from some terpenoids it is possible to derive suitable monomers to produce terpene-based polymers, which is the case for carvone and menthol.[1] Carvone can be transformed into carvomenthone *via* a hydrogenation process and from it carvomenthide is generated through a Baeyer-Villiger oxidation. The Ring Opening Polymerization (ROP) of carvomenthide

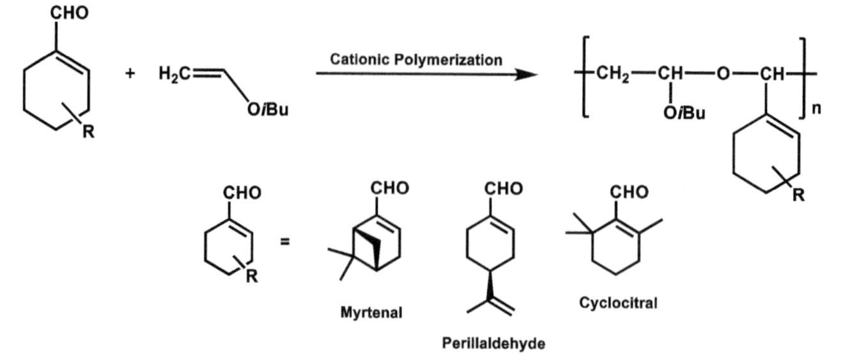

Scheme 14 Controlled copolymerization of natural (1R)-(−)-myrtenal, (S)-(−)-perillaldehyde, and β-cyclocitral with isobutyl vinyl ether *via* a cationic mechanism.[85]

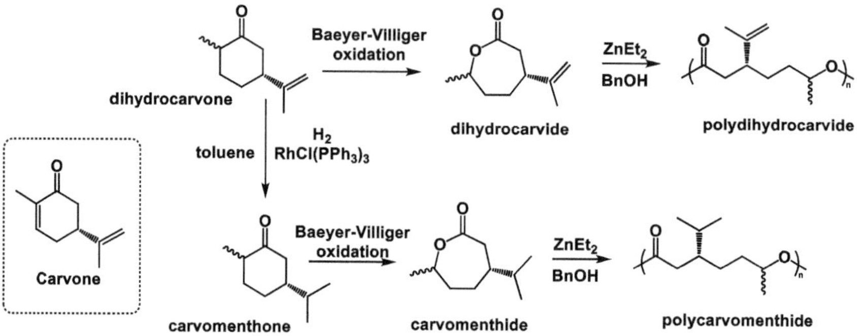

Scheme 15 Production of terpene-based polymers from carvone.[86]

leads to polymers with high molecular weights and low Đ$_M$ (62,000 Da and 1.16, respectively). The intermediate compounds in the hydrogenation process, the dihydrocarvone, can also be polymerized by ROP, although in this case polymers with lower molecular weights are obtained (Scheme 15). The copolymerization of both monomers has also been reported.[86] In addition, from the dihydrocarvone, Hillmyer prepared an epoxidized cyclic lactone that it a suitable monomer for copolymerization with caprolactone,[87] using SnOct$_2$ or ZnEt$_2$ as catalyst. Interestingly, depending on the comonomer ratio, polymers with shape memory properties could be produced.[87]

The ketone form of the menthol, menthone, is easily accessible, and through a Baeyer-Villiger oxidation with *m*-chloroperoxybenzoic acid the menthide lactone can be obtained (Scheme 16A).[88] This oxygenated cyclic monomer has been polymerized *via* ROP with a zinc alkoxide as catalyst at room temperature to give polymers with molecular weights up to 91,000 Da

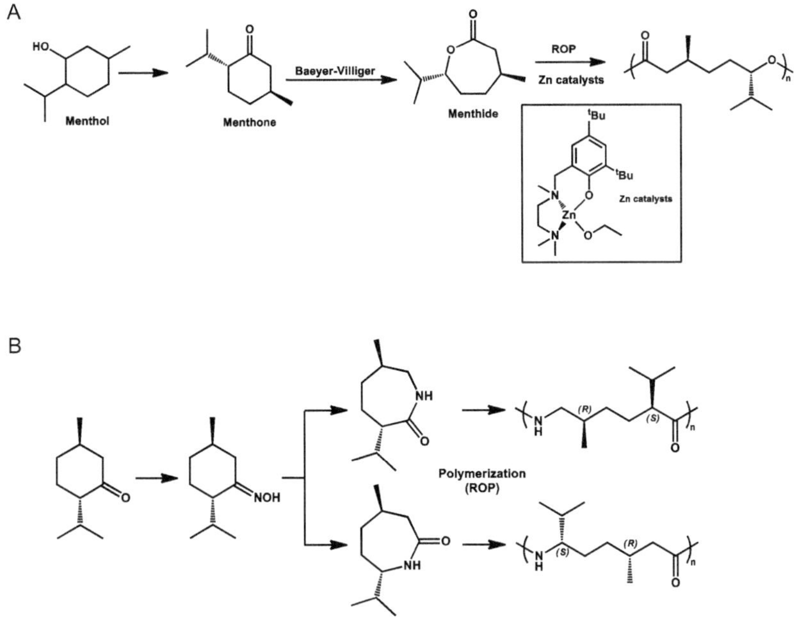

Scheme 16 Production of terpene-based polymers from menthone. (A) Polyesters, (B) Polyamides. *Panel (A) adapted from Zhang D, Hillmyer MA, Tolman WB. Catalytic polymerization of a cyclic ester derived from a "cool" natural precursor. Biomacromolecules 2005;6:2091–2095, with permission from The American Chemical Society. Panel (B) adapted from Winnacker M, Vagin S, Auer V, Rieger B. Synthesis of novel sustainable oligoamides via ring-opening polymerization of lactams based on (−)-menthone. Macromol Chem Phys 2014;215:1654–1660, with permission from Wiley-VCH.*

and moderate molecular-weight dispersities (<1.6) in a controlled polymerization. Another approach is to transform the menthone in lactames that can produce polyamides through a cationic ROP mechanism (Scheme 16B).[89]

A lactame can also be synthesized from β-pinene, as shown in Scheme 17, whereby in this case the polymerization took place *via* an anionic mechanism using NaH or KO^{t}Bu as catalyst and a benzoylated lactam as co-initiator.[90] The anionic polymerization proved to be a more convenient way to obtain these terpene-based polyamides than the cationic route. From α-pinene analogous lactones have been synthesized and again the anionic ROP polymerization leads to polyamides, in this case with molecular weights of up to 30,000 Da. The catalytic system consisted of NaH or potassium as catalyst and BzCl or Ac$_2$O as activator. In this reaction, it has been proposed that the *in situ* generated sodium or potassium lactamates react with the activator to give the corresponding imide that subsequent leads to the formation of the active species.[91]

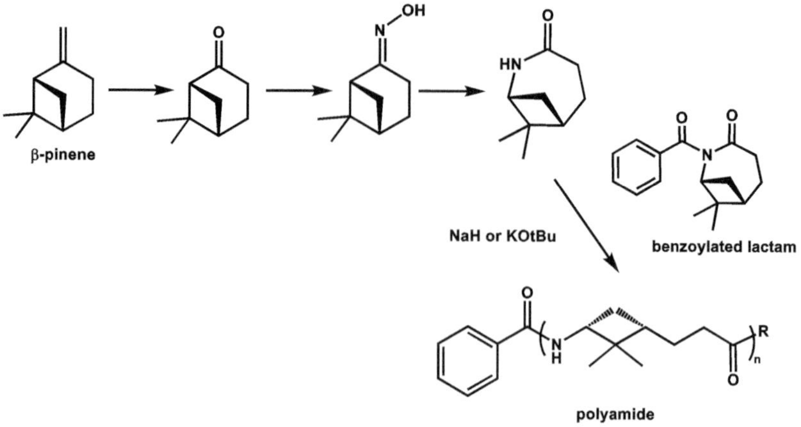

Scheme 17 Synthesis of terpene-based polyamides. *Adapted from Winnacker M, Sag J. Sustainable terpene-based polyamides via anionic polymerization of a pinene-derived lactam. Chem Commun 2018;54:841–844, with permission from The Royal Society of Chemistry.*

The transformation of pinene in other oxygenated cycles such as lactones has also been described; in this case, the cycle can be considered a bio-based version of the caprolactone (Scheme 18).[92] This caprolactone can be polymerized by ROP catalysis using SnOct₂ or ZnEt₂ and benzyl alcohol catalyst systems, where better results were achieved with the Zn catalyst. In addition, the copolymerization with lactide was explored using zirconium phenolate catalysts (Scheme 18) which have shown some heterotactic or isotactic bias in *rac*-LA polymerization. The copolymers produced had low molecular weights, moderate dispersities and low T_g.

Epoxides of terpenes have also great potential to be used in polymerization. In particular, limonene-1,2-oxide is a very interesting candidate to generate terpene-derived polymers and copolymers (Fig. 7).[93] Limonene oxide is easy accessible through the oxidation of limonene and can also be directly produced in biorefinery processes.[94] This unit presents two functional groups that can be polymerized, the C=C double bond and the oxirane. When the polymerization takes place *via* the oxirane a polyether can be generated *via* a ROP process. However, to achieve the ROP of limonene-1,2-oxide the use of efficient catalysts is required to overcome the kinetic barrier for its activation, which is higher than for terminal epoxides, since limonene-1,2-oxide displays an internal, trisubstituted epoxide.[95]

This monomer is quite reluctant to be directly polymerized and there are actually few catalysts able to do so. In 2012, Park described the photoinitiated

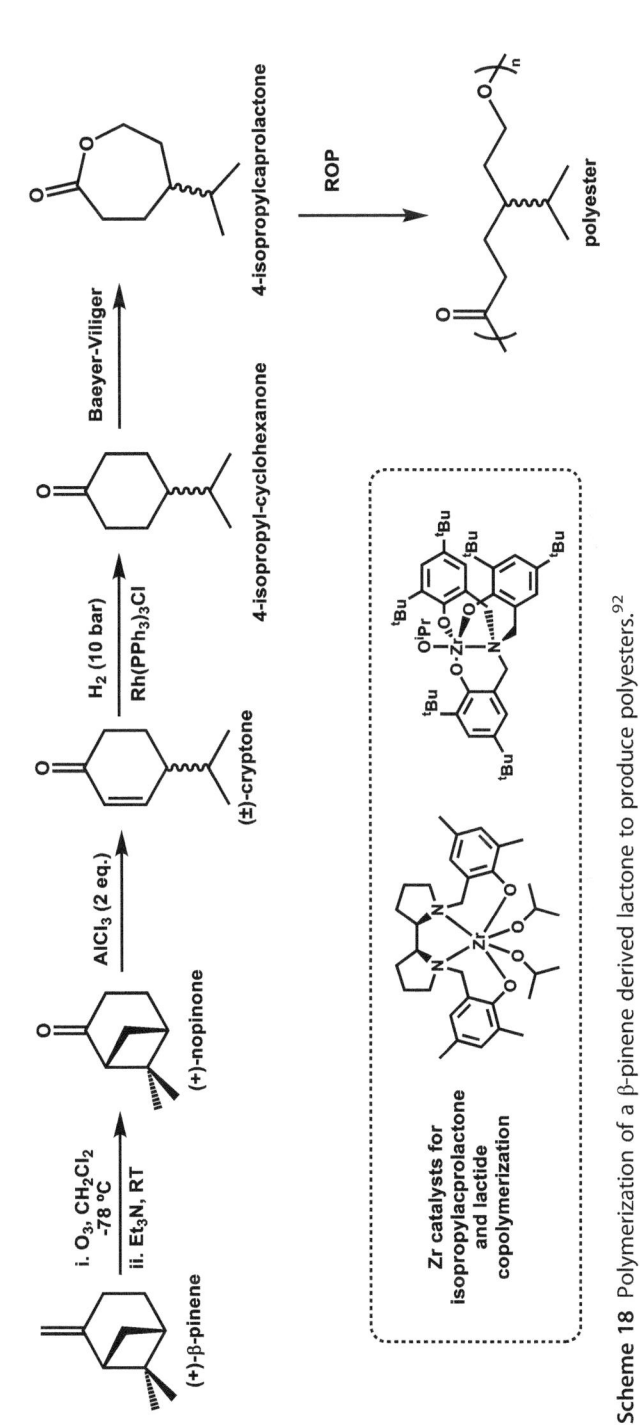

Scheme 18 Polymerization of a β-pinene derived lactone to produce polyesters.[92]

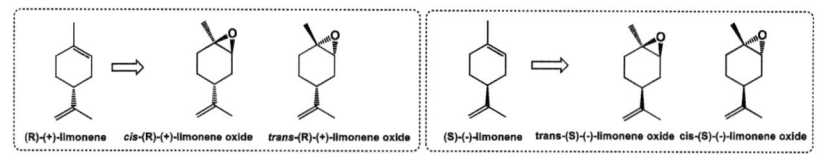

Fig. 7 Isomers of limonene and limonene oxide.

Scheme 19 ROP polymerization of limonene-1,2-oxide with [AlMeCl(OR)] (OR = 2,6-(CHPh$_2$)$_2$-4-^{t}Bu-C$_6$H$_2$O) as catalysts.[97]

cationic ring-opening polymerizations of limonene-1,2-oxide using diaryliodonium salts or triarylsufonium salts as photoinitiators. It should be noted that α-pinene oxide could also be polymerized in this way. However, the polymerization control was poor, secondary and cross-linking reactions were detected, and the products obtained had very low molecular weights and high disperstities.[96] Another example of homopolymerization has been recently reported by ourselves, using a very active aluminum catalyst for ROP, [AlMeCl(OR)] (OR = 2,6-(CHPh$_2$)$_2$-4-^{t}Bu-C$_6$H$_2$O) (Scheme 19).[98] In this case, without the presence of any other initiator or activator, it was possible to efficiently polymerize (+)-limonene oxide in only 30 min. In the reaction, there was a clear preference for *cis*-isomer polymerization. The bio-based polylimonene ether (PLO) produced had good thermal properties, low molecular weights (ca. 1300 Da), and moderate dispersities (1.37–1.42). The limitation in the molecular weight obtained could be attributed to the presence of chain transfer agents (CTA) arising from side reactions. This work also studied the polymers potential as a green additive for other bio-based polymers such as PLA. Indeed, addition of only 10 wt%, led to the modification and improvement of PLA properties in terms of flexibility, thermal stability, and hydrophobicity.[97]

4. Copolymerization of terpenoids

Although the number of reports for the limonene oxide homopolymerization is low, there has been significantly efforts devoted to its copolymerization, particularly with anhydrides and carbon dioxide (CO_2) to produce polyesters and polycarbonates, respectively, particularly over the last 5 years.[99,100]

4.1 Polyesters

The copolymerization with anhydrides is of great interest since it leads to a variety of terpene-based polyesters that could not be achieved by the usual ROP of cyclic esters. With this strategy it would be possible to prepare semi-aromatic or fully-aliphatic polyesters with a large diversity of compositions. The production of polyesters from the copolymerization of anhydrides and epoxides has been described for a significant number of commercial epoxides such as propylene and cyclohexene, among others,[101–107] however, for 1,2-limonene oxide very few examples have been reported.[99,108–113] In those reports, catalysts based on aluminum, zinc, chromium, manganese and cobalt, have been explored for the copolymerization of limonene oxide and various anhydrides (phthalic, naphthalene, succinic, diglycolic and maleic) as highlighted in Scheme 20.

The first 1,2-limonene oxide co-polymerization with anhydrides, diglycolic and maleic, was reported by Coates in 2007 using a highly active diiminate catalysts (Scheme 20), which gave alternating co-polymers with molecular weights up to 36,000 Da and good molecular-weight dispersities (1.1–1.2) when the *trans*-R-limonene oxide isomer was used.[108] Later on, Thomas described a catalytic system composed of a commercial salen catalyst with Al or Co and using *bis*(triphenylphosphine)iminium chloride (PPNCl) as co-catalyst, which was able to achieve the copolymerization of 1,2-limonene oxide and camphoric anhydride to give molecular weights of up to 27,000 Da and narrow molecular-weight dispersities. In this case the monomer used was a mixture of *cis*-and *trans*-R-(+)-limonene oxide. Additionally, this system was able to achieve the co-polymerization of pinene oxide and glutaric anhydride.[109] DuChateau used a similar catalytic system for the copolymerization of 1,2-limonene oxide with phthalic anhydride, using Cr, Al, Mn or Co salophen complexes as catalysts (Scheme 20). The co-catalysts explored were DMAP and PPNCl, where the best results

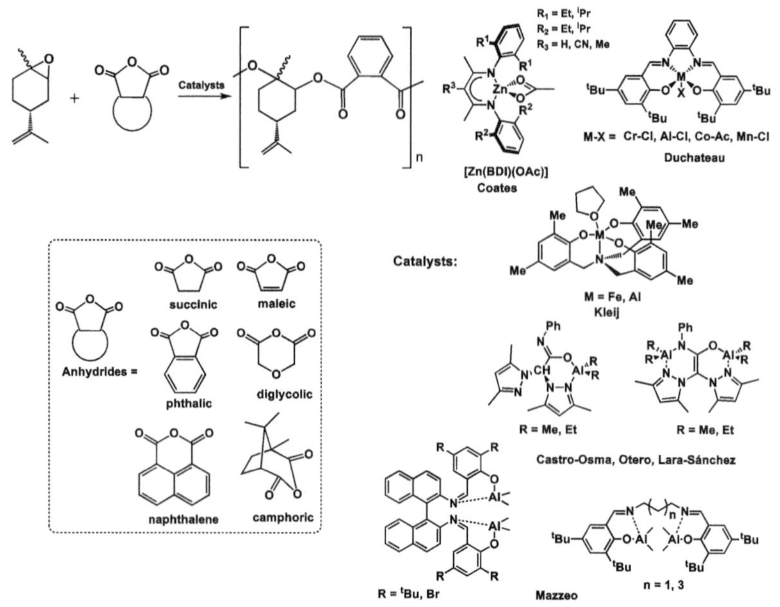

Scheme 20 Copolymerization of 1,2-limonene oxide with anhydrides. Inset: Anhydrides studied.

were obtained with the system formed by the chromium complex and PPNCl as co-catalyst at 130 °C in bulk. The polyesters produced had molecular weights of up 9700 Da and dispersities of 1.4.[110] The low molecular weights obtained were attributed to the existence of side reactions such as transesterifications, irreversible chain terminations or partial catalysts deactivation. The authors also explored the influence of several protic compounds (i.e., alcohols, carboxylic acids, water) as chain transfer agents. In the presence of these agents, a reduction in the molecular weight was observed while the $Đ_M$ remained low, which is in agreement with an immortal catalytic system where the chain transfer is fast in relation to the chain growth.[114] More recently Mazzeo published a study on ROCOP of 1,2-limonene oxide and phthalic anhydride using a bimetallic aluminum salen catalyst in combination with DMAP. The obtained co-polymers presented molecular weight distributions from 5700 to 9200 Da and narrow $Đ_M$'s. Again, the molecular weights were lower than the theoretical values for a living catalytic system which was attributed to the presence of protic impurities.[111,112] In 2018, Lara-Sánchez reported the copolymerization of succinic, maleic and phthalic anhydrides with 1,2-limonene oxide using a catalytic system containing scorpionate aluminum catalysts combined with

TBAB as catalyst. The reaction took place at 80 °C in toluene and the 1,2-limonene oxide conversion was around 50% and resulted in polymers with low molecular weights (up to 3000 Da) and moderate molecular-weight dispersities (~1.36), where again the presence of chain transfer reactions is considered as the reason for lower molecular weights than those expected.[113]

In these aforementioned studies, there was however no mention of the preference for the polymerization of either the *cis* or *trans* isomers of the R-limonene used. In the work reported by Kleij, using a very efficient catalytic system formed by an Fe(III)-based complex and PPNCl as co-catalyst, the authors explored the reaction with the commercial *cis/trans* mixture and just the *cis* isomer.[99] In both cases the *co*-polymerization took place but co-polymers of superior grade, higher molecular weight (up to 16,400 Da) and also higher T_g (141 °C) were obtained when the diastereoisomerically pure *cis*-limonene oxide monomer was used. The amino trisphenolate Fe(III) catalyst described, in combination with PPNCl as co-catalysts, was also interestingly able to achieve the polymerization of other terpenes oxides such as carene oxide, limonene dioxide, and menthene oxide.[99]

Another approach to achieve terpene-based polyesters is the use of terpene-derived anhydrides, which has been explored by both Coates and Kleij using chromium, cobalt and aluminum salen or iron trisphenolate complexes as catalysts to give polyesters with molecular weights up to 32,000 Da and Đ$_M$'s.[115–117] Considering all the different possibilities that the co-polymerization of epoxides and anhydrides can bring to create a wide range of terpene-based polyesters, this field has still much to explore, not only in relation to the different monomers to be used, but also to overcome the low molecular weights of the obtained polymers and to improve the control over the polymerization.

4.2 Polycarbonates

The copolymerization of 1,2-limonene oxide with CO_2 to give polycarbonates has also been explored with particular interest in recent years.[6,22,100,118,119] These terpene-based polycarbonates have shown very good properties, comparable to the commercial ones, with the clear advantage of being completely biobased.

The first contribution on the copolymerization of an epoxide with CO_2 to give polycarbonates appeared in 1969 by Inoue, using diethylzinc as catalysts.[120] After this seminal work, several decades passed until in 2004 Coates published the polymerization of 1,2-limonene oxide and

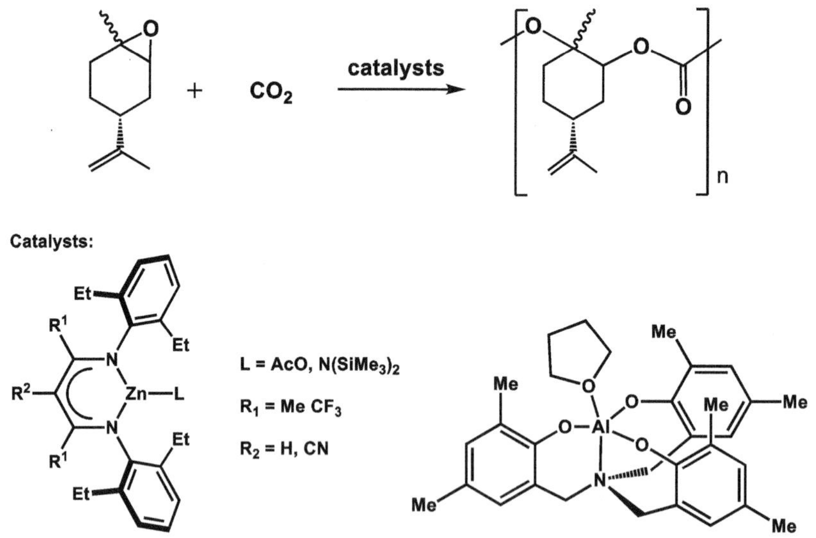

Scheme 21 Polycarbonates derived from limonene oxide.

CO_2 using the β-diiminate zinc acetate complexes catalysts that has also proven to be effective for the copolymerization of 1,2-limonene oxide and anhydrides (Scheme 21).[121] In this case, regioregular and alternating copolymers were obtained with molecular weights, up to 25,000 Da and narrow molecular-weight dispersities (1.16). The catalysts only polymerized the *trans* isomer from the mixture, leaving the *cis* isomer unreacted and recoverable, this was observed for both the S and the R isomers of the 1,2-limonene oxide. Interestingly, the *co*-crystallization poly(1S,2S,4R-limonene carbonate) and poly(1R,2R,4S-limonene carbonate) from a 1:1 solution in hexane, led to the formation of a crystalline stereocomplex.[122] Apart from the initial Zn iminidate complex [Zn(BDI)(AcO)], the modification of the substituent to the more electron withdrawing CF_3 led to a more efficient catalyst that gave TOFs of up to 301 h^{-1}.[123] As well, changing the acetate ligand for an hydrolysable one such as $N(SiMe_3)_2$, [Zn(BDI){N(SiMe_3)_2}], allows the production of polymers with hydroxyl terminated groups.[124]

A decade after Coates's original paper, Greiner optimized the catalytic process and achieved the formation of high molecular weight (>100,000 Da) polycarbonates, at large scale.[125] The strategy used involves copolymerization of limonene oxide with a high *trans* isomer content monomer and also to reduce significantly the presence of hydroxyl impurities in the monomer using ICH_3 to perform the O-methylation. The presence of those impurities is considered to be responsible for the reduction in the molecular weight as they may act as chain transfer agents. The polycarbonate produced showed improved properties, higher T_g (130 °C) and also higher transparency than the polycarbonate prepared from carcinogenic bisphenol A and toxic phosgene. Later on, good properties for gas permeability were revealed for this polymer which make it suitable for the manufacture of new-generation windows.[126]

The copolymerization of both isomers, *cis* and *trans*, was achieved by Kleij using an amino trisphenolate aluminum catalyst in combination with PPNCl as co-catalyst, although the molecular weights attained were lower, 1.3–15.1 kDa.[127–129] In this case, even if the aluminum compound can polymerize both isomers, there is a strong preference for the *cis* isomer which reacts faster and when using a pure *cis*-monomer the molecular weight attained is higher and also the same is true of the stereoregularity. As determined for Coates' β-diiminate zinc catalyst, the polymerization mechanism implies a bimetallic approach.[130]

Further studies have been carried out with the copolymerization and even terpolymerization with other monomers such as cyclohexene oxide, lactide or butyrolactone in order to increase the molecular weight and improve the processability.[123,131–133] In addition, the presence of the pendant double bond opens the possibility for a functionalization post-polymerization. Hence, it is possible to transform it *via* an oxidation processes,[127] and also reactions such as thiol-ene can be applied to provide the polycarbonates with a whole range of new properties, including permanent antibacterial activity.[79,134]

5. Concluding remarks

Terpenes and terpenoids are a varied family of natural products that have strong potential as monomers and comonomers. There is much disperse information about these fascinating molecules, and for many of them, still there are not efficient catalytic systems able to produce polymers in a fast and well-controlled process, hence there is room for improvement to obtain materials able to be competitive with analogues derived from oil.

The discovery of catalytic systems able to provide a good control over polymerization and to avoid chain transfer reactions and other secondary reactions that are the main reason for limiting the control in the polymerization is an important field of research. So far, the polymers obtained lack necessary mechanical properties, although as they are clearly highly functionalized molecules it is evident their potential for obtaining highly functionalized polymers and for post-polymerization modifications. The progress achieved in the copolymerization of 1,2-limonene oxide and CO_2 to prepare polymers with T_g and mechanical properties able to compete with the petrol derived ones is clearly significant and highlights that many new advances can be achieved with terpenes if more efforts are devoted to the study of the polymerization of this family of attractive natural products. In summary, the polymerization of terpenes and terpenoids is an exciting area that is only now awakening.

Acknowledgments

Financial support from Spanish Government (RTI2018-094840-B-C31), Madrid Regional Government (EPU-INV/2020/001, Línea de Actuación de Excelencia del Profesorado Universitario Permanente) and the Alcalá University, Spain (CCG2018/EXP-038) are gratefully acknowledged. D.S.R. and M. P. thanks the Spanish Government and Universidad de Alcalá, respectively, for a Predoctoral Fellowship.

References

1. Yao K, Tang C. Controlled polymerization of next-generation renewable monomers and beyond. *Macromolecules*. 2013;46:1689–1712.
2. Belgacem MN, Gandini A. *Monomers, Polymers and Composites from Renewable Resources*. Elsevier; 2011.
3. Llevot A, Dannecker PK, von Czapiewski M, Over LC, Sçyler Z, Meier MAR. Renewability is not enough: recent advances in the sustainable synthesis of biomass-derived monomers and polymers. *Chem A Eur J*. 2016;22:11510–11521.
4. Jansen DJ, Shenvi RA. Synthesis of medicinally relevant terpenes: reducing the cost and time of drug discovery. *Future Med Chem*. 2014;6:1127–1148.
5. Zhu Y, Romain C, Williams CK. Sustainable polymers from renewable resources. *Nature*. 2016;540:354–362.
6. Ciriminna R, Lomeli-Rodriguez M, Demma Carà P, Lopez-Sanchez J, Pagliaro M. Limonene: a versatile chemical of the bioeconomy. *Chem Commun*. 2014;50:15288–15296.
7. Zhang X, Fevre M, Jones GO, Waymouth RM. Catalysis as an enabling science for sustainable polymers. *Chem Rev*. 2018;118:839–885.
8. Mülhaupt R. Green polymer chemistry and bio-based plastics: dreams and reality. *Macromol Chem Phys*. 2013;214:159–174.
9. Behr A, Johnen L. Myrcene as a natural base chemical in sustainable chemistry: a critical review. *ChemSusChem*. 2009;2:1072–1095.
10. Kühlborn J, Groß J, Opatz T. Making natural products from renewable feedstocks: back to the roots? *Nat Prod Rep*. 2020;37:380–424.
11. Eggersdorfer M. Terpenes. In: *Ullmann's Encyclopedia of Industrial Chemistry*. New York: Wiley-VCH Verlag GmbH & Co; 2000.
12. Mathers RT, Lewis SP. Monoterpenes as polymerization solvents and monomers in polymer chemistry. In: Mathers RT, Meier MAR, eds. *Green Polymerization Methods: Renewable Starting Materials, Catalysis and Waste Reduction*. Weinheim, Germany: Wiley-VCH; 2011:91–128.

13. Tsolakis N, Bam W, Srai JS, Kumar M. Renewable chemical feedstock supply network design: the case of terpenes. *J Clean Prod.* 2019;222:802–822.
14. Croteau R, Kutchan TM, Lewis NG. Natural products (secondary metabolites). *Biochemistry and Molecular Biology of Plants.* 2000;24:1250–1319.
15. Silvestre AJD, Gandini A. Terpenes: major sources, properties and applications. In: *Monomers, Polymers and Composites from Renewable Resources.* Elsevier; 2008:17–38.
16. Wilbon PA, Chu F, Tang C. Progress in renewable polymers from natural terpenes, terpenoids, and rosin. *Macromol Rapid Commun.* 2013;34:8–37.
17. Christmann M. Otto Wallach: founder of terpene chemistry and Nobel Laureate 1910. *Angew Chem Int Ed.* 2010;49:9580–9586.
18. Wang Z, Yuan L, Tang C. Sustainable elastomers from renewable biomass. *Acc Chem Res.* 2017;50:1762–1773.
19. Gandini A. The irruption of polymers from renewable resources on the scene of macromolecular science and technology. *Green Chem.* 2011;13:1061–1083.
20. Kukhta NA, Vasilenko IV, Kostjuk SV. Room temperature cationic polymerization of β-pinene using modified AlCl$_3$ catalyst: toward sustainable plastics from renewable biomass resources. *Green Chem.* 2011;13:2362–2364.
21. Bähr M, Bitto A, Mülhaupt R. Cyclic limonene dicarbonate as a new monomer for non-isocyanate oligo- and polyurethanes (NIPU) based upon terpenes. *Green Chem.* 2012;14:1447–1454.
22. Poland SJ, Darensbourg DJ. A quest for polycarbonates provided via sustainable epoxide/CO$_2$ copolymerization processes. *Green Chem.* 2017;19:4990–5011.
23. Sahu P, Sarkar P, Bhowmick AK. Synthesis and characterization of a terpene-based sustainable polymer: poly-alloocimene. *ACS Sustainable Chem Eng.* 2017;5:7659–7669.
24. Gandini A, Lacerda TM. From monomers to polymers from renewable resources: recent advances. *Prog Polym Sci.* 2015;48:1–39.
25. Sarkar P, Bhowmick AK. Synthesis, characterization and properties of a bio-based elastomer: polymyrcene. *RSC Adv.* 2014;4:61343–61354.
26. Winnacker M, Rieger B. Recent progress in sustainable polymers obtained from cyclic terpenes: synthesis, properties, and application potential. *ChemSusChem.* 2015;8:2455–2471.
27. Swift KAD. Catalytic transformations of the major terpene feedstocks. *Top Catal.* 2004;27:143–155.
28. Valente A, Mortreux A, Visseaux M, Zinck P. Coordinative chain transfer polymerization. *Chem Rev.* 2013;113:3836–3857.
29. Sarkar P, Bhowmick AK. Terpene based sustainable elastomer for low rolling resistance and improved wet grip application: synthesis, characterization and properties of poly(styrene-co-myrcene). *ACS Sustainable Chem Eng.* 2016;4:5462–5474.
30. Zhao J, Schlaad H. Synthesis of terpene-based polymers. In: *Bio-synthetic Polymer Conjugates.* Berlin, Heidelberg: Springer; 2013:151–190.
31. Marvel CS, Hwa CCL. Polymyrcene. *J Polym Sci.* 1960;45:25–34.
32. Loughmari S, Hafid A, Bouazza A, El Bouadili A, Zinck P, Visseaux M. Highly stereoselective coordination polymerization of beta-myrcene from a lanthanide-based catalyst: access to bio-sourced elastomers. *J Polym Sci A Polym Chem.* 2012;50:2898–2905.
33. Breitmaier E. *Terpenes: Flavours, Fragrances, Pharmaca, Pheromones.* Verlag Gmbh, Weinheim: Wiley-VCH; 2006.
34. Puskas JE, Gautriaud E, Deffieux A, Kennedy JP. Natural rubber biosynthesis—a living carbocationic polymerization? *Prog Polym Sci.* 2006;31:533–548.
35. van den Berg KJ, van der Horst J, Boon JJ, Sudeiijer OO. Cis-1,4-poly-β-myrcene; the structure of the polymeric fraction of mastic resin (*Pistacia lentiscus* L.) elucidated. *Tetrahedron Lett.* 1998;39:2645–2648.

36. Runckel WJ, Goldblatt LA. Inhibition of myrcene polymerization during storage. *Ind Eng Chem.* 1946;38:749–751.
37. Rummelsburg AL. *Terpene Resins.* US Patent; 1945. 2383084.
38. Newmark RA, Majumdar RN. ^{13}C-NMR spectra of cis-polymyrcene and cis-polyfarnesene. *J Polym Sci A Polym Chem.* 1988;26:71–77.
39. Grune E, Bareuther J, Blankenburg J, et al. Towards bio-based tapered block copolymers: the behaviour of myrcene in the statistical anionic copolymerisation. *Polym Chem.* 2019;10:1213–1220.
40. Ávila-Ortega A, Aguilar-Vega M, Loría Bastarrachea MI, Carrera-Figueiras C, Campos-Covarrubias M. Anionic synthesis of amine ω-terminated β-myrcene polymers. *J Polym Res.* 2015;22:226.
41. Bolton JM, Hillmyer MA, Hoye TR. Sustainable thermoplastic elastomers from terpene-derived monomers. *ACS Macro Lett.* 2014;3:717–720.
42. Gilbert RG, Hess M, Jenkins AD, Jones RG, Kratochvíl P, Stepto RFT. Dispersity in polymer science. *Pure Appl Chem.* 2009;81:351–353.
43. Cawse JL, Stanford JL, Still RH. Polymers from renewable sources. IV. Polyurethane elastomers based on myrcene polyols. *J Appl Polym Sci.* 1986;31(6):1549–1565.
44. Cawse JL, Stanford JL, Still RH. Polymers from renewable sources: 5. Myrcene-based polyols as rubber-toughening agents in glassy polyurethanes. *Polymer.* 1987;28(3): 368–374.
45. Rummelsburg AL. *Polymerized Acyclic Terpenes and Method of Production.* US Patent; 1945. 2373419.
46. Hulnik MI, Vasilenko IV, Radchenko AV, Peruch F, Ganachaud F, Kostjuk SV. Aqueous cationic homo- and co-polymerizations of β-myrcene and styrene: a green route toward terpene-based rubbery polymers. *Polym Chem.* 2018;9:5690–5700.
47. Georges S, Touré AO, Visseaux M, Zinck P. Coordinative chain transfer copolymerization and terpolymerization of conjugated dienes. *Macromolecules.* 2014;47: 4538–4547.
48. Georges S, Bria M, Zinck P, Visseaux M. Polymyrcene microstructure revisited from precise high-field nuclear magnetic resonance analysis. *Polymer.* 2014;55:3869–3878.
49. Raynaud J, Wu JY, Ritter T. Iron-catalyzed polymerization of isoprene and other 1,3-dienes. *Angew Chem Int Ed.* 2012;51:11805–11808.
50. a) Liu B, Li L, Sun G, Liu D, Li S, Cui D. Isoselective 3,4-(co)polymerization of bio-renewable myrcene using NSN-ligated rare-earth metal precursor: an approach to a new elastomer. *Chem Commun.* 2015;51:1039–1041. b) Liu B, Li L, Sun G, et al. 3,4-polymerization of isoprene by using NSN- and NPN-ligated rare earth metal precursors: switching of stereo selectivity and mechanism. *Macromolecules.* 2014;47: 4971–4978.
51. Kuhlman RL, Wenzel TT. Investigations of chain shuttling olefin polymerization using deuterium labeling. *Macromolecules.* 2008;41:4090–4094.
52. Camara JM, Petros RA, Norton JR. Zirconium-catalyzed carboalumination of α-olefins and chain growth of aluminum alkyls: kinetics and mechanism. *J Am Chem Soc.* 2011;133:5263–5273.
53. Zhang L, Nishiura M, Yuki M, Luo Y, Hou Z. Isoprene polymerization with yttrium amidinate catalysts: switching the regio- and stereoselectivity by addition of AlMe$_3$. *Angew Chem Int Ed.* 2008;47:2642–2645.
54. Laur E, Welle A, Vantomme A, Brusson J, Carpentier J, Kirillov E. Stereoselective copolymerization of styrene with terpenes catalyzed by an ansa-lanthanidocene catalyst: access to new syndiotactic polystyrene-based materials. *Catalysts.* 2017;7:361.
55. Naddeo M, Buonerba A, Luciano E, Grassi A, Proto A, Capacchione C. Stereoselective polymerization of biosourced terpenes β-myrcene and β-ocimene and their copolymerization with styrene promoted by titanium catalysts. *Polymer.* 2017;131:151–159.

56. Radbil AB, Zhurinova TA, Starostina EB, Radbil BA. Preparation of high-melting polyterpene resins from α-Pinene. *Russ J Appl Chem.* 2005;78:1126–1130.
57. Yu P, Li A, Liang H, Lu J. Polymerization of β-pinene with Schiff-base nickel complexes catalyst: synthesis of relatively high molecular weight poly(β-pinene) at high temperature with high productivity. *J Polym Sci A Polym Chem.* 2007;45:3739–3746.
58. Roberts WJ, Day AR. A study of the polymerization of α- and β-Pinene with Friedel—crafts type catalysts. *J Am Chem Soc.* 1950;72:1226–1230.
59. Lu J, Kamigaito M, Sawamoto M, Higashimura T, Deng Y. Cationic polymerization of β-pinene with the AlCl$_3$/SbCl$_3$ binary catalyst: comparison with α-pinene polymerization. *J Appl Polym Sci.* 1996;61:1011–1016.
60. Kennedy JP, Liao T, Guhaniyogi S, Chang VSC. Poly(β-pinenes) carrying one, two, and three functional end groups. II. Syntheses and characterization. *J Polym Sci, Polym Chem Ed.* 1982;20:3229–3240.
61. Winnacker M. Pinenes: abundant and renewable building blocks for a variety of sustainable polymers. *Angew Chem Int Ed.* 2018;57:14362–14371.
62. a) Aoshima S, Kanaoka S. A renaissance in living cationic polymerization. *Chem Rev.* 2009;109:5245–5287. b) Satoh K. Controlled/living polymerization of renewable vinyl monomers into bio-based polymers. *Polym J.* 2015;47:527–536.
63. Lu J, Kamigaito M, Sawamoto M, Higashimura T, Deng Y. Living cationic isomerization polymerization of β-Pinene. 1. Initiation with HCl – 2-chloroethyl vinyl ether adduct/TiCl$_3$(O^{i}Pr) in conjunction with ^{n}Bu$_4$NCl. *Macromolecules.* 1997;30:22–26.
64. Satoh K, Sugiyama H, Kamigaito M. Biomass-derived heat-resistant alicyclic hydrocarbon polymers: poly(terpenes) and their hydrogenated derivatives. *Green Chem.* 2006;8:878–882.
65. Satoh K, Nakahara A, Mukunoki K, Sugiyama H, Saito H, Kamigaito M. Sustainable cycloolefin polymer from pine tree oil for optoelectronics material: living cationic polymerization of β-pinene and catalytic hydrogenation of high-molecular-weight hydrogenated poly(β-pinene). *Polym Chem.* 2014;5:3222–3230.
66. Sawamoto M. Modern cationic vinyl polymerization. *Prog Polym Sci.* 1991;16:111–172.
67. Aoshima S, Yoshida T, Kanazawa A, Kanaoka S. New stage in living cationic polymerization: an array of effective lewis acid catalysts and fast living polymerization in seconds. *J Polym Sci, Part A: Polym Chem.* 2007;45:1801–1813.
68. McCaulay DA, Lien AP. Relative basicity of the methylbenzenes. *J Am Chem Soc.* 1951;73:2013–2017.
69. Laurence C, Graton J, Berthelot M, et al. An enthalpic scale of hydrogen-bond basicity. 4. Carbon π bases, oxygen bases, and miscellaneous second-row, third-row, and fourth-row bases and a survey of the 4-fluorophenol affinity scale. *J Org Chem.* 2010;75:4105–4123.
70. Ulvenlund S, Wheatley A, Bengtsson LA. Spectroscopic investigation of concentrated solutions of gallium(III) chloride in mesitylene and benzene. *J Chem Soc Dalton Trans.* 1995;255–263.
71. Solari E, Floriani C, Chiesi-Villa A, Guastini C. Titanium tetrachloride binding and making arenes from acetylenes: the synthesis and X-ray crystal structure of a titanium(IV)–hexamethylbenzene complex. *J Chem Soc Chem Commun.* 1989;1747–1749.
72. Karasawa Y, Kimura M, Kanazawa A, Kanaoka S, Aoshima S. New initiating systems for cationic polymerization of plant-derived monomers: GaCl$_3$/alkylbenzene-induced controlled cationic polymerization of β-pinene. *Polym J.* 2015;47:152–157.
73. Modena M, Bates RB, Marvel CS. Some low molecular weight polymers of d-limonene and related terpenes obtained by Ziegler-type catalysts. *J Polym Sci A Gen Pap.* 1965;3:949–960.

74. Pietila H, Sivola A, Sheffer H. Cationic polymerization of β-pinene, styrene and α-methylstyrene. *J Polym Sci A-1 Polym Chem*. 1970;8:727–737.
75. Sahu P, Bhowmick AK, Kali G. Terpene based elastomers: synthesis, properties, and applications. *Processes*. 2020;8:553.
76. Thomsett MR, Moore JC, Buchard A, Stockman RA, Howdle SM. New renewably-sourced polyesters from limonene-derived monomers. *Green Chem*. 2019;21:149–156.
77. Crowell PL, Gould MN. Chemoprevention and therapy of cancer by d-limonene. *Crit Rev Oncog*. 1994;5:1–22.
78. Raphael TJ, Kuttan G. Immunomodulatory activity of naturally occurring monoterpenes carvone, limonene, and perillic acid. *Immunopharmacol Immunotoxicol*. 2003;25: 285–294.
79. Firdaus M, Montero de Espinosa L, MAR M. Terpene-based renewable monomers and polymers via thiol–ene additions. *Macromolecules*. 2011;44:7253–7262.
80. de Oliveira ERM, Vieira RP. Synthesis and characterization of poly(limonene) by photoinduced controlled radical polymerization. *J Polym Environ*. 2020;28:2931–2938.
81. Andriotis EG, Achilias DS. Role of polylimonene as a bio-based additive in thermal oxidation of high impact polystyrene. *Macromol Symp*. 2013;331–332:173–180.
82. Doiuchi T, Yamaguchi H, Minoura Y. Cyclocopolymerization of d-limonene with maleic anhydride. *Eur Polym J*. 1981;17:961–968.
83. Sharma S, Srivastava AK. Alternating copolymers of limonene with methyl methacrylate: kinetics and mechanism. *J Macromol Sci A*. 2003;40:593–603.
84. Zhang Y, Dubé MA. Copolymerization of 2-ethylhexyl acrylate and D-limonene. *Polym-Plast Technol Eng*. 2015;54:499–505.
85. Ishido Y, Kanazawa A, Kanaoka S, Aoshima S. New degradable alternating copolymers from naturally occurring aldehydes: well-controlled cationic copolymerization and complete degradation. *Macromolecules*. 2012;45:4060–4068.
86. Lowe JR, Martello MT, Tolman WB, Hillmyer MA. Functional biorenewable polyesters from carvone-derived lactones. *Polym Chem*. 2011;2:702–708.
87. Lowe JR, Tolman WB, Hillmyer MA. Oxidized dihydrocarvone as a renewable multifunctional monomer for the synthesis of shape memory polyesters. *Biomacromolecules*. 2009;10:2003–2008.
88. Zhang D, Hillmyer MA, Tolman WB. Catalytic polymerization of a cyclic ester derived from a "cool" natural precursor. *Biomacromolecules*. 2005;6:2091–2095.
89. Winnacker M, Vagin S, Auer V, Rieger B. Synthesis of novel sustainable oligoamides via ring-opening polymerization of lactams based on (−)-menthone. *Macromol Chem Phys*. 2014;215:1654–1660.
90. Winnacker M, Sag J. Sustainable terpene-based polyamides via anionic polymerization of a pinene-derived lactam. *Chem Commun*. 2018;54:841–844.
91. Stockmann PN, Pastoetter DL, Woelbing M, et al. New bio-polyamides from terpenes: α-Pinene and (+)-3-Carene as valuable resources for lactam production. *Macromol Rapid Commun*. 2019;40:1800903.
92. Quilter HC, Hutchby M, Davidson MG, Jones MD. Polymerisation of a terpene-derived lactone: a bio-based alternative to ε-caprolactone. *Polym Chem*. 2017;8: 833–837.
93. Hosseini Nejad E, van Melis CGW, Vermeer TJ, Koning CE, Duchateau R. Alternating ring-opening polymerization of cyclohexene oxide and anhydrides: effect of catalyst, cocatalyst, and anhydride structure. *Macromolecules*. 2012;45:1770–1776.
94. Parrino F, Fidalgo A, Palmisano L, Ilharco LM, Pagliaro M, Ciriminna R. Polymers of limonene oxide and carbon dioxide: polycarbonates of the solar economy. *ACS Omega*. 2018;3:4884–4890.

95. Castro-Gómez F, Salassa G, Kleij AW, Bo C. A DFT study on the mechanism of the cycloaddition reaction of CO_2 to epoxides catalyzed by Zn(Salphen) complexes. *Chem A Eur J*. 2013;19:6289–6298.
96. Park HJ, Ryu CY, Crivello JV. Photoinitiated cationic polymerization of limonene 1,2-oxide and α-pinene oxide. *J Polym Sci, Part A: Polym Chem*. 2013;51:109–117.
97. Sessini V, Palenzuela M, Damián J, Mosquera MEG. Bio-based polyether from limonene oxide catalytic ROP as green polymeric plasticizer for PLA. *Polymer*. 2020;210: 123003.
98. Muñoz MT, Palenzuela M, Cuenca T, Mosquera MEG. Aluminum aryloxide compounds as very active catalysts for glycidyl methacrylate selective ring-opening polymerization. *ChemCatChem*. 2018;10:936–939.
99. Peña Carrodeguas L, Martín C, Kleij AW. Semiaromatic polyesters derived from renewable terpene oxides with high glass transitions. *Macromolecules*. 2017;50:5337–5345.
100. Neumann S, Leitner L, Schmalz H, Agarwal S, Greiner A. Unlocking the processability and recyclability of biobased poly(limonene carbonate). *ACS Sustainable Chem Eng*. 2020;8:6442–6448.
101. Fischer RF. Polyesters from expoxides and anhydrides. *J Polym Sci*. 1960;44:155–172.
102. Li J, Liu Y, Ren W, Lu X. Asymmetric alternating copolymerization of meso-epoxides and cyclic anhydrides: efficient access to enantiopure polyesters. *J Am Chem Soc*. 2016;138:11493–11496.
103. Longo JM, Sanford MJ, Coates GW. Ring-opening copolymerization of epoxides and cyclic anhydrides with discrete metal complexes: structure–property relationships. *Chem Rev*. 2016;116:15167–15197.
104. Paul S, Zhu Y, Romain C, Brooks R, Saini PK, Williams CK. Ring-opening copolymerization (ROCOP): synthesis and properties of polyesters and polycarbonates. *Chem Commun*. 2015;51:6459–6479.
105. Li J, Ren B, Chen S, et al. Development of highly enantioselective catalysts for asymmetric copolymerization of meso-epoxides and cyclic anhydrides: subtle modification resulting in superior enantioselectivity. *ACS Catal*. 2019;9:1915–1922.
106. Hatazawa M, Takahashi R, Deng J, Houjou H, Nozaki K. Cationic co–salphen complexes bisligated by DMAP as catalysts for the copolymerization of cyclohexene oxide with phthalic anhydride or carbon dioxide. *Macromolecules*. 2017;50:7895–7900.
107. Zhu Y, Romain C, Williams CK. Selective polymerization catalysis: controlling the metal chain end group to prepare block copolyesters. *J Am Chem Soc*. 2015;137:12179–12182.
108. Jeske RC, DiCiccio AM, Coates GW. Alternating copolymerization of epoxides and cyclic anhydrides: an improved route to aliphatic polyesters. *J Am Chem Soc*. 2007;129: 11330–11331.
109. Robert C, de Montigny F, Thomas CM. Tandem synthesis of alternating polyesters from renewable resources. *Nat Commun*. 2011;2:586.
110. Nejad EH, Paoniasari A, van Melis CGW, Koning CE, Duchateau R. Catalytic ring-opening copolymerization of limonene oxide and phthalic anhydride: toward partially renewable polyesters. *Macromolecules*. 2013;46:631–637.
111. Isnard F, Lamberti M, Pellecchia C, Mazzeo M. Ring-opening copolymerization of epoxides with cyclic anhydrides promoted by bimetallic and monometallic phenoxy–imine aluminum complexes. *ChemCatChem*. 2017;9:2972–2979.
112. Santulli F, D'Auria I, Boggioni L, et al. Bimetallic aluminum complexes bearing binaphthyl-based iminophenolate ligands as catalysts for the synthesis of polyesters. *Organometallics*. 2020;39:1213–1220.
113. de Sarasa Buchaca MM, de la Cruz-Martínez F, Martínez J, et al. Alternating copolymerization of epoxides and anhydrides catalyzed by aluminum complexes. *ACS Omega*. 2018;3:17581–17589.

114. Inoue S. Immortal polymerization: the outset, development, and application. *J Polym Sci, Part A: Polym Chem.* 2000;38:2861–2871.
115. Van Zee NJ, Sanford MJ, Coates GW. Electronic effects of aluminum complexes in the copolymerization of propylene oxide with tricyclic anhydrides: access to well-defined, functionalizable aliphatic polyesters. *J Am Chem Soc.* 2016;138:2755–2761.
116. Sanford MJ, Peña Carrodeguas L, Van Zee NJ, Kleij AW, Coates GW. Alternating copolymerization of propylene oxide and cyclohexene oxide with tricyclic anhydrides: access to partially renewable aliphatic polyesters with high glass transition temperatures. *Macromolecules.* 2016;49:6394–6400.
117. Van Zee NJ, Coates GW. Alternating copolymerization of propylene oxide with biorenewable terpene-based cyclic anhydrides: a sustainable route to aliphatic polyesters with high glass transition temperatures. *Angew Chem Int Ed.* 2015;54:2665–2668.
118. Bobbink FD, van Muyden AP, Dyson PJ. En route to CO_2-containing renewable materials: catalytic synthesis of polycarbonates and non-isocyanate polyhydroxyurethanes derived from cyclic carbonates. *Chem Commun.* 2019;55:1360–1373.
119. Anderson TD, Kozak CM. Ring-opening polymerization of epoxides and ring-opening copolymerization of CO_2 with epoxides by a zinc amino-bis(phenolate) catalyst. *Eur Polym J.* 2019;120:109237.
120. Inoue S, Koinuma H, Tsuruta T. Copolymerization of carbon dioxide and epoxide with organometallic compounds. *Makromol Chem.* 1969;130:210–220.
121. Byrne CM, Allen SD, Lobkovsky EB, Coates GW. Alternating copolymerization of limonene oxide and carbon dioxide. *J Am Chem Soc.* 2004;126:11404–11405.
122. Auriemma F, De Rosa C, Di Caprio MR, Di Girolamo R, Ellis WC, Coates GW. Stereocomplexed poly(limonene carbonate): a unique example of the cocrystallization of amorphous enantiomeric polymers. *Angew Chem Int Ed.* 2015;54:1215–1218.
123. Reiter M, Vagin S, Kronast A, Jandl C, Rieger B. A Lewis acid β-diiminato-zinc-complex as all-rounder for co- and terpolymerisation of various epoxides with carbon dioxide. *Chem Sci.* 2017;8:1876–1882.
124. Li C, Sablong RJ, Koning CE. Synthesis and characterization of fully-biobased α,ω-dihydroxyl poly(limonene carbonate)s and their initial evaluation in coating applications. *Eur Polym J.* 2015;67:449–458.
125. Hauenstein O, Reiter M, Agarwal S, Rieger B, Greiner A. Bio-based polycarbonate from limonene oxide and CO_2 with high molecular weight, excellent thermal resistance, hardness and transparency. *Green Chem.* 2016;18:760–770.
126. Hauenstein O, Rahman MM, Elsayed M, et al. Biobased polycarbonate as a gas separation membrane and "breathing glass" for energy saving applications. *Adv Mater Technol.* 2017;2:1700026.
127. Kindermann N, Cristòfol À, Kleij AW. Access to biorenewable polycarbonates with unusual glass-transition temperature (Tg) modulation. *ACS Catal.* 2017;7:3860–3863.
128. Peña-Carrodeguas L, González-Fabra J, Castro-Gómez F, Bo C, Kleij AW. Al(III)-catalysed formation of poly(limonene)carbonate: DFT analysis of the origin of stereoregularity. *Chem A Eur J.* 2015;21:6115–6122.
129. Martín C, Kleij AW. Terpolymers derived from limonene oxide and carbon dioxide: access to cross-linked polycarbonates with improved thermal properties. *Macromolecules.* 2016;49:6285–6295.
130. González-Fabra J, Castro-Gómez F, Kleij AW, Bo C. Mechanistic insights into the carbon dioxide/cyclohexene oxide copolymerization reaction: is one metal center enough? *ChemSusChem.* 2017;10:1233–1240.
131. Kernbichl S, Rieger B. Aliphatic polycarbonates derived from epoxides and CO_2: a comparative study of poly(cyclohexene carbonate) and poly(limonene carbonate). *Polymer.* 2020;205:122667.

132. Stößer T, Li C, Unruangsri J, et al. Bio-derived polymers for coating applications: comparing poly(limonene carbonate) and poly(cyclohexadiene carbonate). *Polym Chem.* 2017;8:6099–6105.
133. Carrodeguas LP, Chen TTD, Gregory GL, Sulley GS, Williams CK. High elasticity, chemically recyclable, thermoplastics from bio-based monomers: carbon dioxide, limonene oxide and ε-decalactone. *Green Chem.* 2020;22:8298–8307.
134. Hauenstein O, Agarwal S, Greiner A. Bio-based polycarbonate as synthetic toolbox. *Nat Commun.* 2016;7:11862.

CHAPTER THREE

Bimetallic frustrated Lewis pairs

Miquel Navarro and Jesús Campos*

Instituto de Investigaciones Químicas (IIQ), Departamento de Química Inorgánica and Centro de Innovación en Química Avanzada (ORFEO-CINQA), Consejo Superior de Investigaciones Científicas (CSIC) and University of Sevilla, Sevilla, Spain
*Corresponding author: e-mail address: jesus.campos@iiq.csic.es

Contents

1. Introduction	95
2. Frustrated Lewis pairs	97
2.1 Main group frustrated Lewis pairs	98
2.2 Transition metal frustrated Lewis pairs	101
3. Bimetallic complexes for bond activation	105
3.1 General concepts	105
3.2 Binuclear orthogonal activation	107
3.3 Homobimetallic activation	111
3.4 Polar heterobimetallic complexes	114
4. Genuine metal-only FLP activation	122
4.1 Transition metal/main group metal frustrated Lewis pairs	122
4.2 Transition metal only frustrated Lewis pairs (TMOFLPs)	123
5. Miscellanea	131
5.1 Cooperative reactivity between main group FLPs and transition metal complexes	131
5.2 Transition metal frustrated Lewis pairs in heterogeneous catalysis	133
6. Conclusions	136
Acknowledgments	136
References	137

1. Introduction

The potential of Lewis frustration in bond activation was first introduced by Stephan in 2006.[1,2] A covalently linked metal-free phosphine-borane revealed the cooperative action between the Lewis base (phosphine) and the Lewis acid (borane) in the heterolytic cleavage of the dihydrogen molecule. More remarkably, the process was shown to be reversible. This discovery, together with the independent work of Power[3] and Bertrand[4] on dihydrogen splitting by multi-bonded and sub-valent main group

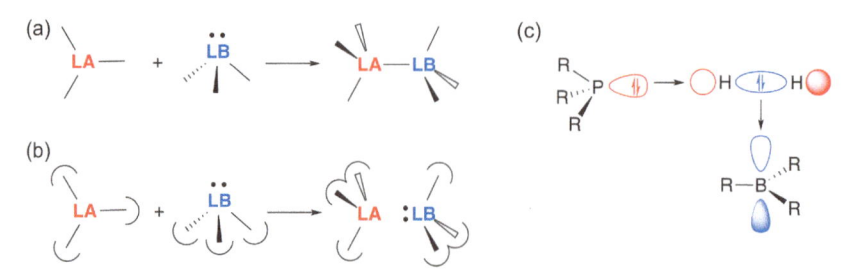

Fig. 1 (A) Formation of a Lewis acid-base adduct through classic Lewis acid-base chemistry; (B) Representation of a Frustrated Lewis Pair (FLP); (C) Frontier orbital interactions for the splitting of H_2 by a P/B frustrated Lewis pair.

systems, respectively, represented a breakthrough in the field of main group chemistry. These studies showed for the first time that a metal is not always required for the cleavage of H_2. In the case of FLPs, although dihydrogen splitting is heterolytic (i.e. R_3P-H^+/Ar_3B-H^-), the initial step, as in metal mediated H_2 activation, involves the synergic donation from the lone pair of the Lewis base (phosphine) to the H_2 σ^*-orbital along with donation from the H_2 σ-orbital to the empty Lewis acid (borane) orbital (Fig. 1). Since these early discoveries, the replacement of the initial phosphine-borane pairs by a manifold of combinations of Lewis bases and Lewis acids, in which steric demands preclude dative bond formation to generate FLPs, has become a hot topic in the field of inorganic chemistry.[5] This growing interest derived not only from the outstanding capacity of FLPs to activate a wide range of small molecules, but also due to its potential in catalysis.[6–10]

Another area of increasing interest in the broader context of cooperative chemistry is that of bimetallic complexes.[11–15] These entities explore the concept of cooperativity from a completely opposite perspective relative to FLPs, where the absence of metals and thus the possibility of effecting metal-free activation and catalysis lies at the heart of their existence. Despite this fact, bimetallic complexes have been proven to present strong similarities with FLP systems, particularly in cases where there is a polarized M–M bond/interaction, as will be discussed in detail throughout this review. Moreover, bimetallic systems in which the two components are based on transition metal fragments with no (or labile) connectivity between the two metal centers exhibit FLP-type bond activation and, as such, their nature can be understood in terms of transition metal-only frustrated Lewis pairs (TMOFLP).

In this chapter we will introduce some general insights within the field of frustrated Lewis pairs, paying particular attention to the potential advantages of introducing transition metals as core components of FLP designs. In addition, we will discuss the broad area of small molecule activation by bimetallic systems, highlighting the diversity of mechanisms for bond activation and their connection to the area of frustration. Once the main aspects of these general areas are discussed we will present recent results on the combination of these two approaches, that is, on the development of bimetallic systems that operate as bimetallic frustrated Lewis pairs. We do not seek for an exhaustive study, particularly in the more general sections dealing with frustrated Lewis pairs and bimetallic complexes. Instead, we rather offer a general approach to the diversity of bond activation mechanisms for these systems with key relevant examples that will be described in more detail, thus serving as a general guide to newcomers and maybe as an inspiration for those interested in the broader field of cooperative chemistry.

2. Frustrated Lewis pairs

Frustrated Lewis pairs are molecular systems comprised of a Lewis acid and a Lewis base that do not form a stable Lewis adduct, generally due to steric or geometric factors (Fig. 1B). FLP systems are able to activate small molecules through a synergic or cooperative action between the Lewis acid and base. In the 1940s the group of Brown studied for the first time the influence of steric strains in the relative stability of different coordination compounds of boron.[16] However, it was not until Stephan first witnessed in 2006 that the phosphino-borane pair p-(Mes$_2$P)C$_6$F$_4$(B(C$_6$F$_5$)$_2$) **1** was able to activate dihydrogen under mild conditions (25 °C, 1 atm H$_2$) that frustrated Lewis pairs could be exploited for synthetic purposes (Scheme 1).[1] Moreover, the dihydrogen splitting observed by Stephan is a reversible process, in which the product **2** derived from the heterolytic cleavage of H$_2$ is capable of regenerating the phosphino-borane p-(Mes$_2$P)C$_6$F$_4$(B(C$_6$F$_5$)$_2$), releasing hydrogen gas above 100 °C.

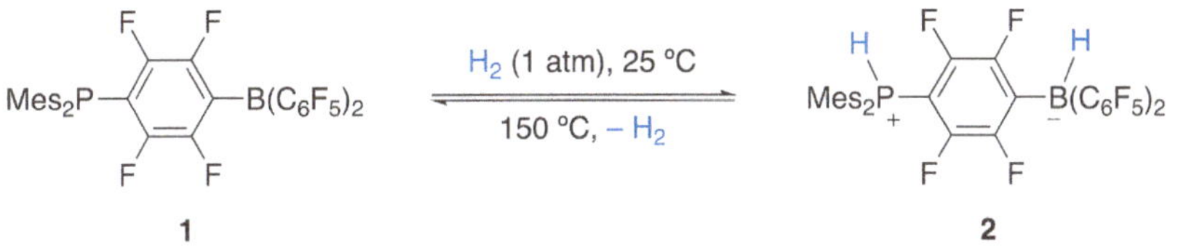

Scheme 1 First example of reversible dihydrogen activation by a FLP system.

After this breakthrough, many other inter- and intramolecular PAr$_3$/B (C$_6$F$_5$)$_3$ based frameworks have been studied in detail as active FLP systems. The interest in this topic has steadily grown in the last decade, and it is

Publications about FLPs

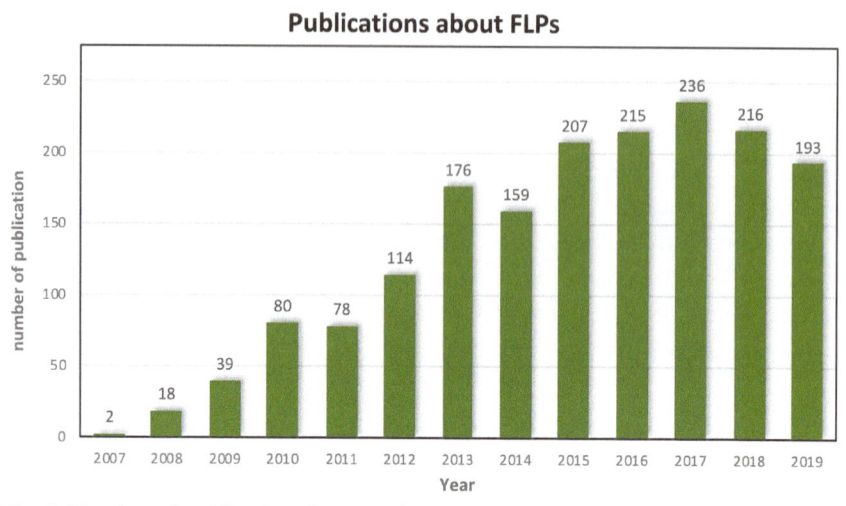

Fig. 2 Number of publications by year that contains the term "Frustrated Lewis Pair" either in the title or in the abstract according to Web of Science.

reflected in the high number of research articles (>1500) published since the pioneer work by Stephan in 2006 (Fig. 2). The potential of frustrated Lewis pairs in the activation of small molecules and as powerful catalysts for homogeneous processes has encouraged several research groups to investigate and expand the use of different Lewis acids and bases based not only on main group elements but also on transition and rare-earth metals. Different reviews gathering the major advances in this field have been published recently, from those describing general aspects of FLP chemistry[17–19] to others dealing with specific applications of FLP systems in homogeneous catalysis or for the functionalization of different molecular oxides, including CO_2.[20–22]

2.1 Main group frustrated Lewis pairs

The absence of metal centers represents a major desire for pharmaceutical and food chemical processes, in which the elimination of metallic traces in the final products is fundamental. For this reason, the use of FLP systems in different catalytic processes has gained great focus during the last years. The widespread interest in FLP systems mainly derives from the feasibility of readily activating small molecules such as dihydrogen under mild conditions. Subsequently, their proficiency as hydrogenation catalysts was efficiently proved providing new metal-free venues for the hydrogenation of imines, enamines, N-heterocycles and carbonyl compounds.[7] Similarly, FLP systems, both inter- and intramolecular versions, have been competently employed as metal-free catalysts for dehydrogenation,[23] hydrosilylation,[24,25] hydroboration[26] and hydroamination[27] transformations (Scheme 2).

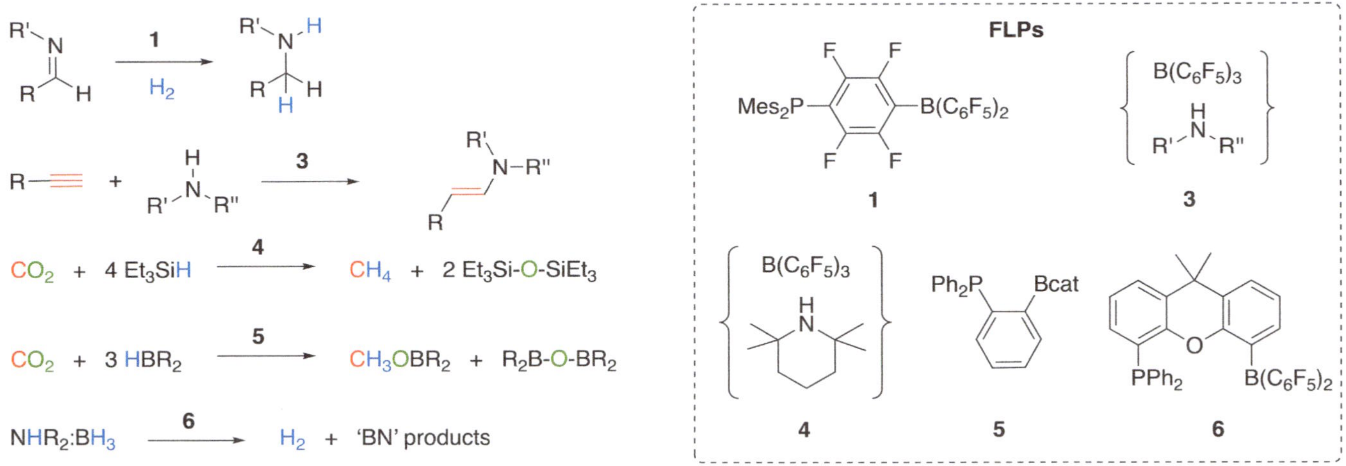

Scheme 2 Representative examples of main group FLP catalyzed reactions.

Likewise, several frustrated Lewis pair-catalyzed asymmetric hydrogenation reactions have been described. Klankermayer described the enantioselective hydrogenation of imines (up to 83% ee's) catalyzed by the intermolecular FLP system **7** based on a chiral borane with inherently chiral terpene groups (Scheme 3).[28] Intramolecular chiral FLP systems have also been described. Appropriately linking the Lewis acid and base to bring them in close proximity, but with bulky substituents that hinder dative bond formation, permits highly efficient FLP reactivity.[29,30] For instance, Pápai and Repo reported the synthesis of a chiral binaphthyl-linked amino-borane **9** with unique reactivity in the enantioselective asymmetric hydrogenation of unhindered imines and enamines (up to 99% ee's).[30] The binaphthyl core fixes the position of both Lewis acid and base next to the asymmetric axis ensuring their close proximity; in addition, the stability of the arene C—B bond impedes retro-hydroboration and decomposition to olefins and amino-boranes.

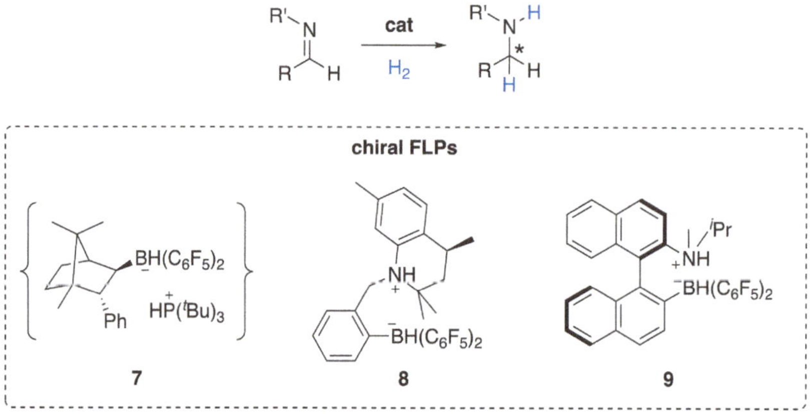

Scheme 3 Chiral frustrated Lewis pair catalysts for asymmetric hydrogenation of imines.

Replacement of the Lewis acidic boron center by main group metals such as aluminum, gallium or indium has provided a variety of FLPs with new unique properties. Stephan, Uhl and Fontaine have reported the use of P/Al frustrated Lewis pairs for the activation of alkynes,[31,32] alkenes,[33,34] dihydrogen[35] and different carbonyl groups,[36–39] as well as the catalytic dehydrocoupling of amine-boranes.[40] Although a number of Al-based FLPs might not be strictly seen as frustrated species in the original sense, since they contain dative Al—P bonds, the mismatch between the hard Lewis acidic aluminum center and the soft Lewis base (phosphorus) in the Al—P bond facilitates access to the independent components. Thus, the resulting pairs show FLP-type reactivity even in cases where the adduct is the resting state. The intermolecular frustrated Lewis pair **10** [P^{t}Bu$_3$][Al(C$_6$F$_5$)$_3$] readily activates dihydrogen with subsequent hydride transfer to inactivated olefins. This reaction involved an Al–olefin activation, a nucleophilic attack by the [HAl(C$_6$F$_5$)$_3$]$^-$ anion followed by a redistribution to aluminate and alane. This mechanism was supported by the isolation of the first Al–olefin complex **11** (Scheme 4A).[35] Alternatively, the dimeric P/Al-based Lewis pair **12** easily activates oxygen (THF and CO$_2$) and nitrogen donor (^{t}BuNCO) molecules by substrate-induced frustration.[36] Under ambient conditions in the presence of CO$_2$ the dimer **13** rapidly dissociates forming the stable CO$_2$ adduct **14** through a cooperative action of two molecules of CO$_2$ to activate the Al—P bond (Scheme 4B).

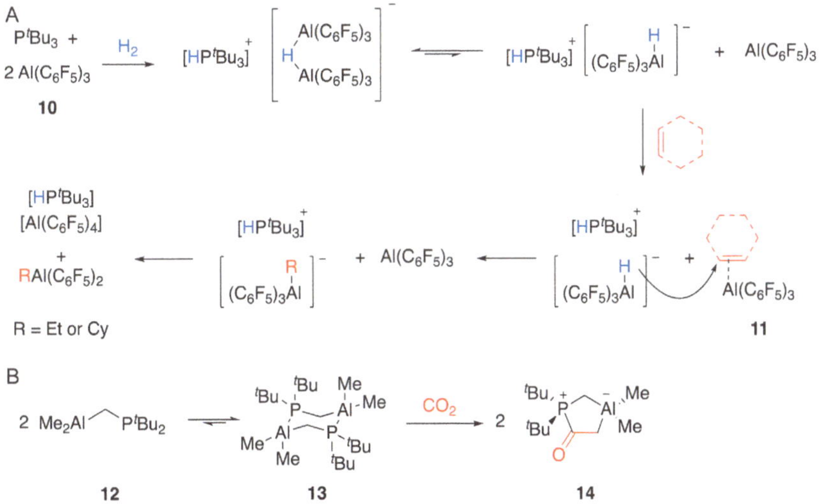

Scheme 4 Dihydrogen and CO$_2$ activation by P/Al-based frustrated Lewis pairs.

2.2 Transition metal frustrated Lewis pairs

During the last recent years, introduction of transition metal centers into FLP frameworks has drawn growing attention inspired by the pioneering work of Wass[41–43] and Erker[44,45] on zirconium/phosphine pairs. At first glance, the incorporation of transition metal centers as acid or basic sites in the design of TM-based FLPs may seem meaningless as the capacity to promote chemical reactions under metal-free conditions is one of the driving forces of the area. However, it provides a whole range of chemical opportunities other than the combinatorial possibilities derived from introducing the three series of transition metals, as well as the rare-earth elements. First, the rich reactivity resulting from the presence of partly occupied d orbitals with accessible energies to participate in elementary reactions such as oxidative addition, reductive elimination or migratory insertion constitutes a tremendously powerful synthetic tool. Besides, organometallic and coordination complexes present a large structural and electronic diversity provided by both the metal center and the ligands surrounding it. The coordination numbers typically range from two to six for the d-block yet they can reach up to nine, leading to a broad variety of structures and geometries. Moreover, the extensive number of described ligands with tunable properties enables unlimited possibilities for the design of virtually any desired transition metal complex. In addition, transition metals exhibit a large diversity of affinities toward specific elements.[46,47] For instance, while main group elements are generally highly oxophilic, the degree of oxophilicity in transition metals is notably wider, from the high tendency of early transition metals such as titanium or hafnium to bind oxygen, to the practically oxophobic character of gold or palladium.[48] Their synthetic amenability may be considered as an additional advantage. The available methodologies to prepare transition metal complexes are vast, and many relatively simple protocols have been described in the last decades.

Electron deficient early transition metal complexes have been widely used as Lewis acid catalysts for a number of transformations.[49] In an early example, Stephan demonstrated that the combination of the acidic titanium compound $[(\eta^5\text{-}C_5H_5)Ti(N=P^tBu_3)][B(C_6F_5)_4]$ with the sterically demanding $P(o\text{-}MeC_6H_4)_3$ phosphine did not lead to ligand coordination due to steric clash, but cooperatively cleaved a C—Cl bond of a dichloromethane solvent molecule instead.[50] In 2011 the group of Wass described likely the first well-characterized transition metal FLP system[41] based on a zirconocene stabilized by two pentamethylcyclopentadienyl groups and an aryloxide phosphine ligand with bulky *tert*-butyl groups that hampers the existence of a Zr—P bond (**15**). As such, this compound consists in a fully frustrated system in which the zirconium center acts as the Lewis acid while the uncoordinated phosphine is the Lewis base. The Zr/P frustrated Lewis pair **15** readily activates a wide range of small molecules such as H_2, CO_2, C_2H_4 and formaldehyde, as well as different C–X bonds (X=Cl, F, O) (Scheme 5). It is worth highlighting the importance of steric effects: while the Zr/P pair activates dihydrogen at ambient temperature, the analogous complex with unsubstituted cyclopentadienyl ligands ($\eta^5\text{-}C_5H_5$) is unreactive under identical conditions.

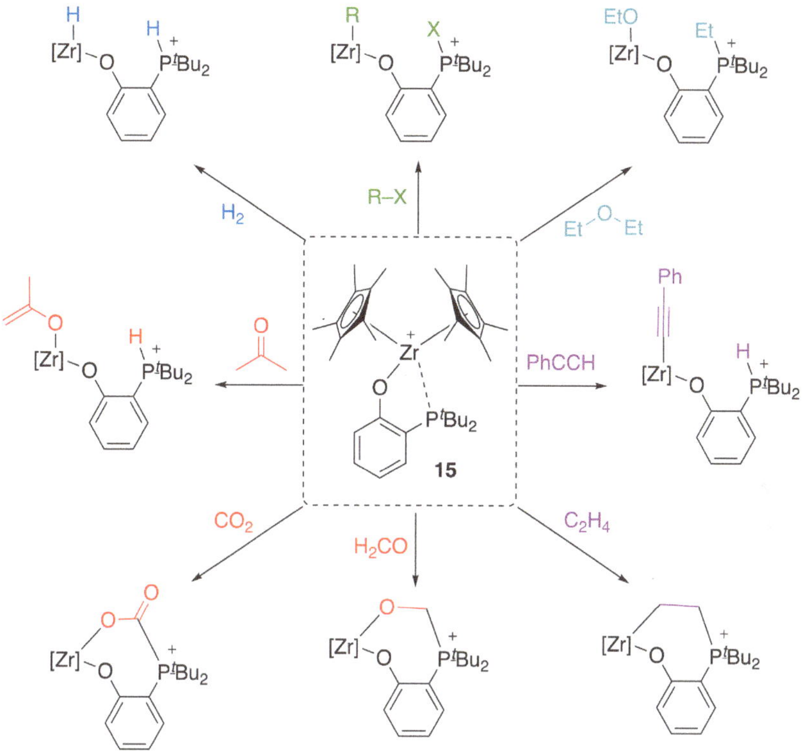

Scheme 5 Reactivity of a Zr/P FLP system toward small molecules where [Zr] denotes the [(η^5-C$_5$Me$_5$)Zr] fragment.

The most appealing profit of the incorporation of transition metals in frustrated Lewis pair systems is their potential for catalysis, mostly due to the aforementioned specific properties of transition metals in comparison with main group elements. Different Zr/P pairs (**15–18**) have been described as efficient homogenous catalysts for the hydrogenation of alkenes, alkynes and imines, as well as the dehydrogenation of amine-boranes.[41,45,51,52] The use of electron-rich transition metals as Lewis bases in cooperation with a boron Lewis acid has also been used for the hydrosilylation of unsaturated systems. The group of Berke combined an electron rich rhenium hydride with the bulky acidic B(C$_6$F$_5$)$_3$ borane (**19**) in order to fixate CO$_2$ in a FLP-type fashion followed by reduction to the hydrosilylated products.[53] Similarly, Peters described the use of a cobalt complex bearing a Z-type ligand with a Lewis acidic boron group (**20**) as a powerful catalyst for the hydrosilylation of carbonyl compounds under mild conditions.[54] More recently, a gold(I) complex with a bulky amino group within the ligand framework (**21**) has been described as an active catalyst for the isomerization of propargylic esters to dienyl esters, which are very valuable entities for several organic synthetic processes. The interaction between the amine nitrogen atom and the gold center is particularly weak and the stereoselectivity of the reaction can be effectively controlled through an appropriate choice of the ligand substituents.[55]

In general, rare-earth elements have been less explored than transition metals, but its choice as Lewis acids to build TMFLPs becomes evident considering their widespread use as strong Lewis acids in catalysis.[56] For instance, Piers and Eisenstein combined the [(η^5-C$_5$Me$_5$)$_2$Sc]$^+$ cation with the hydrido-(perfluorophenyl)-borate anion [HB(C$_6$F$_5$)$_3$]$^-$ (22) to trap small molecules such as CO or CO$_2$ in the polarized Sc$^+$/HB pocket and subsequently activate them in an FLP-type manner by hydride transfer from the borate anion.[57,58] This approach allowed developing an efficient cooperative method for the deoxygenative hydrosilylation of CO$_2$. In addition, Xu examined several rare-earth complexes of type [RE(OAr)$_3$] (RE=Sc, Y, Sm, La; Ar=2,6-^{t}Bu$_2$-C$_6$H$_3$) in combination with commercially available phosphines as Lewis bases. Such rare-earth based frustrated Lewis pairs (23) are capable of initiating the polymerization of polar alkenes through an FLP-like 1,4-addition of the substrate across the intermolecular Lewis pair (Fig. 3).[59,60]

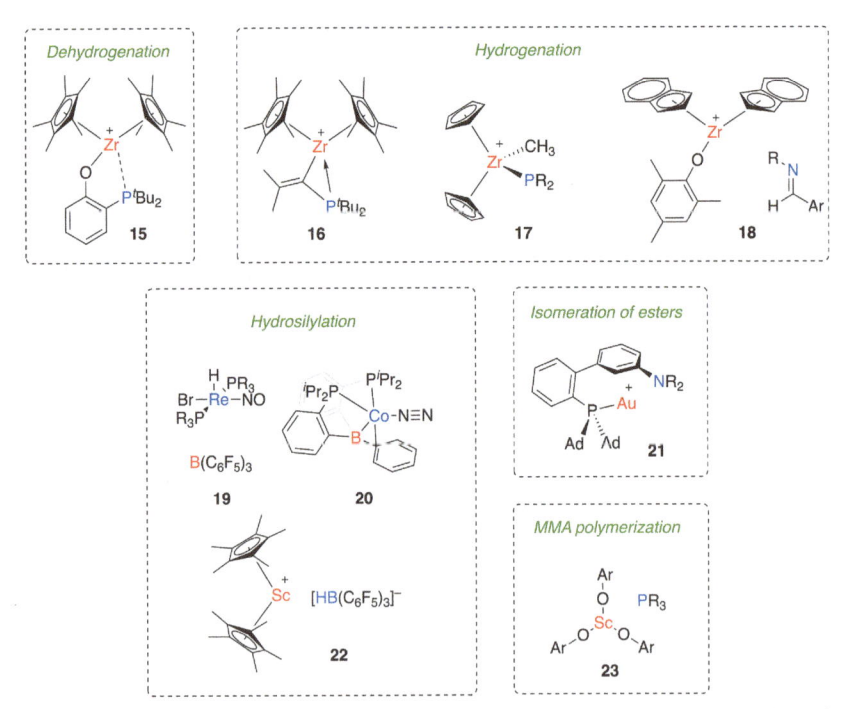

Fig. 3 Representative examples of transition metal frustrated Lewis pairs as catalysts for a variety of catalytic transformations (red=Lewis acid; blue=Lewis base).

3. Bimetallic complexes for bond activation
3.1 General concepts

The elucidation of the structure of $Mn_2(CO)_{10}$ by X-ray diffraction methods in 1957 unambiguously recognized for the first time the existence of M–M bonding.[61] Shortly after, Cotton provided the first example of multiple bonding between metals in $[Re_2Cl_8]^{2-}$, with a quadruple Re—Re bond, thus tearing apart the common belief of a maximum bond order of three.[62] These seminal discoveries paved the way for a rapid development of binuclear and polynuclear chemistry[63–65] which, however, has later witnessed just intermittent periods of progress over the last decades. At variance, the relatively simpler chemistry of monometallic systems has continuously advanced. The fields of organometallic and coordination chemistry have intensively focused on the development of monometallic catalysts, in which the tunable structural parameters are basically the metal center and the ligands (Fig. 4). In this respect, there has been outstanding progress over the last decades on the successful preparation and application of active complexes in a broad scope of homogeneous catalytic transformations.[66]

The renewed interest on the use of Earth-abundant first-row transition metals and the rapid development of cooperative chemistry in the context of catalytic applications, have prompted many research groups to look at nature for inspiration. This is because many inorganic cofactors in biological systems commonly incorporate multiple metals that function in concerted ways to bind/activate chemical substrates. The interest on mimicking the biological inorganic cofactors in order to improve the activity and selectivity in

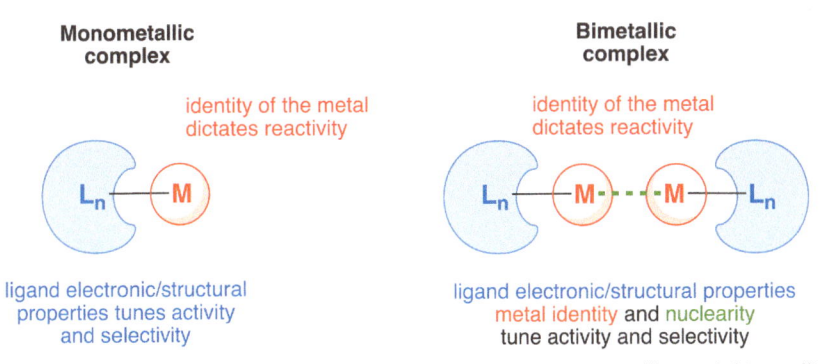

Fig. 4 Structural and electronic differences between monometallic and bimetallic systems.

different catalytic processes has once more encouraged the study of multimetallic, and especially bimetallic, complexes and their application in different homogeneous catalytic transformations.[67,68] The incorporation of a second metal center (mononuclear vs binuclear) presents additional structural and electronic tunability (Fig. 4). In these cases, the active site is largely determined by the cooperative effects between the two metal centers and therefore it can be easily modified providing a wide available space for reaction discovery.[11,69,70]

The interaction of a substrate (e.g. dihydrogen molecule) with a bimetallic active site followed by bond breaking/formation through the cooperative participation of the two metal centers has been a growing case of study.[71,72] Although the amount of possible mechanistic scenarios is broad and delineating clear differences among those is not always obvious, we propose to consider three general scenarios (Fig. 5). These three generic mechanistic proposals are based on the key transition state, in which the bond breaking of the substrate molecule occurs. The first activation mechanism involves the orthogonal reactivity of a metal fragment that initially activates the substrate, followed by subsequent reactivity on the second metal fragment toward the final product (Fig. 5A). Alternatively, the substrate molecule can be added directly across the metal–metal bond in a concerted fashion in what could be considered as a classic bimetallic activation (Fig. 5B). In the third route, the substrate molecule can be cleaved by the cooperative action of two independent metallic fragments by a genuine FLP-like mechanism (Fig. 5C).

Binuclear Bond Activation

	a) Orthogonal Activation	b) Classic Bimetallic Activation	c) FLP-like Activation

	X---┆---Y
	┆
	M----M
		X----Y
		┆ ┆
		M——M
			X----Y---M
			   ┆
			   M

metals involved in bond activation	1	2	2
M–M bond in transition state	Allowed	Yes	No
substrate coordination	Classic	Lateral to M–M	Side-on/End-on

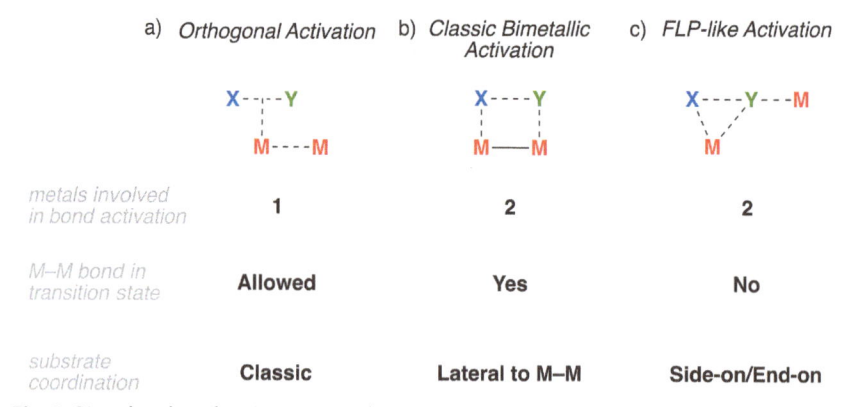

Fig. 5 Binuclear bond activation involving (A) orthogonal activation, (B) classic bimetallic activation or (C) FLP-like activation.

Notwithstanding, these categories merely represent a simplification of a colorful palette of mechanistic options that could even co-exist for a precise bimetallic design and for which the boundaries may be diffuse. We believe, however, that this rationalization may offer some aid in categorizing the extensive variety of bimetallic approaches to bond activation and catalysis and facilitate a better understanding of the grounds on which those systems operate.

3.2 Binuclear orthogonal activation

The use of binuclear systems in the development of efficient catalytic transformations has been exploited for many years. However, the detailed examination of their mechanistic pathways is steadily resurging thanks to many contemporary research groups.[73–76] In the most typical approach, two independent monometallic complexes are mixed up with the substrate(s) and additive(s) in one-pot. Two main mechanistic alternatives can in principle be envisaged. In the first, the two monometallic species operate in an orthogonal independent manner (i.e. orthogonal tandem catalysis), in which the two catalytic cycles are not interconnected. Alternatively, these two fragments may cooperate in an orthogonal co-dependent fashion in which two catalytic cycles are connected as a result of a combination of two intermediate species. Some representative examples for each of those scenarios will be briefly presented in the following sections.

3.2.1 Orthogonal tandem catalysis

Orthogonal tandem catalysis implies performing multiple reactions in a one-pot mode where multiple catalysts and reagents are combined in a single reaction vessel undergoing sequential catalytic processes through two or more functionally distinct catalytic cycles. In addition, it also permits thermodynamic leveraging due to the coupling of the multiple catalytic cycles.[77,78] Pioneer work from Brookhart and Goldman described for the first time alkane metathesis using homogeneous catalysts. To do so they combined olefin hydrogenation/dehydrogenation with olefin metathesis to access medium-weight (C_3–C_8) alkanes desirable for diesel and jet fuels, and also heavier *n*-alkanes for natural gas liquids.[79] For these processes, a combination of iridium pincer catalysts **24** and Schrock-type Mo catalyst **25** was used. In a first step, the iridium pincer complex performs catalytic dehydrogenation of the alkanes, which then are coupled to each other through Mo-catalyzed metathesis. Finally, the heavier olefin is again hydrogenated by the in situ generated iridium dihydride species (Scheme 6).

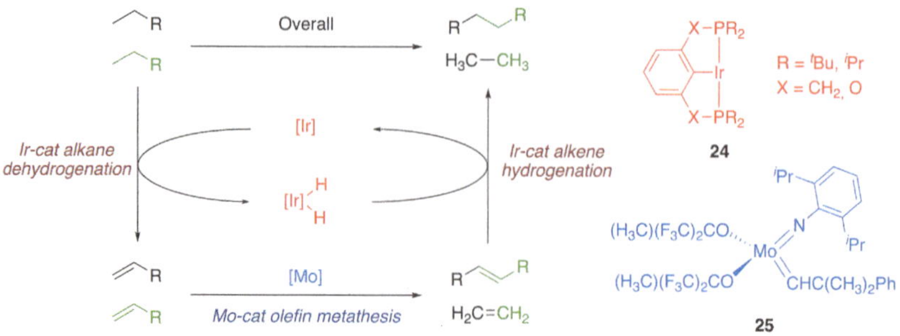

Scheme 6 Alkane metathesis via orthogonal tandem dehydrogenation-olefin metathesis.

As stated above, the two active species in orthogonal binuclear tandem catalysis are in most cases two different independent monometallic complexes. However, Mankad has recently shown that a formally bimetallic complex (NHC)Cu[M$_{CO}$] **26** (NHC = N-heterocyclic carbene; [M$_{CO}$] = [(η^5-C$_5$H$_5$)M(CO)$_n$] where M stands for Fe, Mo, W) with a labile M–M interaction functions as two independent entities in the catalytic deoxygenation of CO$_2$ with pinacolborane (HBpin) (Scheme 7).[80] Pre-activation of the bimetallic complex through cleavage of the Cu–M bond is essential to assess the catalytic transformation. This occurs by initial reaction with HBpin to form the corresponding copper hydride **27** and boryl complex [M$_{CO}$]Bpin **28**. This elementary step consists in the activation of the B–H bond across the Cu–M bond in an event that we described above as "classic bimetallic activation" (see Fig. 5B and Scheme 7). However, in terms of the overall catalytic cycle of CO$_2$ deoxygenation this represents a clear example of orthogonal tandem catalysis, as it will be immediately noted. Formation of the monometallic species is followed by insertion of CO$_2$ onto the (NHC)Cu–hydride bond and subsequent reaction with HBpin to release HCO$_2$Bpin. Finally, the [M$_{CO}$]Bpin unit reacts with the released borylated intermediate, HCO$_2$Bpin, to form CO and Bpin$_2$O. Overall, the roles of the two catalyst components are independent (CO$_2$ reduction and C–O bond cleavage due to copper and its accompanying metal, respectively), while the bimetallic (NHC)Cu[M$_{CO}$] **26** is a dormant off-cycle species.

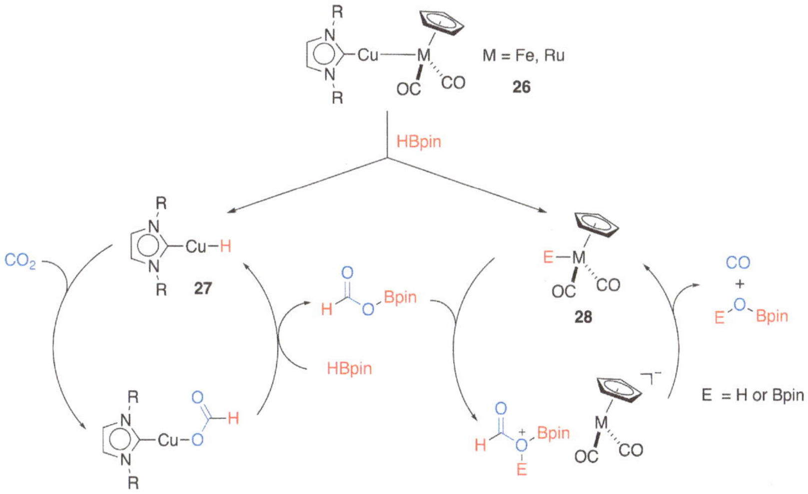

Scheme 7 Catalytic CO_2 deoxygenation by orthogonal tandem catalysis.

In a recent striking work, Hoveyda introduced the concept of "delayed catalysis" for the direct enantioselective conversion of nitriles to NH_2-amines, as a variant of a tandem catalytic transformation.[81] This work involves a copper-based catalytic process with two competing catalytic cycles (Scheme 8). Intriguingly, the incorporation of a nonproductive side catalytic cycle to delay the more active catalyst is key for the production of the desired product. In this particular example, even if it cannot be strictly considered a bimetallic process, the two different catalytically active species (NHC)Cu–H **29** and (NHC)Cu–Bpin **30** are generated concurrently from precatalyst (NHC)CuO*t*Bu **31**. The more reactive Cu—H species hampers the desired initial Cu–Bpin addition to the allene. Thus, in order to favor the Cu–Bpin addition to the allene, which is required for subsequent allyl addition to the nitrile, a strategical feeding of the reaction mixture with excess of inexpensive *t*BuOH and polymethylhydrosiloxane (PHMS) engages the Cu—H complex in a nonproductive side cycle. It is worth noticing that Cu–Bpin did not react in the short term with any of the added additives, but rather added to the allene to accomplish turnover after subsequent nitrile coupling. Finally, protonolysis and CuH-catalyzed ketimine reduction lead to the final desired amine **32**.

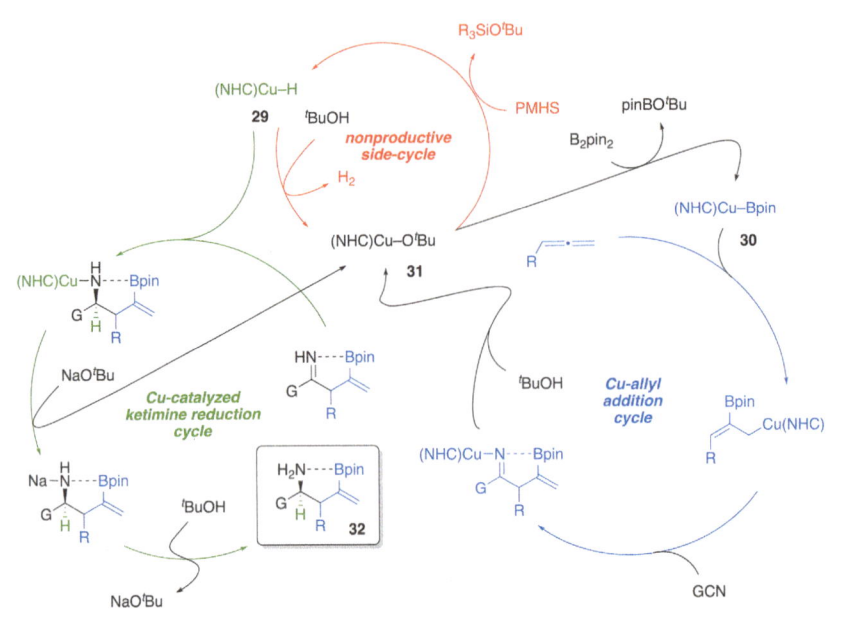

Scheme 8 Binuclear copper-catalyzed enantioselective conversion of nitriles to NH$_2$-amines by incorporation of a nonproductive side-cycle ("delayed catalysis").

3.2.2 Orthogonal co-dependent catalysis

In contrast to orthogonal tandem catalysis, in bimetallic co-dependent catalysis the two catalytic cycles are somehow interconnected. Bimetallic co-dependent catalysis has been intensively explored in the last years to develop efficient binuclear catalytic C—C and C–X coupling transformations.[82–85] A good representative example of such binuclear transformations are the bimetallic catalytic cross-Ullmann reactions envisioned by the group of Waix (Scheme 9).[86] In a one pot procedure, a Pd(0) catalyst **33** bearing the bidentate 1,3-bis(diphenylphosphino)propane (dppp) ligand reacts selectively with aryl triflates through oxidative addition to afford the persistent [(dppp)Pd(II)-Ar]$^+$ intermediate, which accumulates in solution. In parallel, the bipyridine nickel(0) complex **34** reacts preferably with aryl bromides to form a [(bpy)Ni(II)Ar′]$^+$ (bpy=2,2′-bipyridine) transient intermediate. The latter rapidly reacts with the palladium fragment [(dppp)Pd(II)-Ar]$^+$ through a transmetallation process consisting in exchanging the aryl moiety bound to nickel by the weakly coordinating triflate anion. The target coupled biaryl Ar–Ar′ products are formed after reductive elimination from the corresponding Pd(II) intermediate, while Ni(0) is regenerated by the reducing action of Zn(0). It is worth mentioning that this bimetallic catalyzed cross-Ullmann reaction represented a great advance on the development of methods to selectively form unsymmetrical biaryls.

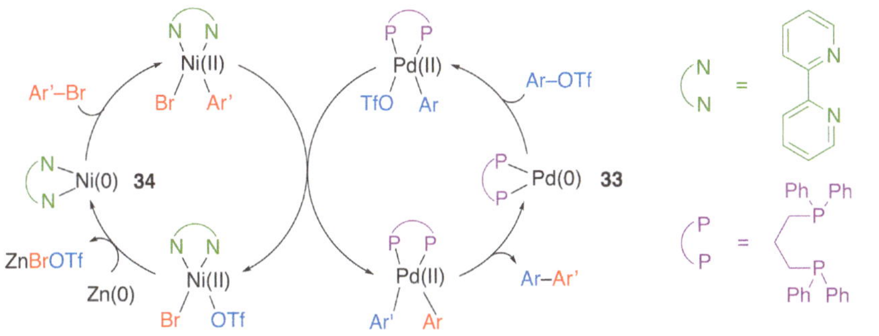

Scheme 9 General orthogonal co-dependent mechanism of the cross-Ullmann reaction catalyzed by combination of nickel and palladium catalysts.

3.3 Homobimetallic activation

Complexes with metal–metal bonds of various bond orders have been long known to participate in binuclear bond activation reactions.[87–93] Homobinuclear species usually rely on the presence of bridging ligands in order to hold the two metal centers in close proximity, as well as on metal–metal bond interactions. It is therefore possible that bond activation reactions take place either through the addition of the substrate molecule to the metal–metal bond, thus altering the bond order, or directly to one or both metal centers without affecting the metal–metal bond.

The coupling of an alkyne, an alkene and CO to yield a cyclopentenone in an overall [2 + 2 + 1] cycloaddition, the so-called Pauson-Khand reaction, was first catalyzed by the bimetallic complex [$Co_2(CO)_8$] **35**, representing a paradigmatic example of bimetallic catalysis in which the integrity of the M–M bond remains.[94–96] Some years later, Nakamura elucidated the mechanism of this transformation by means of DFT calculations.[97] The perpendicular addition of the alkyne to the Co—Co axis through orthogonal π-bonding interactions is crucial for subsequent alkene coordination and insertion into a Co—C bond. Besides, coordination of the alkyne leads to a change of hybridization for the carbon atoms (from sp to sp^3) and an important decrease of the C—C bond order.[98] However, since the addition of the alkyne directly takes place at the cobalt centers there is no apparent alteration on the Co—Co bond. Finally, migratory insertion of a CO ligand and subsequent reductive elimination of the organic product regenerates the [$Co_2(CO)_6$] precatalyst (Scheme 10A). Drawing on the same theme, the group of Uyeda recently reported a detailed study of dinuclear Ni—Ni complexes as intermediates for the Pauson-Khand reaction.[99] In addition, ligand assisted oxidative addition of Br_2 to the well-defined d^9–d^9 dinickel complexes **36** supported by a naphthyridine-diimine pincer ligand (NDI) was

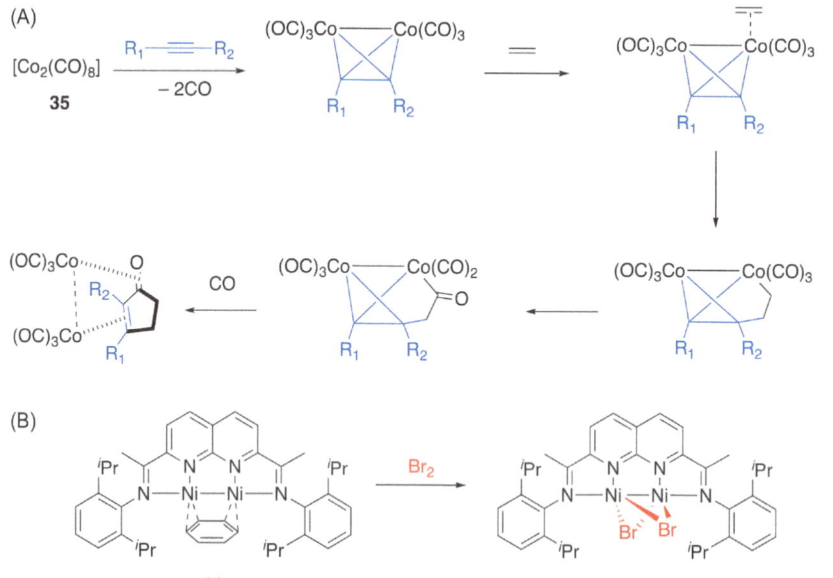

Scheme 10 Pauson-Khand reaction catalyzed by [Co$_2$(CO)$_8$] complex (A) and oxidative addition to a (NDI)dinickel complex (B).

effectively proved, revealing very little changes on the Ni—Ni bond length distances. Once more, these results highlight minimal change in the metal–metal bond order during bimetallic bond activation (Scheme 10B).[100]

In the context of homobimetallic complexes, the first quintuply bonded species, more precisely a dimeric chromium complex stabilized by bulky terphenyl ligands and reported by Power in 2005, represents a landmark discovery.[101] After this outstanding achievement, the study of multiply bonded species has grown and different groups have intensively exploited the chemistry of Cr—Cr and Mo—Mo quintuply bonded complexes stabilized by nitrogen donor ligands showing novel reactivity toward different small molecules.[102–107] Interestingly, the group of Yu and Tsai described that quintuply bonded dimolybdenum amidinates easily activate terminal and internal alkynes. While reaction of [Mo$_2${μ-κ^2-RC(N-2,6-^{i}Pr$_2$C$_6$H$_3$)$_2$}$_2$] **38** with terminal alkynes gave rise to the [2+2+2] cycloaddition products, the reaction with internal alkynes afforded the [2+2] cycloaddition adducts **39** through an oxidative addition transformation, in which the alkyne sits in a symmetric bridging geometry along the metal–metal bond (Scheme 11).[108] As a result, the two amidinate ligands adopt a bent geometry to accommodate the newly formed dimetallacyclobutene, while the Mo—Mo bond order is reduced to four. In

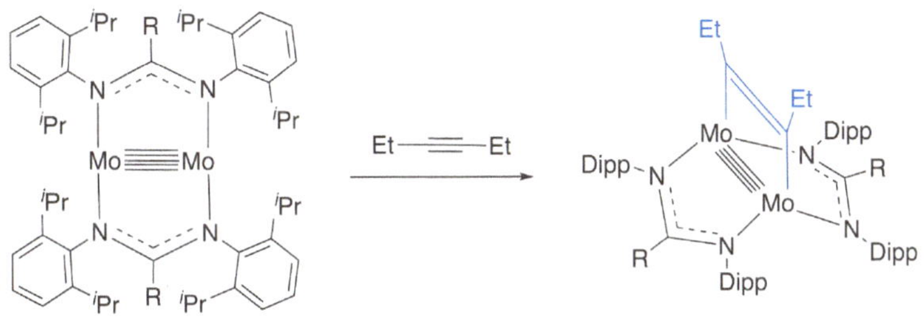

Scheme 11 Reaction of [Mo$_2$\{μ-κ^2-RC(N-2,6-^{i}Pr$_2$C$_6$H$_3$)$_2$\}$_2$] with an internal alkyne to afford the [2+2] cycloaddition product.

addition due to the reduced stability of quintuply bonded metals, the reactivity of quadruple M–M bonds has been investigated to a wider extent.[63,109]

The reactivity of the aforementioned quintuply bonded dimolybdenum complex **38** has been further explored,[105–107] including its application as a catalyst for the synthesis of benzene derivatives by alkyne trimerization.[110] The role of the quintuple bond to accomplish this catalytic transformation was later evaluated by theoretical means by the group of Sakaki.[89] Additionally, in a more recent computational study, the same authors introduced for the first time the notion of multiple metal–metal bonds behaving as frustrated Lewis pairs.[111] To substantiate this idea, the authors suggest that the polarization of the M–M multiple bond parallels the polarization found in FLP systems. As such, the key orbitals involved in σ-bond cleavage during oxidative addition of H—H, C—H and O—H bonds over the quintuply bonded dimolybdenum complex **38**, get polarized in the transition state (Scheme 12). This facilitates the charge transfer from the M–M bond to the antibonding σ*$_{EH}$ orbital while weakening the exchange repulsion between the multiple M–M bond and the E–H substrate. This concept may encourage others to investigate multiply bonded bimetallic complexes from a FLP-like perspective due to the demonstrated facile M–M bond polarization.

In addition, polarization of homobimetallic complexes can be induced through intramolecular ligand rearrangements, which can also lead to mixed-valence complexes.[112–114] Consequently, these systems have the ability to activate different type of substrates due to its polarized electronic properties.[115,116] As an example, Nocera and co-workers described the cooperative activation of dihydrogen in a two-electron mixed-valence diiridium complex (Scheme 13).[117] In a first step, disproportionation of a

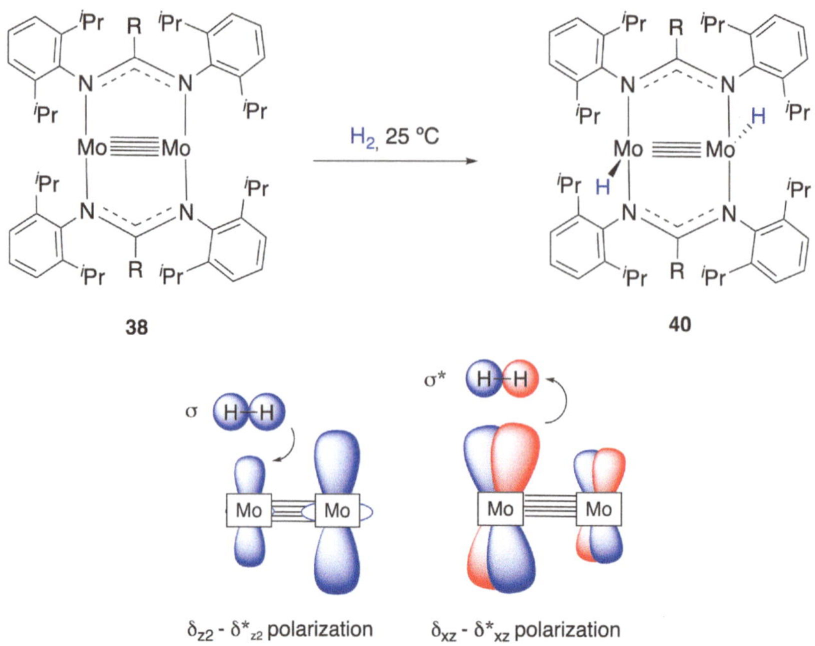

Scheme 12 Addition of dihydrogen to the quintuply bonded Mo₂ complex and simplified representation of the polarized δ_{Mo2} orbitals participating in the E–H bond cleavage reminiscent of FLPs.

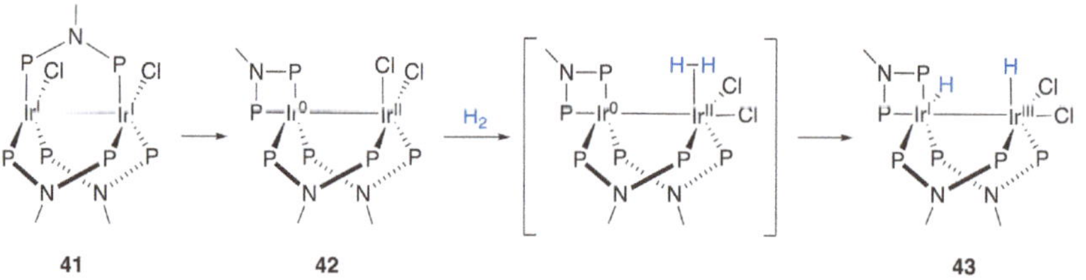

Scheme 13 Dihydrogen activation over a mixed-valence diiridium complex (P=P (OCH₂CF₃)₂).

symmetric $Ir_2^{I,I}$ (d⁸–d⁸ system) **41** into its related mixed-valence $Ir_2^{0,II}$ (d⁷–d⁹ system) **42** results in the generation of an empty orbital at the Ir(II) metal center that permits H₂ coordination. Subsequently, oxidative addition takes place affording an Ir^I-hydride/Ir^{III}-hydride bimetallic complex **43**.

3.4 Polar heterobimetallic complexes

As discussed in the previous section, polarized M–M bonds in homobimetallic complexes can exhibit a non-obvious reactivity that is reminiscent of FLP systems. In turn, heterobimetallic complexes present intrinsic polarization at the M–M bond and, therefore, tend to exhibit a cooperative reactivity that may be sometimes assimilated as FLP-type reactivity.[118–122] However, in accordance to the simplified categorization proposed in Fig. 5, we have here differentiated between these heterobimetallic complexes and those bimetallic entities that could be considered as genuine FLPs. While for the latter there is no M–M bonding in the key transition state involved in a bond activation event, in polar heterobimetallic complexes the integrity of the M–M bond would remain virtually intact in the key transition state or, at least, get cleaved in a concerted fashion along with substrate bond cleavage/formation.

As seen for homobimetallic complexes, a wide range of bridging ligands have been used to bring into proximity the two metal centers in polar heterobimetallic systems. In a pioneer study, Bergman studied the cooperative activation of a variety of small molecules with no or low polarity (e.g. dihydrogen and silanes) as well as more polar ones such as alcohols, amines or sulfides by reaction with the early-late heterobimetallic [(η^5-C$_5$H$_5$)$_2$Zr(μ-N^{t}Bu)Ir(η^5-C$_5$Me$_5$)] complex **44** (Scheme 14).[123,124] Despite the remarkable reactivity of the heterodinuclear complex, mechanistic investigations were not pursued and it is therefore difficult to stablish whether the scission

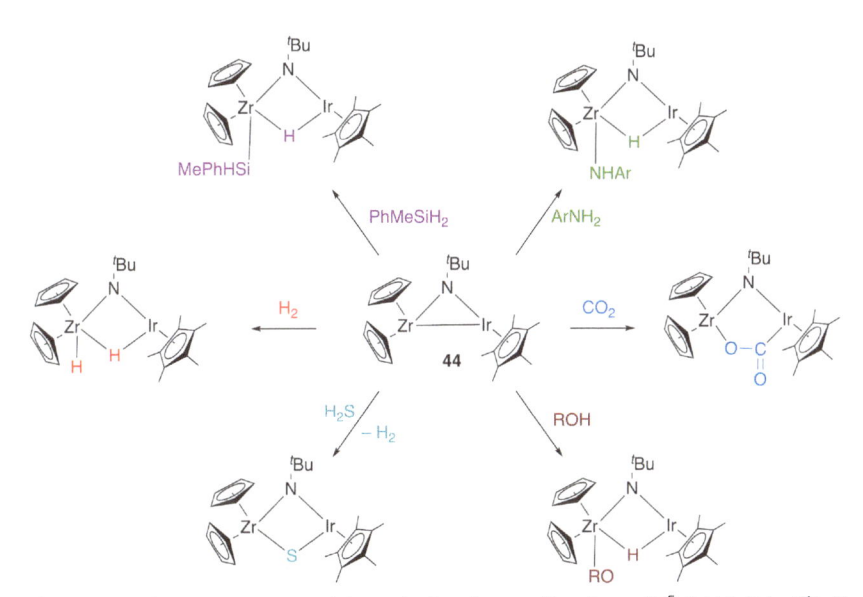

Scheme 14 Cooperative reactivity of the heterodinuclear [(η^5-C$_5$H$_5$)$_2$Zr(μ-N^{t}Bu)Ir(η^5-C$_5$Me$_5$)] complex with small molecules.

of the metal–metal bond precedes the transformation of the substrate or whether it occurs at a later stage. Hence, we favor its discussion in this section since the existence of the Ir—Zr bond in the precursor is beyond any doubt.

In more contemporary studies, Thomas has focused on early-late heterobimetallic complexes in which two transition metals are linked through a bis- or tris-(phosphinoamide) ligand framework. For instance, the system [(THF)Zr(MesNP^{i}Pr$_2$)$_3$CoN$_2$] (Mes=2,4,6-trimethylphenyl) has demonstrated versatile multielectron reactivity toward a wide range of small molecules.[125–127] In most cases, cooperative bond activation involved simultaneous substrate binding to both metal centers. Activation of dihydrogen occurs at room temperature through the addition of two equivalents of H$_2$ to the [(THF)Zr(MesNP^{i}Pr$_2$)$_3$CoN$_2$] complex **45** with concomitant P—N bond cleavage of the phosphinoamide ligand (Scheme 15).[128] Alternatively, the activation of dihydrogen by the related heterobimetallic complex **47** bearing only two bridging phosphinoamide ligands proceeds under mild conditions directly across the metal–metal bond without any ligand rearrangement reaction and only incorporating one equivalent of H$_2$.[129] The versatility of these Zr(IV)/Co(—I) heterobimetallic systems has also allowed for the activation of more challenging substrates such as O$_2$ in a controlled manner.[130]

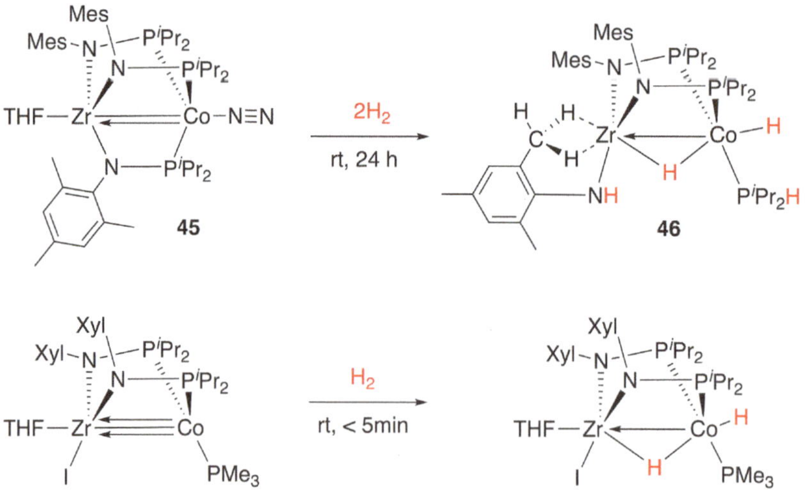

Scheme 15 Hydrogen activation by Zr(IV)/Co(—I) heterobimetallic complexes with tris- and bis-(phosphonamide) bridging ligands.

Despite the utility of bridging ligands to maintain two metals in close proximity, their presence is not always required to access bimetallic structures. Polar heteronuclear bimetallic complexes without bridging ligand frameworks that are only stabilized through metal–metal interactions (M → M' dative bonds between a Lewis basic and a Lewis acidic metal fragment) are better known as metal only Lewis pairs (MOLPs).[131] However, it was Nowell and Russell who first described in 1967 a Lewis acid-base complex of this kind by reporting the X-ray molecular structure of [(η^5-C$_5$H$_5$)(CO)$_2$Co → HgCl$_2$].[132] Since this achievement and owing to the growing interest in bimetallic systems, the investigation of new Lewis pairs constructed around M → M dative bonds has been a topic of great interest in order to further investigate the key role of metal-metal interactions in catalytic transformations.[131,133] Despite the large scope of metal donor Lewis bases examined over the years, linear platinum(0) complexes such as [Pt(PCy$_3$)$_2$] (**50**, Cy=cyclohexyl) or [Pt(P^{t}Bu$_3$)$_2$] have been the most employed as relatively strong Lewis bases. Braunschweig and co-workers have explored the formation of a wide diversity of unsupported dative bonds between transition metal bases and *s*- and *p*-block metal acidic fragments.[134–137] In some of these reports, the exchange of GaCl$_3$ or AlCl$_3$ between different transition metal bases is discussed as a method to gauge Lewis basicity (Scheme 16).[138,139] These exchange studies proved the lability and dynamic behavior of the M → M bond, suggesting strong similarities with main group phosphino-borane adducts, and foreseeing a great potential to act as thermally induced FLP systems.

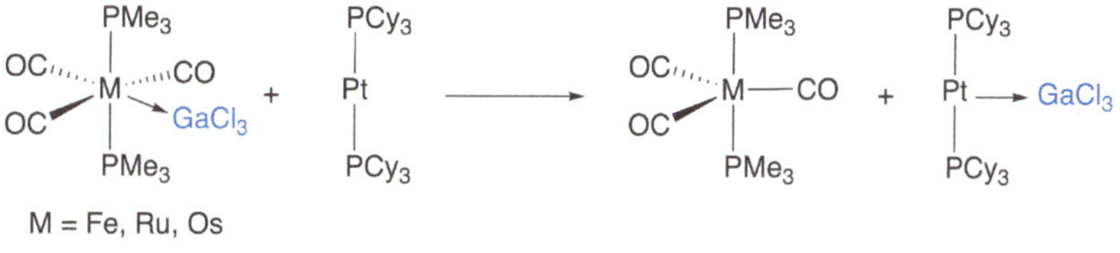

Scheme 16 Transfer of the Lewis acid GaCl$_3$ from M (M=Fe, Ru, Os) carbonyl complexes to [Pt(PCy)$_2$].

Despite the aforesaid potential, the use of MOLPs for the activation of small molecules has not yet been studied in depth. In a pioneer work, Cutler showed that bimetallic [(η^5-C$_5$H$_5$)(CO)$_2$M → Zr(Cl)(η^5-C$_5$H$_5$)$_2$] (M=Fe, Ru) complexes **53** and **54** react with CO$_2$ to yield the corresponding bimetallocarboxylates **55**, yet the FLP-like analogy could not be depicted at that time.[140,141] The bimetallocarboxylates are stabilized by push–pull interactions derived from the Lewis basic group 8 compound and the electrophilic character of the zirconium fragment (Scheme 17). Two plausible mechanistic pathways for the addition of CO_2 onto the M→Zr bond were considered. In the first one, initial ionization of the bimetallic [(η^5-C_5H_5)(CO)$_2$M→Zr(Cl)(η^5-C_5H_5)$_2$] complex takes place to give the ionic [(η^5-C_5H_5)M(CO)$_2$]$^-$ **56** and cationic zirconocene **57**. Subsequently, the electron rich [(η^5-C_5H_5)M(CO)$_2$]$^-$ fragment may intercept the CO_2 molecule by attacking its electrophilic carbon to finally afford the bimetallocarboxylate **55** upon zirconium coordination. This mechanism is highly reminiscent of CO_2 activation by FLPs.[20–22] On the contrary, carbon dioxide could insert directly onto the M→Zr bond *via* a bifunctional cooperative CO_2 activation step leading to the final bimetallocarboxylate product. The authors provided some experimental support for the later activation mechanism, however no computational studies were performed and no information about the M→Zr bond in the suggested transition state was discussed.

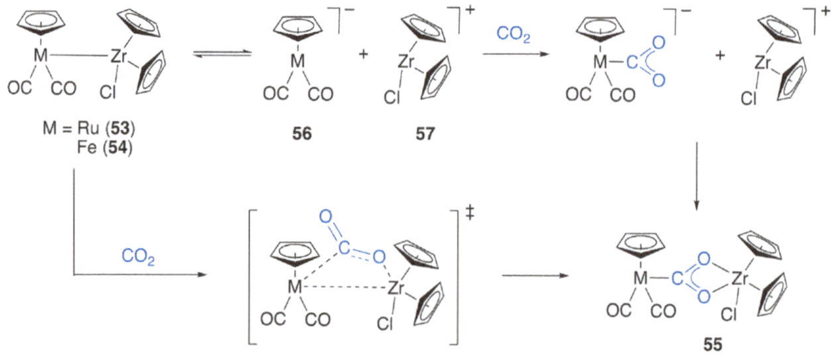

Scheme 17 Bimetallic CO_2 activation by a Zr–M (M=Ru, Fe) metal only Lewis pair.

More recently, Jamali demonstrated the use of the metal-only Lewis adduct **58** [(P^{t}Bu$_3$)$_2$Pt→Cu(NCMe)$_n$] to activate a polar O—H bond in water (by using wet acetone as a solvent) to generate a cationic Pt(II) hydride **59** and copper oxide **60** (Scheme 18A).[142] Three possible reaction pathways were considered during their preliminary mechanistic studies.

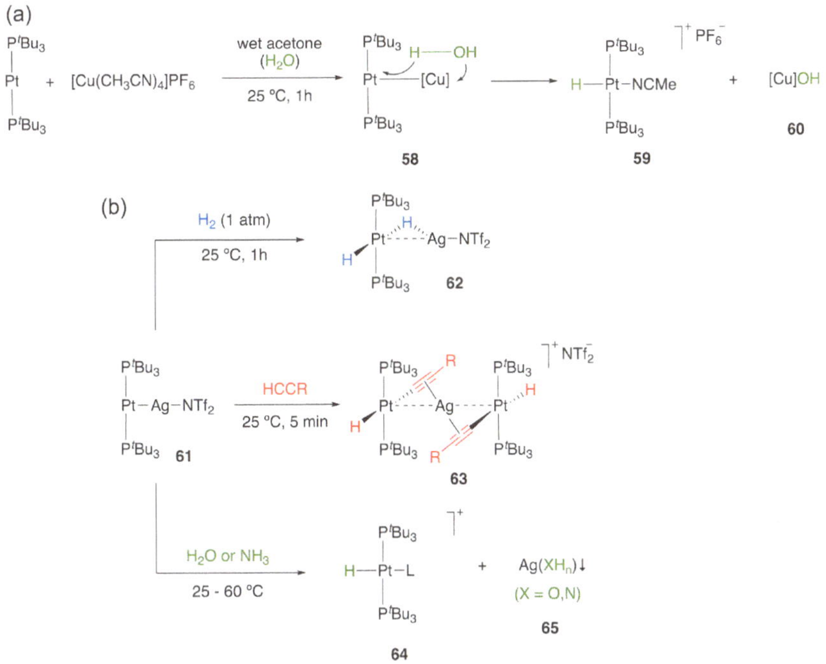

Scheme 18 Activation of water, dihydrogen, alkynes and ammonia by metal-only Lewis pairs based on [Pt (P^{t}Bu$_3$)$_2$].

Direct oxidative addition of water to the platinum(0) complex to form a hydrido,hydroxo-platinum(II) complex followed by transmetallation to copper was ruled out on the basis of NMR spectroscopic observations. Similarly, the acidification of one O—H bond upon coordination to copper, followed by Pt(0) deprotonation was discarded on the same grounds. In turn, the presence of a dative Pt→Cu bond as an intermediate during the progress of the reaction was confirmed by NMR spectroscopy, which made the authors suggest a cooperative pathway implying the bimetallic adduct as the likely active species.

Subsequent work by our group described a [(P^{t}Bu$_3$)$_2$Pt→AgNTf$_2$] adduct **61** with a similar architecture, which actively reacts toward X–H (X=H, C, N, O) bonds (Scheme 18B).[143] In contrast with the monometallic [Pt(P^{t}Bu$_3$)$_2$] and AgNTf$_2$ fragments, the metal only Lewis adduct readily reacts with dihydrogen and phenylacetylene under mild conditions to form the corresponding metal hydride **62** and the uncommon heterobimetallic dibridged bisacetylide **63**, respectively. Moreover, the reactivity toward polar X–H bonds in water and ammonia was additionally investigated, generating the oxidized Pt(II) hydride **64** and the corresponding silver salts **65** (Ag(XH)$_n$), where X=N, O). The activation of the N—H bond in ammonia is particularly appealing since it is often difficult to achieve by monometallic transition metal complexes due to formation of unreactive Werner-type adducts.[144–146]

The group of Mankad has highlighted in different occasions the analogy of a variety of unbridged polarized heterobimetallic systems with frustrated Lewis pairs.[72,147] For example, the bimetallic [CuFe] complex **66** in Scheme 19 reacts with different small molecules such as carbon disulfide,[148] iodomethane,[149] benzyl chlorides[150] and dihydrogen[151] in a FLP-like manner. The mechanism for dihydrogen splitting has been intensively investigated by the group, and computational analysis revealed key orbital interactions that resemble classic FLPs (Scheme 19). For instance, in the activation of dihydrogen, there is donation from the σ_{HH} orbital to a copper valence orbital, with concerted back-donation from the Cu—Fe bond toward the antibonding σ^*_{HH} orbital. This process requires high temperatures (ca 150 °C), however the M–M bond suffers only a smooth dissociation across the coordination reaction and is only partially broken at the transition state. It is only upon dissociation of the two metal hydrides **67** and **68** that the Cu—Ru and H—H bonds are completely broken. Similarly, the bimetallic oxidative addition of H–Bpin to the bimetallic [CuFe] complex **66** proceeds through a concerted mechanism, in which the Cu—Fe bond is only partially cleaved at the key transition state.[151,152] Nevertheless, experimental observations evidenced that Mankad's systems display dynamic equilibrium in solutions from the heterobimetallic species toward the monometallic fragments, thus enabling FLP-like activation pathways in the same fashion as thermally induced FLPs.[153,154] Thus, it may be possible that combining these and related polarized heterobimetallic systems with the concept of frustration could provide new modes of reactivity (activity/selectivity) modulation during bond activation and catalysis. For the later, it is important to remark that Mankad's heterobimetallic complexes have revealed highly active in a variety of applications in catalysis capitalizing on the cooperative reactivity of the two metals in close proximity.[72] Besides, these studies provide computational and experimental evidences for E–H bond heterolytic cleavage that are reminiscent of FLPs systems.

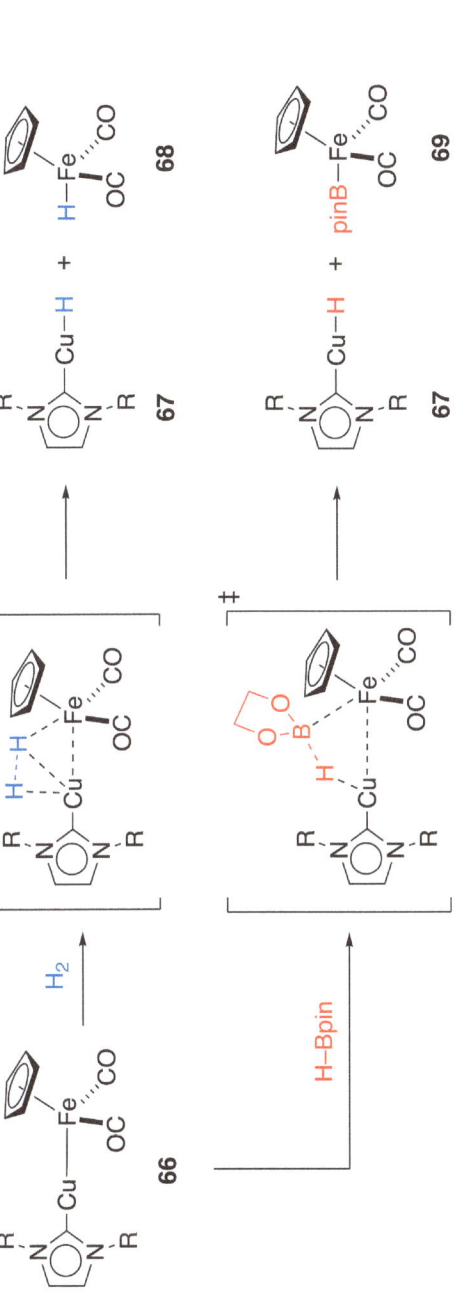

Scheme 19 Heterobimetallic activation of dihydrogen and H–Bpin by [CuFe] complex, highlighting a proposed key transition state.

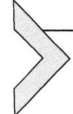

4. Genuine metal-only FLP activation
4.1 Transition metal/main group metal frustrated Lewis pairs

As previously discussed, bimetallic complexes in which only one of the components is based on a transition metal while the other corresponds to a main group metal are well known.[131] The Bourissou group took a step forward and accessed a bimetallic platinum/aluminum complex **70** with a dative Pt→Al bond in a four-membered metallacycle which exhibits a remarkable reactivity.[155] The strain associated to the four-membered metallacycle facilitates insertion of different small molecules along the Pt→Al bond. These reactions proceed under mild conditions in a FLP-manner, that is, the Pt/Al heterobimetallic compound acting as a thermally induced frustrated Lewis pair (Scheme 20).[154] Activation of dihydrogen afforded a trans hydrido-aluminohydride Pt(II) complex **71**, which was formed by irreversible oxidative addition of dihydrogen to platinum and formal insertion of one of the hydrides into the Pt→Al bond. Theoretical studies revealed that dihydrogen is first trapped through a FLP-like manner by the Pt/Al complex via insertion into the Pt→Al bond through an end-on coordination to the basic Pt nucleus and side-one coordination to the acidic Al center, as recurrently proposed for main group based FLPs during dihydrogen activation.[156–160] Subsequent H—H bond cleavage leads to the trans hydrido-aluminohydride Pt(II) complex. Despite the high overall energetic span, each step is facilitated by stabilizing interaction with the Lewis acidic aluminum moicty. It is also worth noticing that in this key transition state the Pt→Al bond is completely broken, thus proving a truly FLP mechanism on the basis of the simplification proposed in Fig. 5. Moreover, activation of other small molecules such as CO_2 (**72**) and CS_2 formed the corresponding adducts which are stabilized by push-pull forces, as occurs in main group FLPs.

Inspired by this work, Brewster and co-workers have recently explored a series of transition metal/aluminum bimetallic complexes with bridging pyridine ligands. The heterobimetallic rhodium-aluminum and iridium-aluminum alkyl complexes were able to activate dihydrogen generating the corresponding alkanes. However, mechanistic investigations indicated that the activation of dihydrogen proceeds through direct oxidative addition at the rhodium or iridium center forming the dihydride species, rather than through a cooperative pathway. The cooperative hydrogenolysis mechanism is energetically unfeasible probably because the transition metal center and the aluminum center are not in close enough proximity.[161,162]

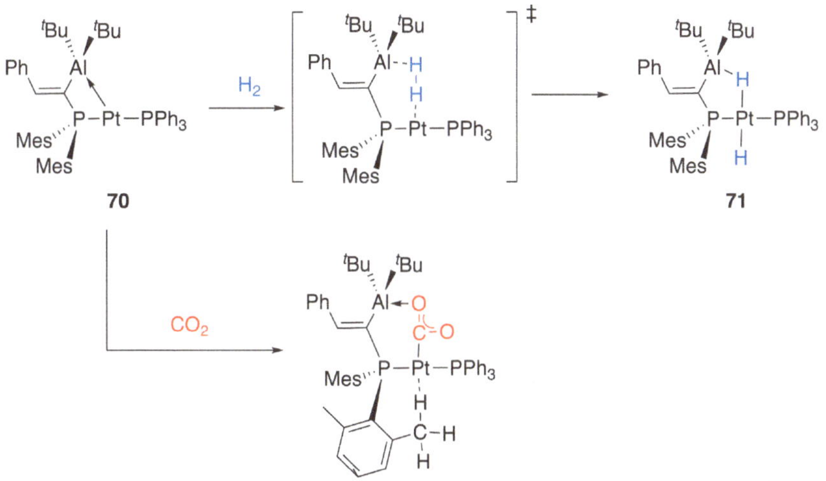

Scheme 20 Activation of dihydrogen and CO_2 in a FLP-type manner along a constraint Pt→Al bond.

4.2 Transition metal only frustrated Lewis pairs (TMOFLPs)

At variance with polarized heterobimetallic complexes that act as FLP-like entities, systems in which the two metal fragments do not present M–M bonds and therefore could be considered as genuine transition metal only Frustrated Lewis pairs (TMOFLPs) are rather scarce. In a first attempt, the group of Wass anticipated the use of a phosphinoaryloxide zirconocene as a suitable framework to coordinate an electron rich Pt(0) center through its pendant phosphine. However the intended Zr(IV)/Pt(0) FLP was not detected, and a new heterobimetallic compound was formed instead by formal insertion of the platinum center into a Zr—C bond.[163] Shortly after, our group described for the first time a truly genuine TMOFLP by combining a gold(I) (**73–75**) and a platinum(0) complex **76** as the Lewis acid and base, respectively.[164] The choice of different bulky phosphine ligands was essential to avoid the formation of a metal-only Lewis pair (MOLP). Modification of the alkyl groups on the terphenyl phosphine ligand that binds the Au(I) fragment had a direct impact on the equilibrium between metal adduct formation and complete frustration (Scheme 21).[165] In addition, the solution equilibria is also strongly affected by solvent effects, in analogy to traditional FLPs.[166] In terms of steric factors, the shielding provided by the three investigated phosphines follows the order $PCyp_2Ar^{Xyl2} > PMe_2Ar^{Dipp2} > PMe_2Ar^{Xyl2}$. NMR spectroscopic studies revealed the capacity to finely control the adduct/frustration equilibrium. Thus, the least congested system

(PMe$_2$Ar^{Xyl2}; Ar^{Xyl2} = C$_6$H$_3$-2,6-(C$_6$H$_3$-2',6'-Me$_2$)$_2$) yielded exclusively the corresponding Lewis pair adduct, while the most hindered (PCyp$_2$Ar^{Xyl2}, Cyp = cyclopentyl) showed complete frustration. In turn, the system with intermediate geometrical constraints (PMe$_2$Ar^{Dipp2}; Ar^{Dipp2} = C$_6$H$_3$-2, 6-(C$_6$H$_3$-2',6'-^{i}Pr$_2$)$_2$) presented an in-between situation in solution, as evinced by broadening of the ^{31}P{^{1}H} NMR signals of the independent monometallic fragments. Although the presence of the monometallic fragments prevails over bimetallic adduct formation, the addition of equivalent amounts of methanol or a simple tetrylene (e.g. germanium dichloride) readily lead to the formation of the corresponding Pt→Au adduct, likely by facilitating triflimide dissociation.[167] Nevertheless, the same did not apply to the bulkier PCyp$_2$Ar^{Xyl2}-based system, where adduct formation is forbidden on steric grounds under all attempted conditions. These results perfectly fit computational studies carried out on this dynamic equilibrium.

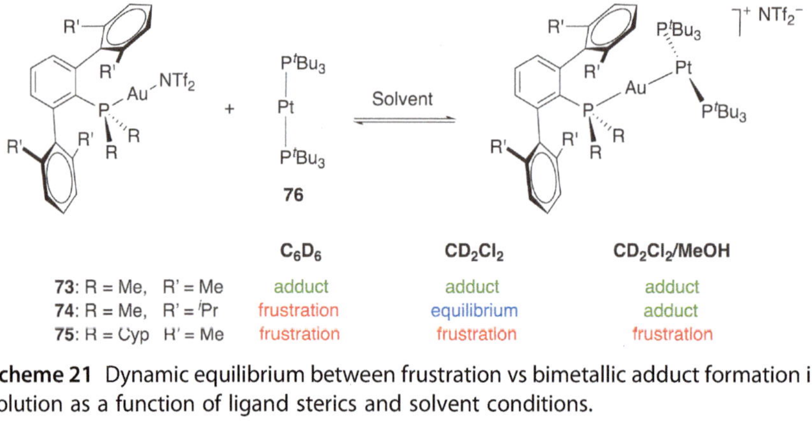

	C$_6$D$_6$	CD$_2$Cl$_2$	CD$_2$Cl$_2$/MeOH
73: R = Me, R' = Me	adduct	adduct	adduct
74: R = Me, R' = ^{i}Pr	frustration	equilibrium	adduct
75: R = Cyp, R' = Me	frustration	frustration	frustration

Scheme 21 Dynamic equilibrium between frustration vs bimetallic adduct formation in solution as a function of ligand sterics and solvent conditions.

The reactivity of these Au(I)/Pt(0) pairs toward small molecules was investigated in order to demonstrate its potential FLP-like reactivity. Activation of dihydrogen[165] and acetylene[168] revealed a strong influence of the equilibrium between Au←Pt adduct and fully frustrated Au(I)/Pt(0) pairs. Thus, the system based on the less congested phosphine (PMe$_2$Ar^{Xyl2}), whose resting state is the bimetallic adduct under all screened conditions, revealed considerably reduced activity for both dihydrogen and acetylene activation. In contrast, the pairs based on PMe$_2$Ar^{Dipp2} and PCyp$_2$Ar^{Xyl2} exhibited a considerably enhanced reactivity. For instance, in the case of dihydrogen activation, the intermediate size phosphine (PMe$_2$Ar^{Dipp2}) effected dihydrogen splitting even at low temperature (−20 °C, $t_{1/2}$ ca 2 h). It is important to state that neither platinum nor gold precursors react with dihydrogen by themselves, thus requiring the cooperative action of the second metal fragment.

The product distribution for the three related systems markedly differs as well (Scheme 22). This is somehow surprising considering dihydrogen as the simplest covalent molecule to activate. Accordingly, dihydrogen splitting with the pair based on [(PMe$_2$Ar^{Xyl2})Au(NTf$_2$)] yields a new heterobimetallic Au/Pt dihydride species, in a process that requires up to 48 h to reach completion at room temperature. In contrast, H—H bond cleavage proceeds immediately for the Pt/Au(PMe$_2$Ar^{Dipp2}) pair to provide an equimolar mixture of the corresponding hydride-bridged digold and [Pt(P^{t}Bu$_3$)$_2$H]$^+$ cation, the former arising from trapping the putative terminal gold hydride by still unreacted gold triflimide. Nonetheless, the foremost mixture evolves over time (ca 12 h) to the analogous heterobimetallic dihydride compound observed for the PMe$_2$Ar^{Xyl2} system. The result is more contrasting for [(PCyp$_2$Ar^{Xyl2})Au(NTf$_2$)], where the only observed gold-containing species is the corresponding hydride-bridged digold, while half of the platinum is converted to the mononuclear dihydride [Pt(P^{t}Bu$_3$)$_2$(H)$_2$]. These differences seem to derive from a set of dynamic equilibrium reactions taking place in solution after dihydrogen splitting that are ultimately dominated by steric factors (Scheme 22).[165]

A series of preliminary experimental observations suggested a truly FLP-type pathway for dihydrogen splitting. First, the reduced activity of the system based on PMe$_2$Ar^{Xyl2} is in agreement with an energetic penalty derived from the need to access the monometallic fragments for H$_2$ splitting to occur, which would not be a prerequisite if traditional heterobimetallic activation was operating. Moreover, the addition of excess amounts (0.5 equiv.) of either gold or platinum precursor, instead of favoring reaction kinetics, had a detrimental effect on the rate of H$_2$ splitting, which once more could be understood in terms of hampering access to monometallic fragments on the basis of the equilibrium depicted in Scheme 21. Computational studies support the notion of a genuine frustrated bimetallic activation, with an overall barrier for the PMe$_2$Ar^{Xyl2} system of $\Delta G^\ddagger = 21.3$ kcal mol^{-1}, in close agreement with experimental observations. The located transition state is represented in Fig. 6, displaying the typical coordination recurrently found for main group FLPs,[156–160] with the basic platinum center laterally approaching dihydrogen, which in turn coordinates to the gold center in a side-on fashion. Electron density analysis confirms bond critical points (bcps) connecting one hydrogen atom to each metal center, as well as another bcp between the two hydrogens. Interaction of the gold center with the triflimide anion proved to

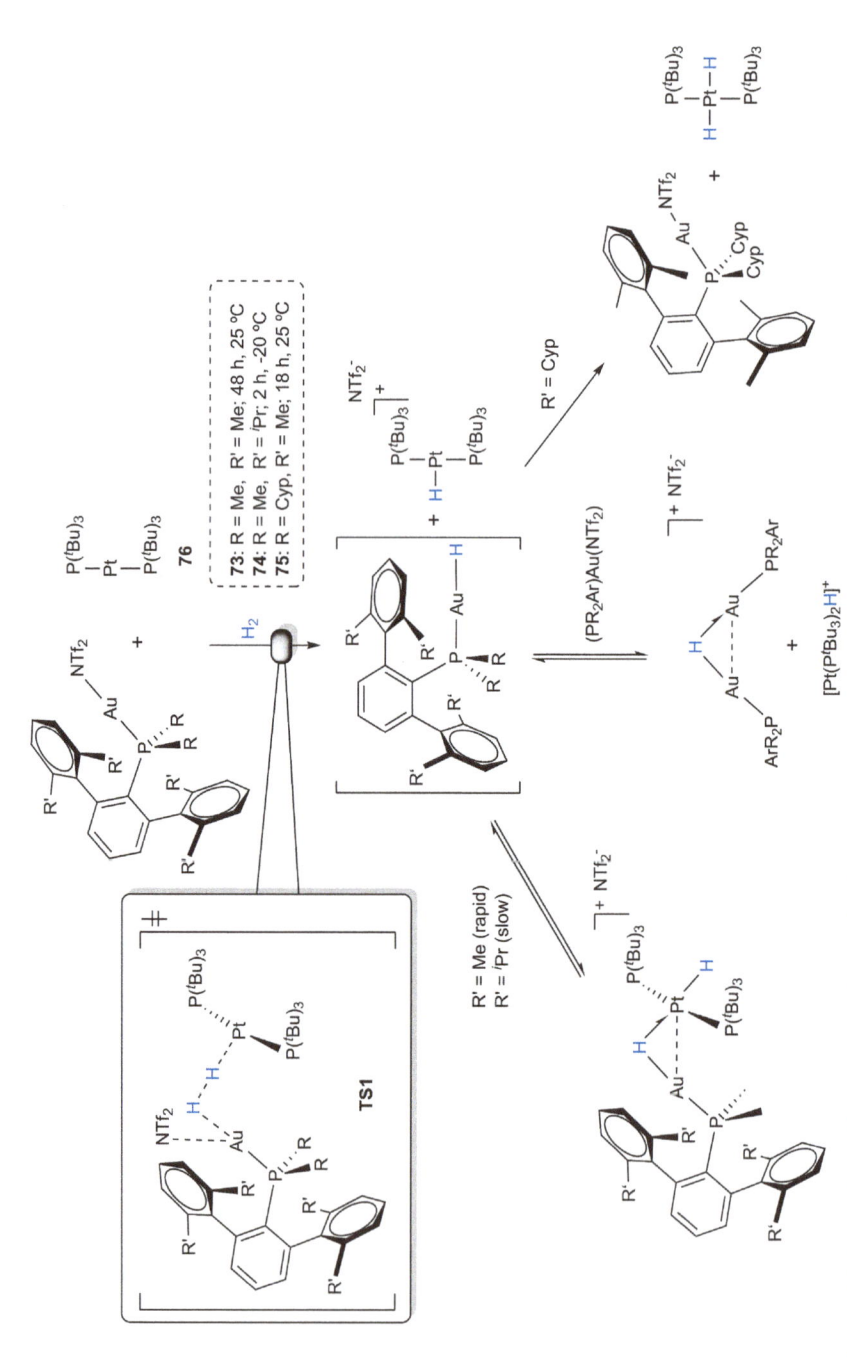

Scheme 22 Heterolytic dihydrogen activation by Au(I)/Pt(0) pairs highlighting the modeled FLP-type transition state and phosphine dependent speciation.

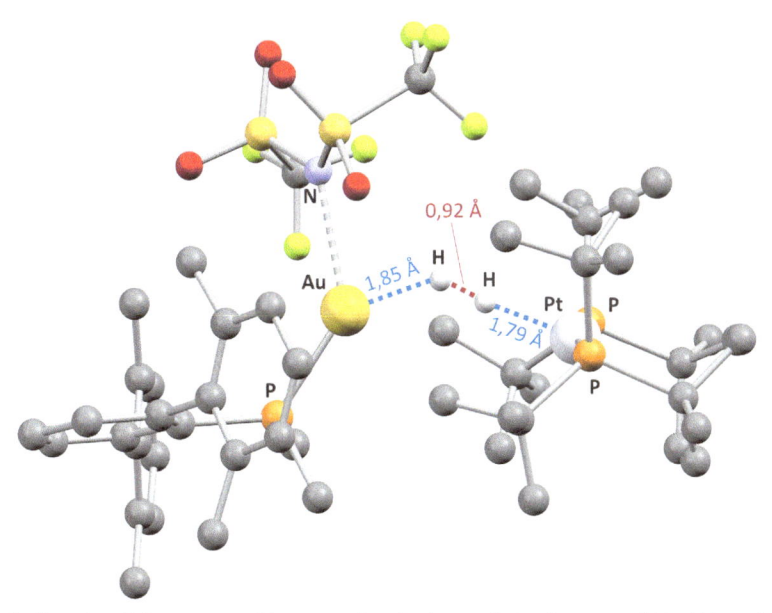

Fig. 6 Genuine FLP-type transition state for the heterolytic cleavage of H—H by a Au(I)/Pt(0) TMOFLP modeled by DFT calculations (ωB97xD). *Figure drawn using data from reference Hidalgo N, Moreno JJ, Pérez-Jiménez M, Maya C, López-Serrano J, Campos J. Evidence for genuine bimetallic frustrated Lewis pair activation of dihydrogen with gold(I)/platinum(0) systems. Chem A Eur J 2020;26(27): 5982–5993. https://doi.org/10.1002/chem.201905793.*

be crucial for stabilizing this transition state, as observed in other related bond activation processes.[169,170] Interestingly, a putative intermediate consisting of a Au(μ-H$_2$)Pt encounter complex forms along the reaction coordinate. Natural bond order analysis of both the encounter complex and the transition state suggests that the two widely postulated mechanistic interpretations for FLP dihydrogen activation,[9,171–173] namely electron transfer (ET) and electric field (EF) pathways, may concurrently operate. In fact, the calculated transfer of electrons from a *d* orbital on platinum to the σ*(H—H) accompanied by an energetically similar donation from σ(H—H) to an empty *s* orbital on gold occurs concomitantly with the polarization of the H—H bond, as estimated by charge analysis. An alternative scenario relying on a traditional heterobimetallic activation across the Au←Pt bond could be envisaged, but a computed higher overall barrier, which poorly fits with experimental observations, rather support the discussed genuine FLP activation.[165,174]

Kinetic studies revealed a strong inverse kinetic isotopic effect.[175] Thus, reaction with D_2 proceeded at a considerably faster rate for both the system based on PMe_2Ar^{Xyl2} (KIE = 0.46) and the one constructed around PMe_2Ar^{Dipp2} (KIE = 0.50). Once more, this analogy between the two systems speaks in favor of the need for adduct dissociation (in PMe_2Ar^{Xyl2}) as a pre-requisite for cooperative bond activation. Besides, further support was found by calculation of the KIE from the zero-point energy differences ($\Delta\Delta ZPE$) for the above mechanism, which is in very good agreement with the experimentally measured inverse KIE. Additionally, this implies that the transition state depicted in Fig. 6 can be considered a late transition state, as anticipated for the calculated side-on geometry of the dihydrogen molecule.

While activation of dihydrogen by the Au(I)/Pt(0) pairs resulted in different phosphine-dependent product distributions, activation of acetylene again evidenced strong selectivity effects derived from subtle ligand modification, and therefore from the metal adduct/full frustration equilibria (Scheme 23A).[168] For the least constrained system a 95:5 mixture of bridging heterobimetallic σ,π-acetylide and a rare bimetallic vinylene (–CH=CH–) was obtained. On the other hand, the fully frustrated system accomplished quantitative formation of the heterobimetallic vinylene. In turn, the intermediate size phosphine led to an in-between situation characterized by a 4:1 ratio of σ,π-acetylide vs vinylene structures.

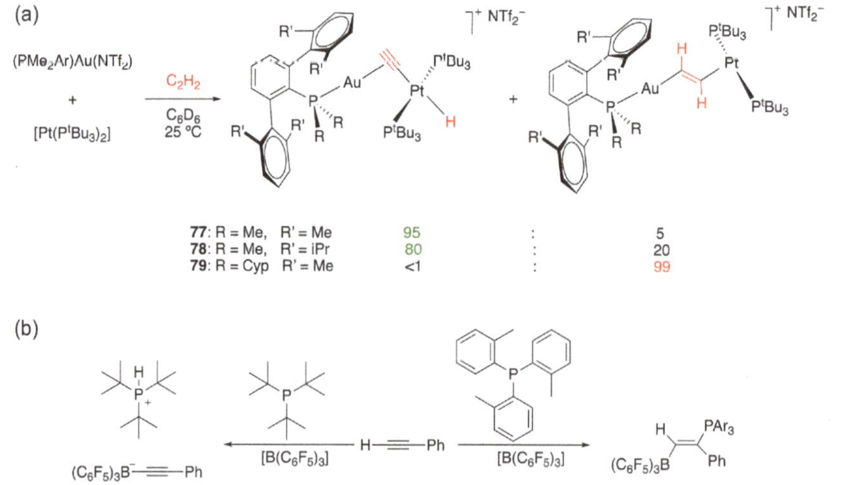

Scheme 23 (A) Regioselectivity effects during acetylene activation by Au(I)/Pt(0) FLPs; (B) Competition between deprotonation and 1,2-addition mechanisms in phosphine/borane FLPs.

It is worth noting that the two heterobimetallic products highly resemble those obtained in the area of main group FLPs, where a competition between deprotonation and addition mechanisms usually takes place, mostly depending on phosphine basicity (Scheme 23B).[176–178] More precisely, while the bimetallic vinylene is identical to the phosphine/borane addition product, the heterobimetallic σ,π-acetylide resembles the corresponding deprotonation isomer. The main difference in this case lies on the existence of metallophilic interactions between the Lewis acidic and basic fragment, not available for main-group designs. Besides, spectroscopic studies and X-ray diffraction analysis revealed that the acetylide bridge is σ-bonded to the Lewis base (platinum), contrasting to what is observed for phosphine/borane pairs. This is due to a transmetallation reaction occurring after the deprotonation event, as supported by computational studies. Those studies also support the interpretation of acetylene (as well as other alkynes) activation by these Au(I)/Pt(0) pairs following an FLP-type mechanism.

Tetrylenes are compounds based on a divalent heavier group 14 element, which possess relatively reduced and tunable HOMO–LUMO gaps and dual nucleophilic (lone electron pair) and electrophilic (empty p orbital) character. Several unstable tetrylene fragments have been characterized owing to the cooperative stabilization by intra- and intermolecular donor and acceptor systems binding to the ambiphilic tetrylene molecule.[179–183] This cooperative stabilization based on electronic push-pull interactions highly resembles the chemistry of frustrated Lewis pairs. Transition metal centers have been used in different occasions to independently prove the ambiphilic character of tetrylenes.[184–186] Directly connected to the area of bimetallic frustration, Rivard has stabilized the tetrylene: $GeCl_2$ in two different metal-only donor-acceptor complexes **80** and **81** based on $W(CO)_5$ as the Lewis acid fragment and $[(\eta^5\text{-}C_5H_5)Rh(PMe_2Ph)_2]$ and $Pt(PCy_3)_2$ as transition metal Lewis bases (Scheme 24A).[187] Encouraged by those results, our group explored the reactivity of the aforementioned Au(I)/Pt(0) FLP toward simple forms of low-valent group 14 elements, particularly: $GeCl_2$ and: $SnCl_2$ (Scheme 24B).[167] Treatment of dichloromethane solutions of the Au(I)/Pt(0) pair with: $GeCl_2$·dioxane cleanly generated the metallic Pt→Au adduct **82**. The tetrylene is key in withdrawing the triflimide anion from gold, likely by formation of $NTf_2 \rightarrow GeCl_2$. On the other hand, tin dichloride promoted a phosphine exchange to yield the heteroleptic compound **83** $[(PMe_2Ar^{Dipp2})Au(P^tBu_3)]^+$. It is also worth noticing that the reaction of : $SnCl_2$ with the mononuclear $Pt(P^tBu_3)_2$ afforded a mixture of complexes, in which the major product is a diplatinum cluster **84** bearing a single tri(*tert*-butyl)phosphine ligand per platinum center and held by three tin chloride units. DFT calculations revealed that the three $SnCl_n$ fragments donate electron density to one of the platinum atoms, which also behaves as a donor toward the other platinum center, in turn providing electron density to the empty *p* orbitals of the $SnCl_2$ bridges (Scheme 24C). Overall, each metal atom (except $SnCl_3^-$) in the highly reduced aggregate exhibits ambiphilic donor–acceptor character.

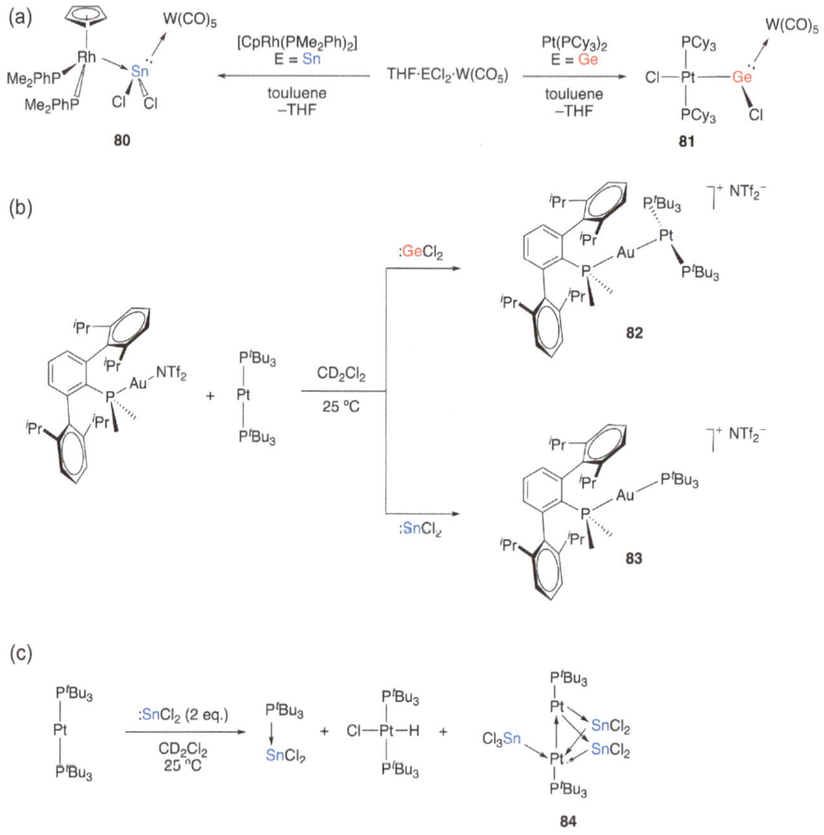

Scheme 24 (A) Metal-only donor-acceptor complexes based on $W(CO)_5$ fragment as Lewis acid and $[(\eta^5\text{-}C_5H_5)Rh(PMe_2Ph)_2]$ and $Pt(PCy_3)_2$ as Lewis bases stabilized by push-pull interactions; (B) reaction of Au(I)Pt(0) FLP toward :$GeCl_2$ and: $SnCl_2$; (C) reaction of $Pt(P^tBu_3)_2$ complex with: $SnCl_2$ toward a diplatinum complex held by three tin chloride units.

5. Miscellanea

As previously seen, Frustrated Lewis pair chemistry is vast and in constant expansion. It is therefore sometimes difficult to construct solid classifications, and hence some FLP related chemistry often falls outside these classifications. Hereafter we will discuss a series of novel examples strongly related to the main topic of this Chapter, "Bimetallic frustrated Lewis pairs" that were considered of importance despite not being able to be classified within the previous sections.

5.1 Cooperative reactivity between main group FLPs and transition metal complexes

Despite the outstanding ability of FLPs to activate a wide range of small molecules, dinitrogen has thus far remained one of its most elusive targets.[188] This contrasts with the isoelectronic CO molecule, which has been readily activated both by main group and transition metal containing FLPs. However, the activation of dinitrogen by genuine FLP systems has not been achieved, which is not surprising considering the high stability of the N_2 molecule and the strength of its triple N≡N bond, though its reluctance to bind p-block elements has surely contributed to this lack of reactivity. In fact, although Braunschweig showed that a B(I) species reacts with dinitrogen, no other p-block systems are known to capture N_2.[189] Once more, introducing transition metals may offer some advantages. Many stable low-valent transition metal complexes have been described to bind N_2, which offered well defined platforms to study new stoichiometric and catalytic transformations. The use of transition metal dinitrogen complexes for the reduction of dinitrogen into NH_3 under mild conditions[190–194] has paved the way for homogenous alternatives to the energy- and resource-intensive Haber-Bosch process.[195] Connected to the concept of frustration, Szymmcsak and Simonneau have independently reported the interaction of boranes with metal–N_2 complexes **85** and **86** and exploited these systems for dinitrogen protonation, borylation and silylation, capitalizing on the FLP behavior of the B—N adduct (Scheme 25A).[196,197] More recently, Liddle has described that the reaction of a chloro titanium triamidoamine complex **87** with

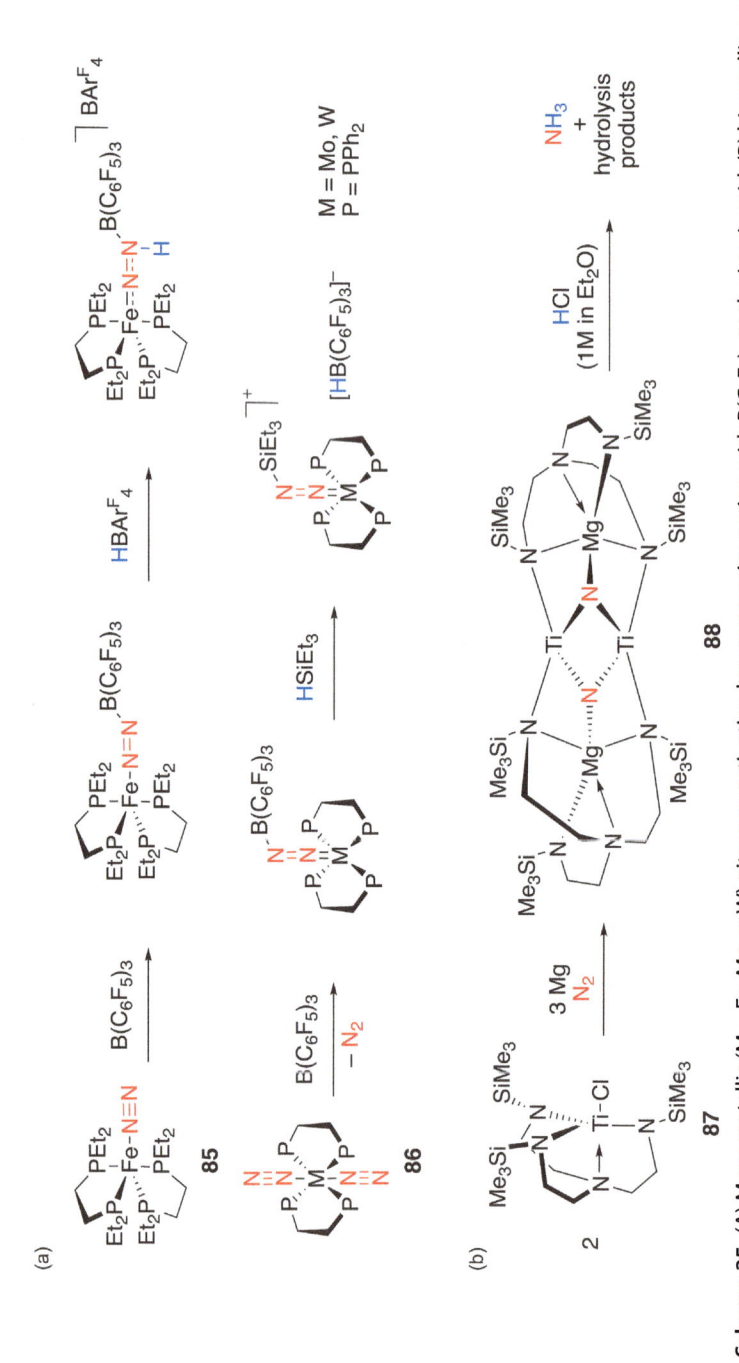

Scheme 25 (A) Monometallic (M = Fe, Mo or W) nitrogen activation by cooperative action with B(C$_6$F$_5$)$_3$ as the Lewis acid; (B) bimetallic cooperative dinitrogen cleavage and tandem frustrated Lewis pair hydrogenation to ammonia.

magnesium effects complete reductive cleavage of dinitrogen to give a dinitride dititanium dimagnesium ditriamidoamine complex **88**. Subsequent stoichiometric tandem H$_2$ splitting by a phosphine-borane FLP shuttles hydrogen atoms to the N^{3-} sites which eventually evolve to ammonia (Scheme 25B).[198] Hence, the combination of transition metal dinitrogen complexes acting as Lewis bases in cooperation with boranes as Lewis acids may serve as a new starting point to strategically overcome the challenging catalytic dinitrogen reduction by an FLP approach. In addition, the use of traditional FLPs in combination with transition metal complexes may provide successful tandem processes for the functionalization of challenging molecules.

5.2 Transition metal frustrated Lewis pairs in heterogeneous catalysis

Most industrial catalytic processes are based on heterogeneous systems, offering important advantages in terms of catalyst stability, product/catalyst separation, and catalyst recycling. For this reason, different efforts to adapt or combine homogeneous FLP systems to heterogeneous catalysts have been made. Guo and co-workers firstly described that a clean gold surface in combination with Lewis basic imines and nitriles activates dihydrogen, achieving the hydrogenation of small unsaturated molecules in a FLP manner.[199] In subsequent studies, several research groups have recently combined Lewis acidic boranes, such as B(C$_6$F$_5$)$_3$ or HB(C$_6$F$_5$)$_2$, with either insoluble bases like α-cyclodextrin or 4 Å molecular sieves, for the effective hydrogenation and reductive deoxygenation of ketones and aldehydes.[200] In addition, immobilized PPh$_3$ on silica supports has been coupled to those boranes for the *Z*-selective reduction of alkynes.[201] In this case the solid-support was successfully recycled, although successive additions of the Lewis acid were required for maintaining the activity of the recovered solid catalyst.

The group of Camp has recently reported a robust molecular heterobimetallic complex **89** featuring a Ta=Ir double bond by the stoichiometric reaction of the tantalum tri-neopentyl neopentylidene complex [Ta(=CH^{t}Bu)(CH$_2^t$Bu)$_3$] with the iridium tetrahydride complex [(η^5-C$_5$Me$_5$)IrH$_4$].[202] The resulting heterobimetallic complex [{Ta(CH$_2^t$Bu)$_3$}{[IrH$_2$(η^5-C$_5$Me$_5$)] presents the shortest Ta—Ir bond reported to date, with a calculated formal shortness ratio below unity ($r=0.905$), supporting the notion of multiple metal–metal bond. This complex was subsequently immobilized on SBA-15$_{700}$ silica (Santa Barbara Amorphous type material; mesoporous silica support dehydroxylated under high-vacuum at 700 °C) through a Si–O–Ta linker, which implied the loss of a neopentyl unit. The atomic-scale structure of the Ta/Ir supported catalyst was characterized in detail by high-resolution solid-state NMR spectroscopy and DFT calculations. The supported Ta/Ir species **90** exhibits drastically increased catalytic performances in H/D catalytic exchange reactions in comparison to the molecular [{Ta(CH$_2^t$Bu)$_3$}{IrH$_2$(η^5-C$_5$Me$_5$)}] complex. In particular, the supported material promotes the H/D exchange between fluorobenzene and C$_6$D$_6$ and D$_2$ as deuterium sources under mild conditions (25 °C, sub-atmospheric D$_2$ pressure) without any external additive (Scheme 26).

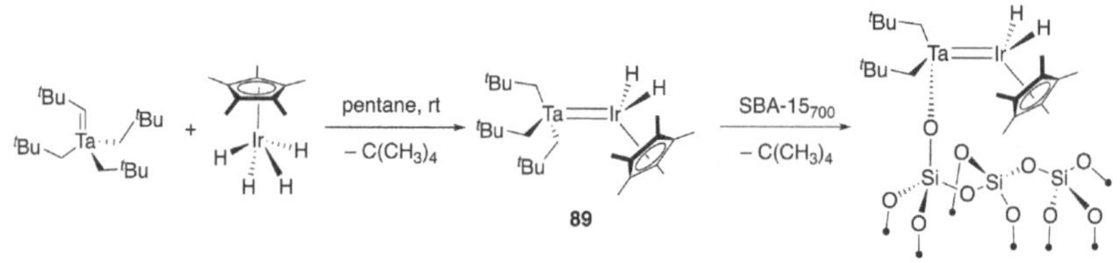

Scheme 26 Synthesis of [{Ta(CH$_2^t$Bu)}{IrH$_2$(η^5-C$_5$Me$_5$)}] complex and its silica-supported derivative.

Metal organic frameworks (MOFs) have also been envisioned as supports for FLP systems after several computational studies.[203–205] In subsequent recent work, Ma and co-workers have prepared the first stable and recyclable MOF-supported FLP by combination of the Lewis acidic B(C$_6$F$_5$)$_3$ and the Lewis basic DABCO in the MOF MIL-101(Cr) (Fig. 7).[206] This heterogeneous supported catalyst was effectively used for the reduction of several imines and alkylidene malonates, with excellent substrate size selectivity owing to the MOF pore size. Somehow related, other materials such as graphene with boron and nitrogen doped FLP sites on its surfaces[207] or CeO$_2$ porous nanorods[208,209] with high concentration of surface defects have been proved efficient for the hydrogenation of alkenes and alkynes.

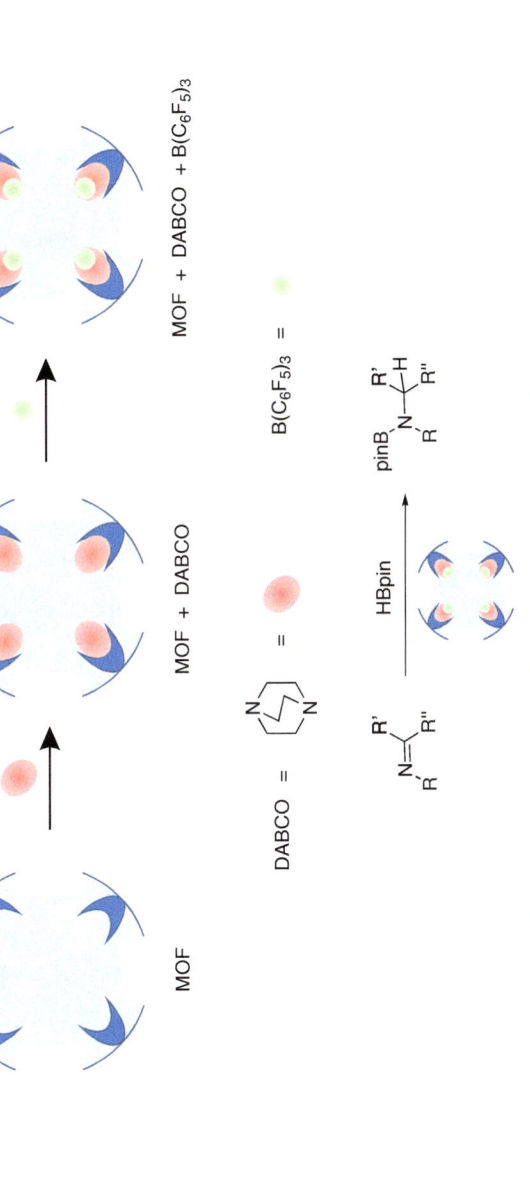

Fig. 7 Schematic representation of a mixed metal organic framework (MOF)/frustrated Lewis pair (FLP) material.

6. Conclusions

The concept of frustration triggered a revolution in the field of main group chemistry and offered countless possibilities in the broad area of cooperative chemistry. The introduction of transition metals into frustrated systems was early recognized as a fruitful avenue of investigation, mainly due to key advantages associated to catalysis. Those include the comprehensive assortment of tunable stereoelectronic properties and structural diversity accounted by transition metal complexes, as well as their capacity to mediate fundamental elementary reactions on behalf of accessible and partly occupied *d* orbitals. However, despite being an obvious subsequent synthetic target, the incorporation of a second transition metal fragment to achieve bimetallic frustration has been barely investigated.

We have discussed herein the main available examples of bimetallic frustrated Lewis pairs, as well as other related bimetallic designs with who they share many common features. In this regard, polarized heterobimetallic compounds are clearly the most paradigmatic example. In fact, it is possible that, in not few occasions, the mechanisms by which these bimetallic species activate small molecules could be understood in terms of bimetallic frustration rather than by means of traditional bimetallic activation across a M—M bond. Thus, the concept of frustration in bimetallic complexes might have been repeatedly overseen, though we foresee more often references to this notion in the near future.

Considering the broader context of cooperativity in bimetallic complexes, which involves orthogonal activation processes, tandem catalysis, multiply bonded bimetallic compounds, delayed processes and so on, it becomes clear that a better understanding of the factors governing bimetallic frustration may provide a whole new range of chemical options. Myriad opportunities for small molecule activation, molecular recognition and catalytic applications should emerge by bringing together inspiration from two seemingly independent areas, namely those of transition metal bimetallic complexes and main group chemistry.

Acknowledgments

This work has been supported by the European Research Council (ERC Starting Grant, CoopCat, Project 756575) and by the Spanish Ministry of Economy and Competitiveness (Project CTQ2016-75193-P [AEI/FEDER, UE]). M. N. thanks the Spanish Ministry of Science and Innovation for a Juan de la Cierva Fellowship (FJC2018-035514-I). We gratefully acknowledge Dr. Marta Roselló for helpful discussions.

References

1. Welch GC, San Juan RR, Masuda JD, Stephan DW. Reversible, metal-free hydrogen activation. *Science*. 2006;314(5802):1124–1128. https://doi.org/10.1126/science.1134230.
2. Stephan DW. The broadening reach of frustrated Lewis pair chemistry. *Science*. 2016;354(6317):7229. https://doi.org/10.1126/science.aaf7229.
3. Spikes GH, Fettinger JC, Power PP. Facile activation of dihydrogen by an unsaturated heavier main group compound. *J Am Chem Soc*. 2005;127(35):12232–12233. https://doi.org/10.1021/ja053247a.
4. Frey GD, Lavallo V, Donnadieu B, Schoeller WW, Bertrand G. Facile splitting of hydrogen and ammonia by nucleophilic activation at a single carbon center. *Science*. 2007;316(5823):439–441. https://doi.org/10.1126/science.1141474.
5. Fontaine FG, Stephan DW. On the concept of frustrated Lewis pairs. *Phil Trans R Soc A*. 2017;375:2017004. https://doi.org/10.1098/rsta.2017.0004.
6. Stephan DW. Frustrated Lewis pair hydrogenations. *Org Biomol Chem*. 2012;10(30):5740–5746. https://doi.org/10.1039/C2OB25339A.
7. Scott DJ, Fuchter MJ, Ashley AE. Designing effective 'frustrated Lewis pair' hydrogenation catalysts. *Chem Soc Rev*. 2017;46(19):5689–5700. https://doi.org/10.1039/C7CS00154A.
8. Lam J, Szkop KM, Mosafer E, Stephan DW. FLP catalysis: main group hydrogenations of organic unsaturated substrates. *Chem Soc Rev*. 2019;48(13):3592–3612. https://doi.org/10.1039/C8CS00277K.
9. Liu L, Lukose B, Jaque P, Ensing B. Reaction mechanism of hydrogen activation by frustrated Lewis pairs. *Green Energy Environ*. 2019;4(1):20–28. https://doi.org/10.1016/j.gee.2018.06.001.
10. Jupp AR, Stephan DW. New directions for frustrated Lewis pair chemistry. *Trends Chem*. 2019;1(1):35–48. https://doi.org/10.1016/j.trechm.2019.01.006.
11. Buchwalter P, Rosé J, Braunstein P. Multimetallic catalysis based on heterometallic complexes and clusters. *Chem Rev*. 2015;115(1):28–126. https://doi.org/10.1021/cr500208k.
12. Pye DR, Mankad NP. Bimetallic catalysis for C–C and C–X coupling reactions. *Chem Sci*. 2017;8(3):1705–1718. https://doi.org/10.1039/C6SC05556G.
13. Mankad NP. Catalysis with multinuclear complexes. In: Klein Gebbink RJM, Moret M-E, eds. *Non-Noble Metal Catalysis: Molecular Approaches and Reactions*. 1st ed. Weinheim, Germany: Wiley-VCH; 2019:49–68.
14. Rej S, Tsutugi H, Mashima K. Multiply-bonded dinuclear complexes of early-transition metal as minimum entities of metal cluster catalysts. *Coord Chem Rev*. 2018;355:223–239. https://doi.org/10.1016/j.ccr.2017.08.016.
15. Knorr M, Jourdain I. Activation of alkynes by diphosphine- and µ-phosphido-spanned heterobimetallic complexes. *Coord Chem Rev*. 2017;350(1):217–247. https://doi.org/10.1016/j.ccr.2017.07.001.
16. Brown HC, Schlesinger HI, Cardon SZ. Studies in stereochemistry. I. Steric strains as a factor in relative stability of some coordination compounds of boron. *J Am Chem Soc*. 1942;64(2):325–329. https://doi.org/10.1021/ja01254a031.
17. Erker G, Stephan DW. Frustrated Lewis pairs I: uncovering and understanding. *Top Curr Chem*. 2013. Springer.
18. Erker G, Stephan DW. Frustrated Lewis Pairs II: expanding the scope. *Top Curr Chem*. 2013. Springer.
19. Stephan DW, Erker G. Frustrated Lewis pair chemistry: development and perspectives. *Angew Chem Int Ed*. 2015;54(22):6400–6441. https://doi.org/10.1002/anie.201409800.
20. Stephan DW. Frustrated Lewis pairs: from concept to catalysis. *Acc Chem Res*. 2015;48(2):306–316. https://doi.org/10.1021/ar500375j.

21. Kehr G, Erker G. Frustrated Lewis pair chemistry: searching for new reactions. *Chem Rec*. 2017;17(8):803–815. https://doi.org/10.1002/tcr.201700010.
22. Fontaine F-G, Courtemanche M-A, Légaré M-A, Rochette E. Design principles in frustrated Lewis pair catalysis for the functionalization of carbon dioxide and heterocycles. *Coord Chem Rev*. 2017;334:124–135. https://doi.org/10.1016/j.ccr.2016.05.005.
23. Mo Z, Rit A, Campos J, Kolychev EL, Aldrige S. Catalytic B–N dehydrogenation using frustrated Lewis pairs: evidence for a chain-growth coupling mechanism. *J Am Chem Soc*. 2016;138(10):3306–3309. https://doi.org/10.1021/jacs.6b01170.
24. Houghton AY, Hurmalainen J, Mansikkamäki A, Piers WE, Tuononen HK. Direct observation of a borane-silane complex involved in frustrated Lewis-pair-mediated hydrosilylations. *Nat Chem*. 2014;6:983–988. https://doi.org/10.1038/nchem.2063.
25. Berkefeld A, Piers WE, Parvez M. Tandem frustrated Lewis pair/tris(pentafluorophenyl)borane-catalyzed deoxygenative hydrosilylation of carbon dioxide. *J Am Chem Soc*. 2010;132(31):10660–10661. https://doi.org/10.1021/ja105320c.
26. Courtemanche M-A, Légare M-A, Maron L, Fontaine F-G. A highly active phosphine-borane organocatalyst for the reduction of CO_2 to methanol using hydroboranes. *J Am Chem Soc*. 2013;135(25):9326–9329. https://doi.org/10.1021/ja404585p.
27. Mahdi T, Stephan DW. Frustrated Lewis pair catalyzed hydroamination of terminal alkynes. *Angew Chem Int Ed*. 2013;52(47):12418–12421. https://doi.org/10.1002/anie.201307254.
28. Chen D, Wang Y, Klankermayer J. Enantioselective hydrogenation with chiral frustrated Lewis pairs. *Angew Chem Int Ed*. 2010;49(49):9475–9478. https://doi.org/10.1002/anie.201004525.
29. Sumerin V, Chernichenko NM, Leskelä M, Rieger B, Repo T. Highly active metal-free catalysts for hydrogenation of unsaturated nitrogen-containing compounds. *Adv Synth Catal*. 2011;353(11):2093–2110. https://doi.org/10.1002/adsc.201100206.
30. Lindqvist M, Borre K, Axenov K, et al. Chiral molecular tweezers: synthesis and reactivity in asymmetric hydrogenation. *J Am Chem Soc*. 2015;137(12):4038–4041. https://doi.org/10.1021/ja512658m.
31. Appelt C, Westenberg H, Bertini F, et al. Geminal phosphorus/aluminum-based frustrated Lewis pairs: C–H versus C≡C activation and CO_2 fixation. *Angew Chem Int Ed*. 2011;50(17):3925–3928. https://doi.org/10.1002/anie.201006901.
32. Holtrichter-Rössmann T, Rösener C, Hellmann J, et al. Generation of weakly bound Al–N Lewis pairs by hydroalumination of ynamines and the activation of small molecules: phenylethyne and dicyclohexylcarbodiimide. *Organometallics*. 2012;31(8):3272–3283. https://doi.org/10.1021/om3001179.
33. Ménard G, Tran L, McCahill JS, Lough AJ, Stephan DW. Contrasting the reactivity of ethylene and propylene with P/Al and P/B frustrated Lewis pairs. *Organometallics*. 2013;32(22):6759–6763. https://doi.org/10.1021/om400222w.
34. Ménard G, Stephan DW. C–H activation of isobutylene using frustrated Lewis pairs: aluminum and boron σ-allyl complexes. *Angew Chem Int Ed*. 2012;51(18):4409–4412. https://doi.org/10.1002/anie.201200328.
35. Ménard G, Stephan DW. H_2 activation and hydride transfer to olefins by $Al(C_6F_5)_3$-based frustrated Lewis pairs. *Angew Chem Int Ed*. 2012;51(33):8272–8275. https://doi.org/10.1002/anie.201203362.
36. Bertini F, Hoffmann F, Appelt C, et al. Reactivity of dimeric P/Al-based Lewis pairs toward carbon dioxide and *tert*-butyl isocyanate. *Organometallics*. 2013;32(22):6764–6769. https://doi.org/10.1021/om3011382.
37. Boudreau J, Courtemanche M-A, Fontaine F-G. Reactivity of Lewis pairs $(R_2PCH_2AlMe_2)_2$ with carbon dioxide. *Chem Commun*. 2011;47(39):11131–11133. https://doi.org/10.1039/C1CC14641F.

38. Ménard G, Gilbert TM, Hatnean JA, Kraft A, Krossing I, Stephan DW. Stoichiometric reduction of CO_2 to CO by phosphine/AlX_3-based frustrated Lewis pairs. *Organometallics*. 2013;32(15):4416–4422. https://doi.org/10.1021/om400619y.
39. Uhl W, Appelt C. Reactions of an Al–P-based frustrated Lewis pair with carbonyl compounds: dynamic coordination of benzaldehyde, activation of benzoyl chloride, and Al–C bond cleavage with benzamide. *Organometallics*. 2013;32(18):5008–5014. https://doi.org/10.1021/om400620h.
40. Appelt C, Slooetweg JC, Lammersma K, Uhl W. Reaction of a P/Al-based frustrated Lewis pair with ammonia, borane, and amine-boranes: adduct formation and catalytic dehydrogenation. *Angew Chem Int Ed*. 2013;52(15):4256–4259. https://doi.org/10.1002/anie.201208746.
41. Chapman AM, Haddow MF, Wass DF. Frustrated Lewis pairs beyond the main group: cationic zirconocene-phosphinoaryloxide complexes and their application in catalytic dehydrogenation of amine boranes. *J Am Chem Soc*. 2011;133(23):8826–8829. https://doi.org/10.1021/ja201989c.
42. Chapman AM, Haddow MF, Wass DF. Frustrated Lewis pairs beyond the main group: synthesis, reactivity, and small molecule activation with cationic zirconocene-phosphinoaryloxide complexes. *J Am Chem Soc*. 2011;133(45):18463–18478. https://doi.org/10.1021/ja207936p.
43. Chapman AM, Haddow MF, Wass DF. Cationic group 4 metallocene-(*o*-phosphanylaryl)oxido complexes: synthetic routes to transition-metal frustrated Lewis pairs. *Eur J Inorg Chem*. 2012;(9):1546–1554. https://doi.org/10.1002/ejic.201100968.
44. Xu X, Fröhlich R, Daniliuc CG, Kehr G, Erker G. Reactions of a methylzirconocene cation with phosphinoalkynes: an alternative pathway for generating $Cp_2Zr(II)$ systems. *Chem Commun*. 2012;48(49):6109–6111. https://doi.org/10.1039/C2CC31883K.
45. Xu X, Kehr G, Daniliuc CG, Erker G. Reactions of a cationic germinal Zr^+/P pair with small molecules. *J Am Chem Soc*. 2013;135(17):6465–6476. https://doi.org/10.1021/ja3110076.
46. Hartwig J. *Organotransition Metal Chemistry*. Mill Valey, CA: University Science Books; 2010.
47. Crabtree RH. *The Organometallic Chemistry of the Transition Metals*. 7th ed. New York, NY: Wiley; 2019.
48. Kepp KP. A quantitative scale of oxophilicity and thiophilicity. *Inorg Chem*. 2016;55(18):9461–9470. https://doi.org/10.1021/acs.inorgchem.6b01702.
49. Yamamoto H. *Lewis Acids in Organic Synthesis*. Weinheim: Wiley-VCH; 2000. https://doi.org/10.1002/aoc.201.
50. McCahill JSJ, Welch GC, Stephan DW. Reactivity of "frustrated Lewis pairs": three-component reactions of phosphines, a borane, and olefins. *Angew Chem Int Ed*. 2007;119(26):5056–5059. https://doi.org/10.1002/ange.200701215.
51. Normand AT, Daniliuc CG, Wibbeling B, Kehr G, Le Gendre P, Erker G. Phosphido- and amidozirconocene cation-based frustrated Lewis pair chemistry. *J Am Chem Soc*. 2015;137(33):10796–10808. https://doi.org/10.1021/jacs.5b06551.
52. Flynn SR, Metters OJ, Manners I, Wass DF. Zirconium-catalyzed imine hydrogenation via a frustrated Lewis pair mechanism. *Organometallics*. 2016;35(6):847–850. https://doi.org/10.1021/acs.organomet.6b00027.
53. Jiang Y, Blacque O, Fox T, Berke H. Catalytic CO_2 activation assisted by rhenium hydride/$B(C_6F_5)_3$ frustrated Lewis pairs—metal hydrides functioning as FLP bases. *J Am Chem Soc*. 2013;135(20):7751–7760. https://doi.org/10.1021/ja402381d.
54. Nesbit MA, Suess DLM, Peters JC. E–H bond activations and hydrosilylation catalysis with iron and cobalt metalloboranes. *Organometallics*. 2015;34(19):4741–4752. https://doi.org/10.1021/acs.organomet.5b00530.

55. Wang Z, Ying A, Fan Z, Hervieu C, Zhang L. Tertiary amino group in cationic gold catalyst: tethered frustrated Lewis pairs that enable ligand-controlled regiodivergent and stereoselective isomerizations of propargylic esters. *ACS Catal.* 2017;7(5):3676–3680. https://doi.org/10.1021/acscatal.7b00626.
56. Anwander R, Kobayashi S. *Tanthanides: Chemistry and Use in Organic Synthesis*. Berlin, Heidelberg: Springer; 1999.
57. Berkefeld A, Piers WE, Parvez M, Castro L, Maron L, Eisenstein O. Carbon monoxide activation via O-bound CO using decamethylscandocinium–hydridoborate ion pairs. *J Am Chem Soc.* 2012;134(26):10843–10851. https://doi.org/10.1021/ja300591v.
58. Berkefeld A, Piers WE, Parvez M, Castro L, Maron L, Eisenstein O. Decamethylscandocinium-hydrido-(perfluorophenyl)borate: fixation and tandem tris(perfluorophenyl)borane catalyzed deoxygenative hydrosilation of carbon dioxide. *Chem Sci.* 2013;4(5):2152–2162. https://doi.org/10.1039/C3SC50145K.
59. Xu P, Xu X. Homoleptic rare-earth aryloxide based Lewis pairs for polymerization of conjugated polar alkenes. *ACS Catal.* 2018;8(1):198–202. https://doi.org/10.1021/acscatal.7b02875.
60. Xu P, Wu L, Dong L, Xu X. Chemoselective polymerization of polar divinyl monomers with rare-earth/phosphine Lewis pairs. *Molecules.* 2018;23(2):360–369. https://doi.org/10.3390/molecules23020360.
61. Dahl LF, Ishishi E, Rundle RE. Polynuclear metal carbonyls. I. Structure of $Mn_2(CO)_{10}$ and $Re_2(CO)_{10}$. *J Chem Phys.* 1957;26:1750–1751. https://doi.org/10.1063/1.1743615.
62. Cotton FA, Curtis NF, Harris CB, et al. Mononuclear and polynuclear chemistry of rhenium(III): its pronounced homophilicity. *Science.* 1964;145(3638):1305–1307. https://doi.org/10.1126/science.145.3638.1305.
63. Cotton FA, Murillo CA, Walton RA. *Multiple Bonds Between Metal Atoms*. Springer; 2005.
64. Parkin G. *Metal-Metal Bonding*. Springer; 2010.
65. Fackler Jr JP. *Metal-Metal Bonds and Cluster in Chemistry and Catalysis*. Springer; 1990.
66. Cornils B, Herrmann WA, Beller M, Paciello R. *Applied Homogeneous Catalysis with Organometallic Compounds: A Comprehensive Handbook in Four Volumes*. 3rd ed. New York: Wiley; 2017.
67. Kim J, Rees DC. Structural models for the metal centers in the nitrogenase molybdenum-iron protein. *Science.* 1992;257(5077):1677–1682. https://doi.org/10.1126/science.1529354.
68. Dobbek H, Svetlitchnyi V, Gremer L, Huber R, Meyer O. Crystal structure of a carbon monoxide dehydrogenase reveals a [Ni-4Fe-5S] cluster. *Science.* 2001;293(5533): 1281–1285. https://doi.org/10.1126/science.1061500.
69. Mankad NP. Selectivity effects in bimetallic catalysis. *Chem A Eur J.* 2016;22(17): 5822–5829. https://doi.org/10.1002/chem.201505002.
70. Powers IG, Uyeda C. Metal-metal bonds in catalysis. *ACS Catal.* 2017;7(2):936–958. https://doi.org/10.1021/acscatal.6b02692.
71. Kalck P. Homo- and heterobimetallic complexes in catalysis. *Top Organomet Chem.* 2016. Springer.
72. Mankad NP. Diverse bimetallic mechanisms emerging from transition metal Lewis acid/base pairs: development of co-catalysis with metal carbenes and metal carbonyl anions. *Chem Commun.* 2018;54(11):1291–1302. https://doi.org/10.1039/C7CC09675E.
73. Hirner JJ, Shi Y, Blum SA. Organogold reactivity with palladium, nickel, and rhodium: transmetalation, cross-coupling, and dual catalysis. *Acc Chem Res.* 2011;44(8):603–613. https://doi.org/10.1021/ar200055y.
74. Hruszkewycz DP, Balcells D, Guard LM, Hazari N, Tilset M. Insight into the efficiency of cinnamyl-supported precatalysts for the Suzuki-Miyaura reaction: observation of Pd(I) dimers with bridging allyl ligands during catalysis. *J Am Chem Soc.* 2014;136(2):7300–7316. https://doi.org/10.1021/ja412565c.

75. Garden JA, Saini PK, Williams CK. Greater than the sum of its parts: a heterodinuclear polymerization catalyst. *J Am Chem Soc.* 2015;137(48):15078–15081. https://doi.org/10.1021/jacs.5b09913.
76. Dürr AB, Fisher DC, Kalvet I, Troung K-N, Schoenebeck F. Divergent reactivity of dinuclear (NHC)nickel(I) versus nickel(0) enables chemoselective trifluoromethylselenolation. *Angew Chem Int Ed.* 2017;129(43):13616–13620. https://doi.org/10.1002/ange.201706423.
77. Fogg DE, dos Santos EN. Tandem catalysis: a taxonomy and illustrative review. *Coord Chem Rev.* 2004;248(21):2365–2379. https://doi.org/10.1016/j.ccr.2004.05.012.
78. Lohr T, Marks T. Orthogonal tandem catalysis. *Nat Chem.* 2015;7:477–482. https://doi.org/10.1038/nchem.2262.
79. Goldman AS, Roy AH, Huang Z, Ahuja R, Schinski W, Brookhart M. Catalytic alkane metathesis by tandem alkane dehydrogenation-olefin metathesis. *Science.* 2006;312(5771):257–261. https://doi.org/10.1126/science.1123787.
80. Bagherzadeh S, Mankad NP. Catalyst control of selectivity in CO_2 reduction using a tunable heterobimetallic effect. *J Am Chem Soc.* 2015;137(34):10898–10901. https://doi.org/10.1021/jacs.5b05692.
81. Zhang S, del Pozo J, Romiti F, Mu Y, Torker S, Hoveyda AH. Delayed catalyst function enables direct enantioselective conversion of nitriles to NH_2-amines. *Science.* 2019;346(6435):45–51. https://doi.org/10.1126/science.aaw4029.
82. Jacobsen EN. Asymmetric catalysis of epoxide ring-opening reactions. *Acc Chem Res.* 2000;33(6):421–431. https://doi.org/10.1021/ar960061v.
83. García-Domínguez P, Nevado C. Au–Pd bimetallic catalysis: the importance of anionic ligands in catalyst separation. *J Am Chem Soc.* 2016;138(10):3266–3269. https://doi.org/10.1021/jacs.5b10277.
84. Friis SD, Pirnot MT, Buchwald SL. Asymmetric hydroarylation of vinylarenes using a synergistic combination of CuH and Pd catalysis. *J Am Chem Soc.* 2016;138(27):8372–8375. https://doi.org/10.1021/jacs.6b04566.
85. Green SA, Matos JLM, Yagi A, Shenvi RA. Branch-selective hydroarylation: iodoarene-olefin cross-coupling. *J Am Chem Soc.* 2016;138(39):12779–12782. https://doi.org/10.1021/jacs.6b08507.
86. Ackerman L, Lovel M, Weix D. Multimetallic catalyzed cross-coupling of aryl bromides with aryl triflates. *Nature.* 2015;524:454–457. https://doi.org/10.1038/nature14676.
87. Farley CM, Uyeda C. Organic reactions enabled by catalytically active metal–metal bonds. *Trends Chem.* 2019;1:497–509. https://doi.org/10.1016/j.trechm.2019.04.002.
88. Broussard ME, Juma B, Train SG, Peng W-G, Laneman SA, Stanley GG. A bimetallic hydroformylation catalyst: high regioselctivity and reactivity through homobimetallic cooperativity. *Science.* 1993;260(5115):1784–1788. https://doi.org/10.1126/science.260.5115.1784.
89. Chen Y, Sakaki S. The important role of the Mo–Mo quintuple bond in catalytic synthesis of benzene from alkynes. A theoretical study. *Dalton Trans.* 2014;43(39):11478–11492. https://doi.org/10.1039/C4DT00595C.
90. Tsurugi H, Hayakawa A, Kando S, Sugino Y, Mashima K. Mixed-ligand complexes of paddlewheel dinuclear molybdenum as hydrodehalogenation catalysts for polyhaloalkanes. *Chem Sci.* 2015;6:3434–3439. https://doi.org/10.1039/C5SC00721F.
91. Inatomi T, Koga Y, Matsubara K. Dinuclear nickel(I) and palladium(I) complexes for highly active transformations of organic compounds. *Molecules.* 2018;23(1):140. https://doi.org/10.3390/molecules23010140.
92. Pérez-Jiménez M, Campos J, López-Serrano J, Carmona E. Reactivity of a *trans*-[H–Mo≡Mo–H] unit towards alkenes and alkynes: bimetallic migratory insertion, H-elimination and other reactions. *Chem Commun.* 2018;59(66):9186–9189. https://doi.org/10.1039/C8CC04945A.

93. Hübner O, Himmel H-J. Oxidative addition of dihydrogen to divanadium in solid Ne: multiple-bonded triplet HVVH and singlet $V_2(\mu\text{-H})_2$. *Angew Chem Int Ed.* 2020;59 (29):12206–12212. https://doi.org/10.1002/anie.202004241.
94. Khand IU, Knox GR, Pauson PL, Watts WE. A cobalt induced cleavage reaction and a new series of arenecobalt carbonyl complexes. *J Chem Soc D.* 1971;(1):36a. https://doi.org/10.1039/C2971000036A.
95. Khand IU, Knox GR, Pauson PL, Watts WE. Organocobalt complexes. Part I. Arene complexes derived from dodecarbonyltetracobalt. *J Chem Soc Perkin Trans.* 1973;1:975–977. https://doi.org/10.1039/P19730000975.
96. Khand IU, Knox GR, Pauson PL, Watts WE, Foreman MI. Organocobalt complexes. Part II. Reactions of acetylenehexacarbonyldicobalt complexes, $(R^1C_2R^2)Co_2(CO)_6$, with norbornene and its derivatives. *J Chem Soc Perkin Trans.* 1973;1:977–981. https://doi.org/10.1039/P19730000977.
97. Yamanaka M, Nakamura E. Density functional studies on the Pauson-Khand reaction. *J Am Chem Soc.* 2001;123(8):1703–1708. https://doi.org/10.1021/ja005565.
98. Dickson RS, Frase PJ. Compounds derived from alkynes and carbonyl complexes of cobalt. *Adv Organomet Chem.* 1974;12:323–377. https://doi.org/10.1016/S0065-3055(08)60454-2.
99. Hartline DR, Zeller M, Uyeda C. Well-defined models for the elusive dinuclear intermediates of the Pauson-Khand reaction. *Angew Chem Int Ed.* 2016;55(20):6084–6087. https://doi.org/10.1002/anie.201601784.
100. Zhou Y-Y, Hartline DR, Steiman TJ, Fanwick PE, Uyeda C. Dinuclear nickel complexes in five states of oxidation using a redox-active ligand. *Inorg Chem.* 2014;53 (21):11770–11777. https://doi.org/10.1021/ic5020785.
101. Nguyen T, Sutton AD, Brynda M, Fettinger JC, Long GJ, Power PP. Synthesis of a stable compound with a fivefold bonding between two chromium(I) centers. *Science.* 2005;310(5749):844–847. https://doi.org/10.1126/science.1116789.
102. Noor A, Glatz G, Müller R, Kaupp M, Demeshko S, Kempe R. Carboalumination of a chromium–chromium quintuple bond. *Nat Chem.* 2009;1:322–325. https://doi.org/10.1038/nchem.255.
103. Ni C, Ellis BD, Long GL, Power PP. Reactions of Ar'CrCrAr' with N_2O or N_3(1-Ad): complete cleavage of the Cr–Cr quintuple interaction. *Chem Commun.* 2009;(17): 2332–2334. https://doi.org/10.1039/B901494B.
104. Schwarzmaier C, Noor A, Glatz G, et al. Formation of cyclo-E_4^{2-} units ($E_4 = P_4$, As_4, AsP_3) by a complex with a Cr–Cr quintuple bond. *Angew Chem Int Ed.* 2011;50(32): 7283–7286. https://doi.org/10.1002/anie.201102361.
105. Tsai Y-C, Chen H-Z, Chang C-C, et al. Journey from Mo–Mo quadruple bonds to quintuple bonds. *J Am Chem Soc.* 2009;131(35):12534–12535. https://doi.org/10.1021/ja905035f.
106. Lu D-Y, Chem PP-Y, Kuo T-S, Tsai Y-C. The Mo–Mo quintuple bond as a ligand to stabilize transition-metal complexes. *Angew Chem Int Ed.* 2015;127(31):9234–9238. https://doi.org/10.1002/ange.201504414.
107. Harisomayajula NVS, Nair AK, Tsai Y-C. Discovering complexes containing a metal–metal quintuple bond: from theory to practice. *Chem Commun.* 2014;50 (26):3391–3412. https://doi.org/10.1039/C3CC48203K.
108. Chen H-Z, Liu S-C, Yen C-H, et al. Reactions of metal–metal quintuple bonds with alkynes: [2 + 2 + 2] and [2 + 2] cycloadditions. *Angew Chem Int Ed.* 2012;128 (41):10488–10492. https://doi.org/10.1002/ange.201205027.
109. Liddle ST. *Molecular Metal–Metal Bonds: Compounds, Synthesis, Properties.* Weinheim: Wiley-VCH Verlag GmbH & Co; 2014.
110. Chen H-Z, Liu S-C, Yen C-H, et al. Reactions of metal–metal quintuple bonds with alkynes: [2 + 2 + 2] and [2 + 2] cycloadditions. *Angew Chem Int Ed.* 2012;51(41): 10342–10346. https://doi.org/10.1002/anie.201205027.

111. Chen Y, Sakaki S. Mo–Mo quintuple bond is highly reactive in H–H, C–H, and O–H σ-bond cleavages because of the polarized electronic structure in transition state. *Inorg Chem*. 2017;56(7):4011–4020. https://doi.org/10.1021/acs.inorgchem.6b03103.
112. Rosenthal J, Bachman J, Dempsey JL, et al. Oxygen and hydrogen photocatalysis by two-electron mixed-valence coordination compounds. *Coord Chem Rev*. 2005; 249(13):1316–1326. https://doi.org/10.1016/j.ccr.2005.03.034.
113. Weller AS, McIndoe JS. Reversible binding of dihydrogen in multimetallic complexes. *Eur J Inorg Chem*. 2007;(28):4411–4423. https://doi.org/10.1002/ejic.200700661.
114. Powers DC, Chambers MB, Teets TS, Elgrishi N, Anderson BL, Nocera DG. Halogen photoelimination from dirhodium phosphazane complexes *via* chloride bridged intermediates. *Chem Sci*. 2013;4(7):2880–2885. https://doi.org/10.1039/C3SC50462J.
115. Esswein AJ, Veige A, Piccole PMB, Schultz AJ, Nocera DG. Intramolecular C–H bond activation and redox isomerization across two-electron mixed valence diiridium cores. *Organometallics*. 2008;27(6):1073–1083. https://doi.org/10.1021/om7007748.
116. Teets TS, Cook TR, McCarthy BD, Nocera DG. Oxygen reduction to water mediated by dirhodium hydrido-chloride complex. *J Am Chem Soc*. 2011;133(21): 8114–8117. https://doi.org/10.1021/ja201972v.
117. Gray TG, Veige AS, Nocera DG. Cooperative bimetallic reactivity: hydrogen activation in two-electron mixed-valence compounds. *J Am Chem Soc*. 2004;124 (31):9760–9778. https://doi.org/10.1021/ja0491432.
118. Stephan DW. Early-late heterobimetallics. *Coord Chem Rev*. 1989;95(1):41–107. https://doi.org/10.1016/0010-8545(89)80002-5.
119. Herberhold M, Jin G-X. Heterodimetallic complexes with an unbridged, polar metal–metal bond. *Angew Chem Int Ed*. 1994;44(9):946–966. https://doi.org/10.1002/anie.199409641.
120. Wheatley N, Kalck P. Structure and reactivity of early-late heterobimetallic complexes. *Chem Rev*. 1999;99(12):3379–33420. https://doi.org/10.1021/cr980325m.
121. Gade LH. Highly polar metal–metal bonds in "early-late" heterodimetallic complexes. *Angew Chem Int Ed*. 2000;39(15):2658–2678. https://doi.org/10.1002/1521-3773(20000804)39:15<2658::AID-ANIE2658>3.0.CO;2-C.
122. Cooper BG, Napoline JW, Thomas CM. Catalytic applications of early/late heterobimetallic complexes. *Catal Rev*. 2012;54(1):1–40. https://doi.org/10.1080/01614940.2012.619931.
123. Baranger AM, Bergman RG. Cooperative reactivity in the interactions of the X–H bonds with zirconium-iridium bridging imido complex. *J Am Chem Soc*. 1994;116(9): 3822–3835. https://doi.org/10.1021/ja00088a019.
124. Hanna TA, Baranger AM, Bergman RG. Reaction of carbon dioxide with heterocumulenes with an unsymmetrical meta–metal bond. Direct addition of carbon dioxide across a zirconium-iridium bond and stoichiometric reduction of carbon dioxide to formate. *J Am Chem Soc*. 1995;117(45):11363–11364. https://doi.org/10.1021/ja00150a044.
125. Marquard SL, Bezpalko MW, Foxman BM, Thomas CM. Stoichiometric C=O bond oxidative addition of benzophenone by a discrete radical intermediate to form a cobalt(I) carbene. *J Am Chem Soc*. 2013;135(16):6018–6021. https://doi.org/10.1021/ja4022683.
126. Wu B, Hernández-Sánchez R, Bezpalko MW, Foxman BM, Thomas CM. Formation of heterobimetallic zirconium/cobalt diimido complexes via a four-electron transformation. *Inorg Chem*. 2014;53(19):10021–10023. https://doi.org/10.1021/ic501490e.
127. Zhang H, Wu B, Marquard SL, et al. Investigation of ketone C=O bond activation processes by heterobimetallic Zr/Co and Ti/Co tris(phosphinoamide) complexes. *Organometallics*. 2017;36(18):3498–3507. https://doi.org/10.1021/acs.organomet.7b00445.

128. Thomas CM, Napoline JW, Rowe GT, Foxman BM. Oxidative addition across Zr/Co multiple bonds in early/late heterobimetallic complexes. *Chem Commun.* 2010;46 (31):5790–5792. https://doi.org/10.1039/C0CC01272F.
129. Gramigna KM, Dickie DA, Foxman BM, Thomas CM. Cooperative H_2 activation across a metal–metal multiple bond and hydrogenation reactions catalyzed by a Zr/Co heterobimetallic complex. *ACS Catal.* 2019;9(4):3153–3164. https://doi.org/10.1021/acscatal.8b04390.
130. Zhang H, Hatzis GP, Moore CE, et al. O_2 activation by a heterobimetallic Zr/Co complex. *J Am Chem Soc.* 2019;141(24):9516–9520. https://doi.org/10.1021/jacs.9b04215.
131. Bauer J, Braunschweig H, Dewhurst RD. Metal-only Lewis pairs with transition metal Lewis bases. *Chem Rev.* 2012;112(8):4320–4346. https://doi.org/10.1021/cr3000048.
132. Nowell IN, Russell DR. The crystal structure of the dicarbonylcyclopentadienylcobalt-mercuric chloride complex. *Chem Commun.* 1967;(16):817. https://doi.org/10.1039/C19670000817.
133. Ma M, Sidiropoulos A, Lalrempuia R, Stasch A, Jones C. Metal-only Lewis pairs featuring unsupported Pt→M (M = Zn or Cd) dative bonds. *Chem Commun.* 2013;49 (1):48–50. https://doi.org/10.1039/C2CC37442K.
134. Braunschweig H, Gruss K, Radacki K. Complexes with dative bonds between d- and s-block metals: synthesis and structure of [(Cy$_3$P)$_2$Pt–Be(Cl)] (X = Cl, Me). *Angew Chem Int Ed.* 2009;48(23):4239–4241. https://doi.org/10.1002/anie.200900521.
135. Braunschweig H, Gruss K, Radacki K. Interaction between d- and p- block metals: synthesis and structure of platinum–alane adducts. *Angew Chem Int Ed.* 2007;46 (41):7782–7784. https://doi.org/10.1002/anie.200702726.
136. Braunschweig H, Gruss K, Radacki K. Reactivity of Pt^0 complexes towards gallium(III) halides: synthesis of a platinum gallane complex and oxidative addition of gallium halides to Pt^0. *Inorg Chem.* 2008;47(19):8595–8597. https://doi.org/10.1021/ic801293e.
137. Braunschweig H, Damme A, Dewhurst RD, Hupp F, Jimenez-Halla JOC, Radacki K. σ-Donor-σ-acceptor plumbylene ligands: synergic σ-donation between ambiphilic Pt^0 and Pb^{II} fragments. *Chem Commun.* 2012;48(84):10410–10412. https://doi.org/10.1039/C2CC35777A.
138. Braunschweig H, Dewhurst RD, Hupp F, et al. Gauging metal Lewis basicity of zerovalent iron complexes *via* metal-only Lewis pairs. *Chem Sci.* 2014;5(10): 4099–4104. https://doi.org/10.1039/C4SC01539H.
139. Bissert R, Braunschweig H, Dewhurst RD, Schneider C. Metal-only Lewis pairs based on zerovalent osmium. *Organometallics.* 2016;35(15):2567–2573. https://doi.org/10.1021/acs.organomet.6b00495.
140. Pinkes JR, Steffy BD, Vites JC, Cutler AR. Carbon dioxide insertion into the iron-zirconium and ruthenium-zirconium bonds of the heterobimetallic complexes Cp(CO)$_2$MZr(cl)Cp$_2$: direct production of the μ^1-η^1(C):η^2(O,O')-CO2 compounds Cp(CO)$_2$M–CO$_2$–Zr(cl)Cp2. *Organometallics.* 1994;13(1):21–23. https://doi.org/10.1021/om00013a009.
141. Gambarotta S, Arena F, Floriani C, Zanazzi PF. Carbon dioxide fixation: bifunctional complexes containing acidic and basic sites working as reversible carriers. *J Am Chem Soc.* 1982;194(19):5082–5092. https://doi.org/10.1021/ja00383a015.
142. Jamali S, Abedanzadeh S, Khaledi NK, et al. A cooperative pathway for water activation using a bimetallic Pt^0–Cu^I system. *Dalton Trans.* 2016;45(44):17644–17651. https://doi.org/10.1039/c6dt03305a.
143. Hidalgo N, Maya C, Campos J. Cooperative activation of X–H (X = H, C, O, N) bonds by a Pt(0)/Ag(I) metal-only Lewis pair. *Chem Commun.* 2019;55(60): 8812–8815. https://doi.org/10.1039/C9CC03008E.

144. Zhao J, Goldman AS, Hartwig JF. Oxidative addition of ammonia to form a stable monomeric amido hydride complex. *Science*. 2005;307(5712):1080–1082. https://doi.org/10.1126/science.1109389.
145. Fafard CM, Adhikari D, Foxman BM, Mindiola DJ, Ozerov OV. Addition of ammonia, water, and dihydrogen across a single Pd–Pd bond. *J Am Chem Soc*. 2007;129(34): 10318–10319. https://doi.org/10.1021/ja0731571.
146. Scheibel MG, Abbenseth J, Kinauer M, et al. Homolytic N–H activation of ammonia: hydrogen transfer of parent iridium ammine, amide, imide, and nitride species. *Inorg Chem*. 2015;54(19):9290–9302. https://doi.org/10.1021/acs.inorgchem.5b00829.
147. Leon NJ, Yu H-C, Mazzacano TJ, Mankad NP. Pursuit of C–H borylation reactions with non-precious heterobimetallic catalysts: hypothesis-driven variation on a design theme. *Synlett*. 2020;31(02):125–132. https://doi.org/10.1055/s-0039-1691504.
148. Jayarathne U, Parmelee SR, Mankad NP. Small molecule activation chemistry of Cu–Fe heterobimetallic complexes toward CS_2 and N_2O. *Inorg Chem*. 2014;53(14): 7730–7737. https://doi.org/10.1021/ic501054z.
149. Karunananda MK, Vázquez FX, Alp EE, et al. Experimental determination of redox cooperativity and electronic structures in catalytically active Cu–Fe and Zn–Fe heterobimetallic complexes. *Dalton Trans*. 2014;43(36):13661–13671. https://doi.org/10.1039/C4DT01841A.
150. Karunananda MK, Parmelee SR, Waldhart GW, Mankad NP. Experimental and computational characterization of the transition state for C–X bimetallic oxidative addition at a Cu–Fe reaction center. *Organometallics*. 2015;34(15):3857–3864. https://doi.org/10.1021/acs.organomet.5b00476.
151. Karunananda MK, Mankad NP. Heterobimetallic H_2 addition and alkene/alkane elimination reactions related to the mechanism of E-selective alkyne semihydrogenation. *Organometallics*. 2017;36(1):220–227. https://doi.org/10.1021/acs.organomet.6b00356.
152. Zhang Y, Karunananda MK, Yu H-C, et al. Dynamically bifurcating hydride transfer mechanism and origin of inverse isotope effect for heterodinuclear AgRu-catalyzed alkyne semihydrogenation. *ACS Catal*. 2019;9(3):2657–2663. https://doi.org/10.1021/acscatal.8b04130.
153. Parmelee SR, Mazzacano TJ, Zhu Y, Mankad NP, Keith J. A heterobimetallic mechanism for the C–H borylation elucidated from experimental and computational data. *ACS Catal*. 2015;5(6):3689–3699. https://doi.org/10.1021/acscatal.5b00275.
154. Rokob TA, Hamza A, Stirling A, Pápai I. On the mechanism of $B(C_6F_5)_3$-catalyzed direct hydrogenation of imines: inherent and thermally induced frustration. *J Am Chem Soc*. 2009;131(5):2029–2036. https://doi.org/10.1021/ja809125r.
155. Devillard M, Declercq R, Nicolas E, et al. A significant but constrained geometry Pt→Al interaction: fixation of CO_2 and CS_2, activation of H_2 and $PhCONH_2$. *J Am Chem Soc*. 2016;138(14):4917–4926. https://doi.org/10.1021/jacs.6b01320.
156. Holschumacher D, Bannenberg T, Hrib CG, Jones PG, Tamm M. Heterolytic dihydrogen activation by a frustrated carbene-borane Lewis pair. *Angew Chem Int Ed*. 2008;47(39):7428–7432. https://doi.org/10.1002/anie.200802705.
157. Grimme S, Kruse H, Goerigk L, Erker G. The mechanism of dihydrogen activation by frustrated Lewis pairs revisited. *Angew Chem Int Ed*. 2010;49(8):1402–1405. https://doi.org/10.1002/anie.200905484.
158. Lu Z, Cheng Z, Chen Z, Weng L, Li ZH, Wang H. Heterolytic cleavage of dihydrogen by "frustrated Lewis pairs" comprising bis(2,4,6-tris(trifluoromethyl)-phenyl)borane and amines: stepwise versus concerted mechanism. *Angew Chem Int Ed*. 2011;50(51):12227–12231. https://doi.org/10.1002/anie.201104999.

159. Könczöl L, Makkos E, Bourissou D, Szieberth D. Computational evidence for a new type of η^2-H$_2$ complex: when main-group elements act in concert to emulate transition metals. *Angew Chem Int Ed.* 2012;51(38):9521–9524. https://doi.org/10.1002/anie.201204794.
160. Daru J, Bakó I, Stirling A, Pápai I. Mechanism of heterolytic hydrogen splitting by frustrated Lewis pairs: comparison of static and dynamic models. *ACS Catal.* 2019;9(7):6049–6057. https://doi.org/10.1021/acscatal.9b01137.
161. Brewster TP, Nguyen TH, Li Z, Eckenhoff WT, Schley ND, DeYonker NJ. Synthesis and characterization of heterobimetallic iridium-aluminum and rhodium-aluminum complexes. *Inorg Chem.* 2018;57(3):1148–1157. https://doi.org/10.1021/acs.inorgchem.7b02601.
162. Carlles III RM, Yokley TW, Schley ND, DeYonker NJ, Brewster TP. Hydrogen activation and hydrogenolysis facilitated by late-transition-metal-aluminum heterobimetallic complexes. *Inorg Chem.* 2019;58(19):12635–12645. https://doi.org/10.1021/acs.inorgchem.9b01359.
163. Chapman AM, Flynn SR, Wass DF. Unexpected formation of early late heterobimetallic complexes from transition metal frustrated Lewis pairs. *Inorg Chem.* 2016;55(3):1017–1021. https://doi.org/10.1021/acs.inorgchem.5b01424.
164. Campos C. Dihydrogen and acetylene activation by a gold(I)/platinum(0) transition metal only frustrated Lewis pair. *J Am Chem Soc.* 2017;139(8):2944–2947. https://doi.org/10.1021/jacs.7b00491.
165. Hidalgo N, Moreno JJ, Pérez-Jiménez M, Maya C, López-Serrano J, Campos J. Evidence for genuine bimetallic frustrated Lewis pair activation of dihydrogen with gold(I)/platinum(0) systems. *Chem A Eur J.* 2020;26(27):5982–5993. https://doi.org/10.1002/chem.201905793.
166. Dang LX, Schenter GK, Chang T-M, Kathmann SM, Autrey T. Role of solvents on the thermodynamics and kinetics of forming frustrated Lewis pairs. *J Phys Chem Lett.* 2012;3(22):3312–3319. https://doi.org/10.1021/jz301533a.
167. Hidalgo N, Bajo S, Moreno JJ, Navarro-Gilabert C, Mercado BQ, Campos J. Reactivity of a gold(I)/platinum(0) frustrated Lewis pair with germanium and tin dihalides. *Dalton Trans.* 2019;48(25):9127–9138. https://doi.org/10.1039/C9DT00702D.
168. Hidalgo N, Moreno JJ, Pérez-Jiménez M, Maya C, López-Serrano J, Campos J. Tuning activity and selectivity during alkyne activation by gold(I)/platinum(0) frustrated Lewis pairs. *Organometallics.* 2020;39(13):2534–2544. https://doi.org/10.1021/acs.organomet.0c00330.
169. Gunsalus NJ, Park SH, Hashiguchi BG, et al. Selective N functionalization of methane and ethane to aminated derivatives by main-group-directed C–H activation. *Organometallics.* 2019;38(11):2319–2322. https://doi.org/10.1021/acs.organomet.9b00246.
170. Das S, Mondal S, Pati SK. Mechanistic insights into hydrogen activation by frustrated N/Sn Lewis pairs. *Chem A Eur J.* 2018;24(11):2575–2579. https://doi.org/10.1002/chem.201705861.
171. Rokob TA, Pápai I. Hydrogen activation by frustrated Lewis pairs: insights from computational studies. *Top Curr Chem.* 2013;332:157–211. https://doi.org/10.1007/128_2012_399.
172. Paradies J. Mechanisms in frustrated Lewis pair-catalyzed reactions. *Eur J Org Chem.* 2019;2:283–293. https://doi.org/10.1002/ejoc.201800944.
173. Rocchigiani L. Experimental insights into the structure and reactivity of frustrated Lewis pairs. *Isr J Chem.* 2015;55(2):134–149. https://doi.org/10.1002/ijch.201400139.
174. Zhang L, Hu S, Yang L, et al. H$_2$ activation by heterobimetallic gold(I)/platinum(0) complex: theoretical understanding of electronic processes and prediction on more active species. *J Phys Chem C.* 2020;124(8):4525–4533. https://doi.org/10.1021/acs.jpcc.9b09452.
175. Gómez-Gallego M, Sierra MA. Kinetic isotope effects in the study of organometallic reaction mechanisms. *Chem Rev.* 2011;111(8):4857–4963. https://doi.org/10.1021/cr100436k.

176. Dureen MA, Stephan DW. Terminal alkyne activation by frustrated and classical Lewis acid/phosphine pairs. *J Am Chem Soc*. 2009;131(24):8396–8397. https://doi.org/10.1021/ja903650w.
177. Fukazawa A, Yamada H, Yamaguchi S. Phosphonium- and borate-bridged zwitterionic ladder stilbene and its extended analogues. *Angew Chem Int Ed*. 2008;47(30): 5582–5585. https://doi.org/10.1002/anie.200801834.
178. Geier SK, Dureen MA, Ouyang EY, Stephan DW. New strategies to phosphino-phosphonium cations and zwitterions. *Chem A Eur J*. 2010;16(3):988–993. https://doi.org/10.1002/chem.200902369.
179. Swarnakar AK, McDonald SM, Deutsch KC, et al. Application of the donor-acceptor concept to intercept low oxidation state group 14 element hydrides using a Wittig reagent as a Lewis base. *Inorg Chem*. 2014;53(16):8662–8671. https://doi.org/10.1021/ic501265k.
180. Al-Rafia SMI, Malcolm AC, McDonald R, Fegruson MJ, Rivard E. Efficient generation of stable adducts of Si(II) dihydride using a donor-acceptor approach. *Chem Commun*. 2012;48(9):1308–1310. https://doi.org/10.1039/C2CC17101E.
181. Ghadwal RS, Roesky HW, Merkel S, Stalke D. Ambiphilicity of dichlorosilylene in a single molecule. *Chem A Eur J*. 2010;16(1):85–88. https://doi.org/10.1002/chem.200902930.
182. Abraham MY, Wang Y, Xie Y, et al. Cleavage of carbene-stabilized disilicon. *J Am Chem Soc*. 2011;133(23):8874–8876. https://doi.org/10.1021/ja203208t.
183. Azhakar R, Tavcar G, Roesky HW, Hey J, Stalke D. Facile synthesis of a rare chlorosilylene–BH$_3$ addcut. *Eur J Inorg Chem*. 2011;4:475–477. https://doi.org/10.1002/ejic.201001188.
184. Al-Rafia SMI, Malcolm AC, Liew SK, Ferguson MJ, Rivard E. Stabilization of the heavy methylene analogues, GeH$_2$ and SnH$_2$, within the coordination sphere of a transition metal. *J Am Chem Soc*. 2011;133(4):777–779. https://doi.org/10.1021/ja1106223.
185. Eisenhut C, Szilvási T, Dübek G, Breit NC, Inoue S. Systematic study of N-heterocyclic carbene coordinate hydrosilylene transition-metal complexes. *Inorg Chem*. 2017;56(16):10061–10069. https://doi.org/10.1021/acs.inorgchem.7b01541.
186. Feng Z, Jiang Y, Ruan H, et al. A diamidinatogermylene as a Z-type ligand in a nickel(0) complex. *Dalton Trans*. 2019;48(40):14975–14978. https://doi.org/10.1039/C9DT03803E.
187. Swarnakar AK, Ferguson MJ, McDonald R, Rivard E. Transition metal-mediated donor-acceptor coordination of low-oxidation state group 14 element halides. *Dalton Trans*. 2015;45(14):6071–6078. https://doi.org/10.1039/C5DT03018H.
188. Melen R. A step closer to metal-free dinitrogen activation: a new chapter in the chemistry of frustrated Lewis pairs. *Angew Chem Int Ed*. 2018;57(4):880–882. https://doi.org/10.1002/anie.201711945.
189. Légaré M-A, Bélanger-Chabot G, Dewhurst RD, et al. Nitrogen fixation and reduction at boron. *Science*. 2018;359(6378):896–900. https://doi.org/10.1126/science.aaq1684.
190. Tanabe Y, Nishibayashi Y. Catalytic dinitrogen fixation to form ammonia at ambient reaction conditions using transition metal-dinitrogen complexes. *Chem Rec*. 2016;16(3): 1549–1577. https://doi.org/10.1002/tcr.201600025.
191. Nishibayashi Y. Recent progress in transition metal-catalyzed reduction of molecular dinitrogen under ambient reaction conditions. *Inorg Chem*. 2015;54(19):9234–9247. https://doi.org/10.1021/acs.inorgchem.5b00881.
192. Jia H-P, Quadrelli EA. Mechanistic aspects of dinitrogen cleavage and hydrogenation to produce ammonia in catalysis and organometallic chemistry: relevance of metal hydride bonds and dihydrogen. *Chem Soc Rev*. 2014;43(2):547–564. https://doi.org/10.1039/C3CS60206K.
193. MacLeod KC, Holland PL. Recent developments in the homogeneous reduction of dinitrogen by molybdenum and iron. *Nat Chem*. 2013;5:559–565. https://doi.org/10.1038/nchem.1620.

194. Hölscher M Leitner W. Catalytic NH$_3$ synthesis using N$_2$/H$_2$ at molecular transition metal complexes: concepts for lead structure determination using computational chemistry. *Chem A Eur J*. 2017;23(50):11992–12003. https://doi.org/10.1002/chem.201604612.
195. Jennings JR. *Catalytic Ammonia Synthesis: Fundamentals and Practice*. New York: Springer; 1991.
196. Geri JB, Shanahan JP, Szymczak NK. Testing the push-pull hypothesis: Lewis acid augmented N$_2$ activation at iron. *J Am Chem Soc*. 2017;139(16):5952–5956. https://doi.org/10.1021/jacs.7b01982.
197. Simmonneau A, Turrel R, Vendier L, Etienne M. Group 6 transition-metal/boron frustrated Lewis pair templates activate N$_2$ and allow its facile borylation and silylation. *Angew Chem Int Ed*. 2017;56(40):12268–12272. https://doi.org/10.1002/anie.201706226.
198. Doyle LR, Wooles AJ, Liddle ST. Bimetallic cooperative cleavage of dinitrogen to nitride and tandem frustrated Lewis pair hydrogenation to ammonia. *Angew Chem Int Ed*. 2019;58(20):6674–6677. https://doi.org/10.1002/anie.201902195.
199. Lu G, Zhang P, Sun D, et al. Gold catalyzed hydrogenations of small imines and nitriles: enhanced reactivity of Au surface toward H$_2$ *via* collaboration with a Lewis base. *Chem Sci*. 2014;5(3):1082–1090. https://doi.org/10.1039/C3SC52851K.
200. Mahdi T, Stephan DW. Facile protocol for catalytic frustrated Lewis pair hydrogenation and reductive deoxygenation of ketones and aldehydes. *Angew Chem Int Ed*. 2015;54(29):8511–8514. https://doi.org/10.1002/anie.201503087.
201. Szeto KC, Sahyoun W, Merle N, et al. Development of silica-supported frustrated Lewis pairs: highly active transition metal-free catalysts for the Z-selective reduction of alkynes. *Cat Sci Technol*. 2016;6(3):882–889. https://doi.org/10.1039/C5CY01372K.
202. Lassalle S, Jabbour R, Schiltz P, et al. Metal–metal synergy in well-defined surface tantalum–iridium heterobimetallic catalysts for H/D exchange reactions. *J Am Chem Soc*. 2019;141(49):19321–19335. https://doi.org/10.1021/jacs.9b08311.
203. Ye J, Johnson K. Design of Lewis pair-functionalized metal organic frameworks for CO$_2$ hydrogenation. *ACS Catal*. 2015;5(5):2921–2929. https://doi.org/10.1021/acscatal.5b00396.
204. Ye J, Johnson K. Screening Lewis pair moieties for catalytic hydrogenation of CO$_2$ in functionalized UiO-66. *ACS Catal*. 2015;5(10):6219–6229. https://doi.org/10.1021/acscatal.5b01191.
205. Ye J, Johnson K. Catalytic hydrogenation of CO$_2$ to methanol in a Lewis pair functionalized MOF. *Cat Sci Technol*. 2016;6(24):8392–8405. https://doi.org/10.1039/C6CY01245K.
206. Niu Z, Bhagya Gunatilleke WDC, Sun Q, et al. Metal-organic framework anchored with a Lewis pair as a new paradigm for catalysis. *Chem*. 2018;4(11):2587–2599. https://doi.org/10.1016/j.chempr.2018.08.018.
207. Sun X, Li B, Liu T, Song J, Su DS. Designing graphene as a new frustrated Lewis pair catalyst for hydrogen activation by co-doping. *Phys Chem Chem Phys*. 2016;18(16):11120–11124. https://doi.org/10.1039/C5CP07969A.
208. Zhang S, Huang Z-Q, Ma Y, et al. Solid frustrated-Lewis-pair catalysts constructed by regulations on surface defects of porous nanorods of CeO$_2$. *Nat Commun*. 2017;8:15266. https://doi.org/10.1038/ncomms15266.
209. Huang Z-Q, Liu L-P, Qi S, Zhang S, Qu Y, Chang C-R. Understanding all-solid frustrated-Lewis-pair sites on CeO$_2$ from theoretical perspectives. *ACS Catal*. 2018;8(1):546–554. https://doi.org/10.1021/acscatal.7b02732.

CHAPTER FOUR

Metallic-based magnetic switches under confinement

Alejandro López-Moreno and Maria del Carmen Giménez-López*

Centro Singular de Investigación en Química Biolóxica e Materiais Moleculares (CIQUS), Universidade de Santiago de Compostela, Santiago de Compostela, Spain
*Corresponding author: e-mail address: maria.gimenez.lopez@usc.es

Contents

1. Introduction	149
2. Endohedral fullerenes: Zero-dimensional host cages	153
2.1 Dysprosium-containing clusterfullerenes molecular magnets	155
2.2 Dimetallofullerenes molecule magnets: Single-electron lanthanide-lanthanide bonds	160
3. Carbon nanotubes: One-dimensional host nanocontainers	162
3.1 Strategies for filling CNT	163
3.2 EMF-SMM as guest molecules in carbon nanotubes: One-dimensional peapod structures	164
3.3 Other molecule magnets confined in carbon nanotubes	165
4. Metal-organic frameworks: As three-dimensional platforms	167
4.1 Spin crossover into three-dimensional structures	168
4.2 Single molecule magnets in three-dimensional structures	173
5. Conclusions	181
Acknowledgments	183
References	183

1. Introduction

Magnetic switching phenomena[1,2] have become key for the development of a plethora of different technological applications in spintronics and quantum computing for storing and processing information faster and in an energy-efficient manner. Strategies for advancement in this field have been focusing on molecular sizing. Thus, magnetic bistability in molecular magnets, essential for data storage applications, arises from a combination of two factors: a large spin ground state (as result of intramolecular superexchange interactions) and a high axial magnetic anisotropy (due to a negative and large zero-field splitting parameter, leading to a preferential magnetization direction).[3] These two factors combined result in a large energy barrier for magnetization reversal, which gives rise to slow relaxation of the magnetization and concurrent magnetic bistability, showing hysteresis below a threshold temperature (T_B, blocking temperature). Molecular magnets can also exhibit properties unique to this class of materials, including quantum tunneling of magnetization,[4,5] and quantum coherence.[6,7] Since the discovery by Sessoly et al. of the quantum tunneling of magnetization in a manganese polynuclear cluster almost three decades ago,[2,8] the ongoing race for obtaining molecular magnets with slower relaxation of magnetization at ever increasing temperatures started (Fig. 1).[9] The first single-molecule magnets (SMM) were based on clusters of transition metals aiming to increase spin ground state values, however in the last decade synthetic strategies involving lanthanides instead have been put forward because of their strong single-ion anisotropy.[10,11] Thus, other systems exhibiting a SMM-like behavior are single-ion magnets (SIM), polyoxometallates (POM) and endohedral fullerenes. Among all the lanthanides, Dy has become the most popular metal for creating new SMM, including efforts in endohedral metallofullerenes (EMF).[12] The highest temperature of magnetic hysteresis in SMM exceeding the liquid nitrogen temperature has recently been achieved in organometallic compounds involving Dy-metallocenium salts thanks to amazing efforts from Layfield, Mills and Long groups.[13–16]

A special class of bistable molecule-based materials are spin-crossover systems in which d orbitals play a fundamental role.[17,18] The spin-crossover (SCO) process comprises the rearrangement of electrons in metal ions, from

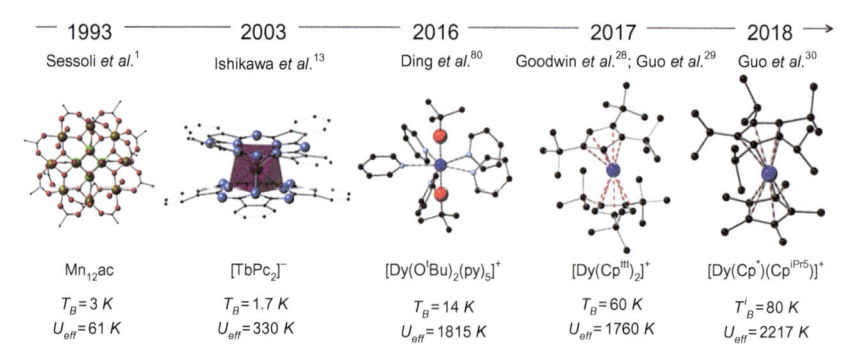

Fig. 1 Rising values of the effective barriers (U_{eff}) and blocking temperatures (T_B) in molecular nanomagnets. *Reproduced from Escalera-Moreno L, Baldovi J, Gaita-Ariño A, Coronado E. Exploring the high-temperature frontier in molecular nanomagnets: from lanthanides to actinides.* Inorg Chem. *2019;58:11883–11892.*

a low-spin (LS) to a high-spin (HS) state, corresponding to the distributions of electrons within metal orbital giving rise to a minimum and maximum number of unpaired electrons, respectively. The switching phenomenon is particularly popular in iron chemistry and can take place in any phase of matter, however in the solid state is the most usual. Thus, the magnetic state of Fe(II) bistable materials can be tuned from a low-spin configuration (LS, S = 0) to a high-spin configuration (HS, S = 2) through external stimuli: thermally, by means of light irradiation, or under pressure.[19–25] Attempts to control the spin states on SCO complexes through electrical and mechanical stretching have also been reported.[26,27] Changes in SCO materials are accompanied by a change in their magnetic moment, color, dielectric constant and electrical resistance. Besides SCO shows pronounced hysteresis (it is a *thermal hysteresis* when the variable causing it is temperature), which often is concomitant with a structural phase change during the transition. Within the hysteresis loop, the SCO molecule-based materials are simply bistable switches as they can be either LS or HS depending on the history.

Bistability is undoubtedly one of the most desired properties when designing new technological materials.[28,29] One particular challenge for harnessing the magnetic bistability of these molecular switches is however their coupling to the macroscopic world through the integration into functional devices, while controlling their functional properties[30] (that is essential for read and write purposes) and retaining their chemical integrity and magnetic properties. In the last decade, efforts have been done in controlling organization of these switches so that they can be addressed individually, but it has proven to be a difficult task, mainly because of the limited stability of some magnetic switches during the nanostructuration process (Fig. 2). Most of the efforts were initially focused on the archetypical Mn_{12} SMM. Pioneering work in this area focused on the two-dimensional nanostructuration of these system has been done by Coronado and co-workers using the Langmuir-Blodgett technique.[31] Subsequently, other groups reported on the two-dimensional assembly of SMM using both inorganic and organic two-dimensional supports[32–35] and the methods and techniques used are collected in several detailed reviews.[36,37] It is worth noting that SMM behaviors were preserved in only a few of the reported materials, and exhibited similar magnetic features when compared to the unsupported magnetic switches. More recent investigations have demonstrated that different confinement strategies involving hollow and layered supports,[3] such as mesoporous silicas,[38] carbon nanotubes,[39] metallo supra-molecular cages,[40] chalcogenide and oxalate networks[41,42] and coordination polymer networks,[43–45]

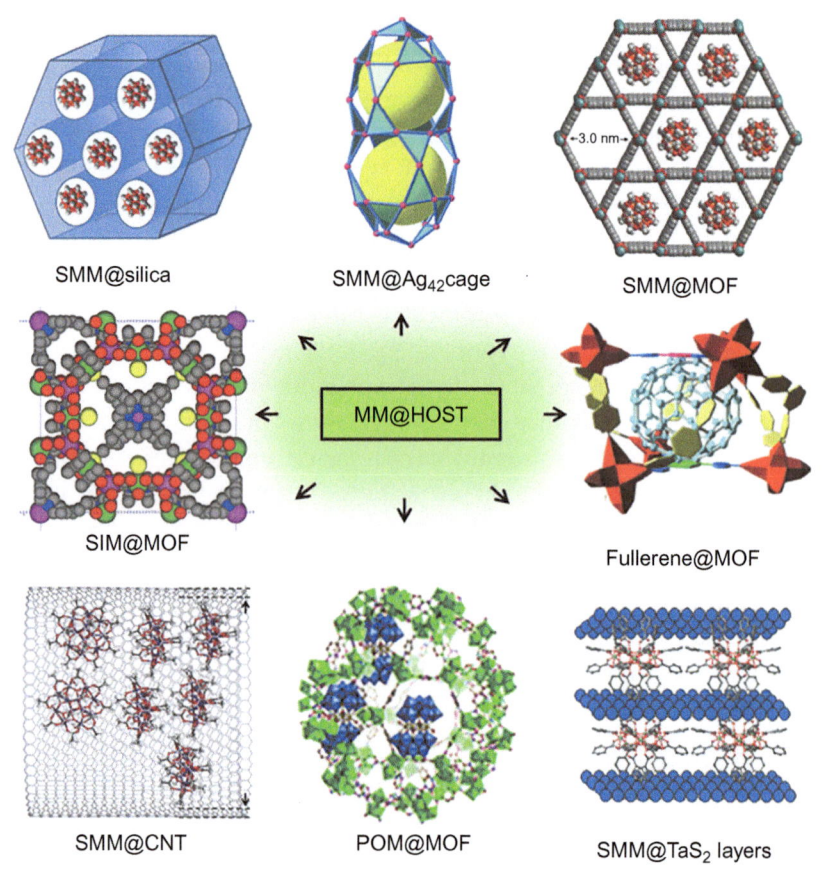

Fig. 2 Schematic representation of the different strategies for the nanostructuration of SMM in different hollow and layered supports. *Reproduced from Aulakh D, Bilan H, Wriedt M. Porous substrates as platforms for the nanostructuring of molecular magnets.* Cryst Eng Comm *2018;20:1011–1030.*

may be the way forward for the nanostructuration of magnetic switches with an increasing blocking temperature (Fig. 2). As the goal is to minimize the effect of the local vibrations on the spin relaxation of the molecular switches, as it has been shown recently,[15] it is expected that the interactions between the guest nanoscale switches with the host structures will lead to significant changes in the magnetic properties of the confined magnetic switches, such as the blocking temperature[46] and the relaxation times of the magnetization.[47] Although significant progress has been done in this field of nanostructuration, articles providing a comprehensive review on the confinement effects of nanoscale switches in void cages and pore nanostructures are needed in the area. Among the most intriguing systems, we will review the current literature on fullerene, carbon nanotubes and metal-organic frameworks, as zero-dimensional host-cages, one-dimensional nanocontainers and three-dimensional templates, respectively. We will conclude with a discussion of current challenges and future perspectives in this newly emerging field on the nanostructuring of molecular switches.

2. Endohedral fullerenes: Zero-dimensional host cages

The unique ability of the fullerene cages to act as host cages to stabilize species that would otherwise exist has been exploited to create families of endohedral metallofullerenes (EMF).[48–50] Thus, the confinement of single atoms, atomic clusters, or even small molecules inside the sub-nanometer sized void of the carbon fullerene cages has yielded to the formation of very exotic endohedral fullerene structures[51] with unique magnetic properties, among others. Unlike metallofullerenes (only containing metal ions inside the cage), endohedral fullerenes contain clusters comprising multiple metallic ions (from one to four, and typically in their three-valent state) along with electronegative atoms and species (i.e., N^{3-}, C_2^{2-}, S^{2-}, O^{2-}), which attain a negative charge, compensating the Coulombic repulsion of the positively charged metal ions inside the cage. Since the discovery of the first clusterfullerene $Sc_3N@C_{80}$ in 1999, a variety of EMF have been reported,[48–50] including di- and trimetallic clusters,[52,53] carbides,[54,55] carbon nitrides,[56] nitrides,[57] sulfides,[58] and oxide clusters[59] in a variety of C_n cages. Arc-discharge methods involving the evaporation of graphite electrodes impregnated with a metal or a metal oxide precursor are utilized for their production. And the addition of organic compounds or reactive gases may be needed to tune the composition and selectivity of the desired system. Investigations on the formation mechanism of EMF[60] suggest that the endohedral species may trigger the formation of the fullerene cage acting as templates. Thus, the production of nitride clusterfullerenes has been referred as the "trimetallic nitride template" (TNT) process.[57]

Single molecule magnetism in clusterfullerenes containing a 4f-lanthanide metal ($DySc_2N@C_{80}$) was observed for the first time by Popov et al. in 2012.[61] This breakthrough in EMF marked a turning point in the area and since then, in only few years, SMM behavior has been demonstrated for other EMF (Fig. 3).[12,61–63] Magnetic anisotropy in lanthanide EMF SMM arises by negatively charged nonmetallic species inside the cage and the carbon atoms of the negatively charged fullerene cages themselves. Magnetic anisotropy in

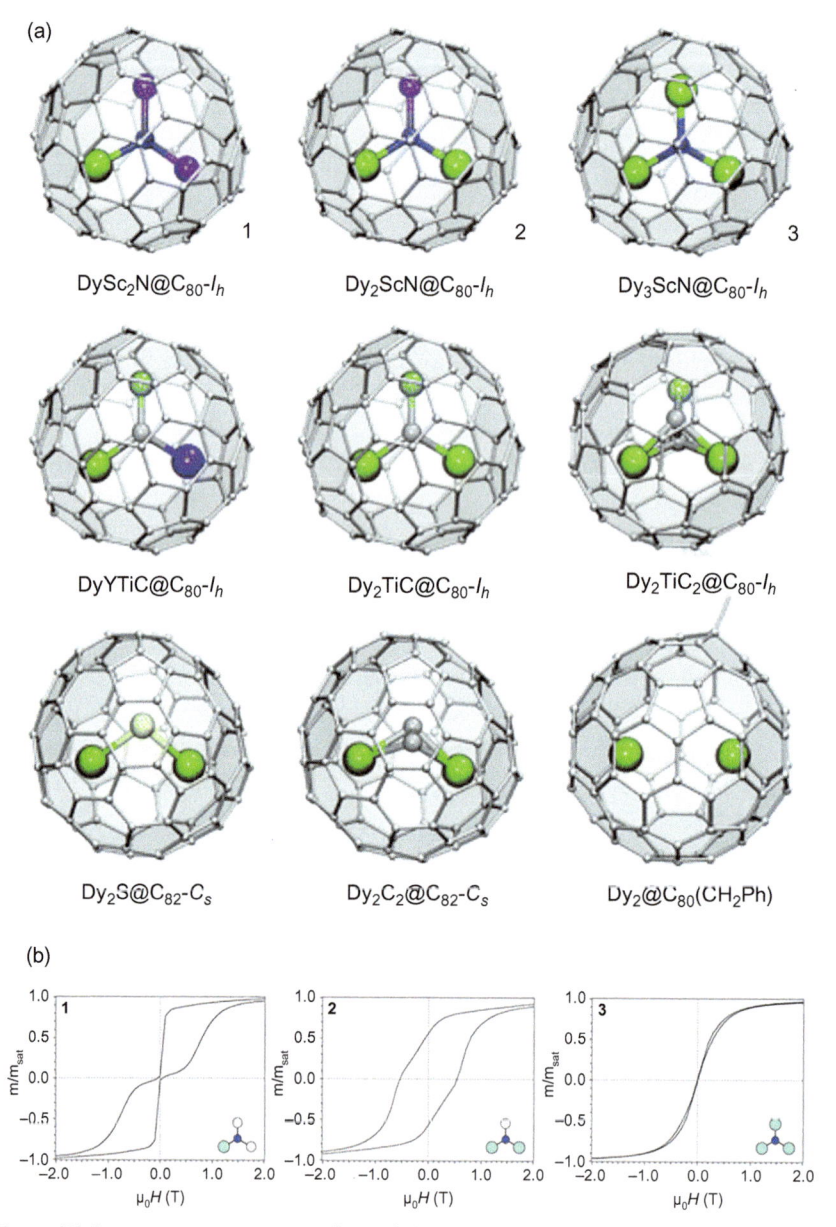

Fig. 3 (A) Dysprosium-containing clusterfullerenes exhibiting SMM behavior. Coulour code: Dy is shown in green, Sc – magenta, Ti – cyan, Y – violet, N – blue, C – gray, and S – yellow. Note: the structure of the benzyl group of $Dy_2@C_{80}(CH_2Ph)$ is not completed. (B) Magnetic hysteresis curves at 2 K (field sweep rate of 0.8 mT s^{-1}) for the $Dy_xSc_{3-x}N@C_{80-Ih}$ family (from left to right: x = 1, 2, and 3). Panel A: reproduced from Spree L, Popov A. Recent advances in single molecule magnetism of dysprosium-metallofullerenes. Dalton Trans. 2019;48:2861–2871. Panel B: adapted from Westerström R, Dreiser J, Piamonteze C, Muntwiler M, Weyeneth S, Krämer K, Liu S, Decurtins S, Popov A, Yang S, Dunsch L, Greber T. Tunneling, remanence, and frustration in dysprosium-based endohedral single-molecule magnets. Phys Rev B 2014;89:060406.

EMF has also shown to have effect on the increasing proton relaxivities (r_1) of water-soluble derivatives of $Gd_3N@C_{80}$ as magnetic resonance imaging agents compared to that reported for $Gd@C_{82}$ under the same conditions.[64]

The host fullerene cages with high thermal stability enable the stabilization and protection from ambient detrimental conditions of unstable variety of exotic molecular configurations that would otherwise not exist. Moreover, EMF display a very rich addition chemistry, allowing the functionalization of the fullerene cage with different pendant groups without disrupting the structure of the endohedral species.[65] Thus, the combination of these unique physical and chemical attributes makes EMF a perfect zero-dimensional object for developing bistable functional materials. There exist reports on the theoretical prediction of hetero dimers of EMF that may act as single-molecule spin switches[66,67] or experimental data designing metallic complexes bridging two fullerene cages to obtain dimers in which the spin transition is accompanied by a metal-to-fullerene charge transfer.[68] Nevertheless, in the present review we will focus on providing recent advances in single molecule magnetism of different families of clusterfullerenes containing Dy, especially in the oxide clustersfullerenes and the dimetallofullerenes, showing very intriguing SMM features caused by the fullerene cage. Recent theoretical advances in Co-encapsulated metallofullerenes are however out of the scope of this review.[69]

2.1 Dysprosium-containing clusterfullerenes molecular magnets

The combination of a bistable magnetic ground state with a strong single-ion magnetic anisotropy is the key for single molecule magnetism in Dy-containing clusterfullerenes. The short distance between Dy ions and negatively charged nonmetal atom gives rise to a strong axial ligand field, and thus a strong magnetic anisotropy for the dysprosium ions in the clusterfullerenes.[70-72] Moreover, Dy(III) with $4f^9$ electrons is a Kramers ion with a doubly degenerate $\pm m_j$ ground state, for which quantum tunneling of the magnetization (QTM) should be forbidden in the absence of magnetic field (QTM usually prevents the observation of magnetic hysteresis at zero magnetic field in lanthanide complexes). Indeed, a long relaxation of magnetization with a butterfly-shaped hysteresis due to QTM was first proven for the nitrile clusterfullerene $DySc_2N@C_{80}$-I_h with a blocking temperature of $T_B = 7$ K.[61] To prove that the SMM behavior observed was indeed a single molecule phenomenon instead of a collective effect, $DySc_2N@C_{80}$-I_h was diluted with nonmagnetic C_{60}. Magnetic studies for the dinuclear Dy$_2$ScN@C$_{80}$-I$_h$ showed an increase of the blocking temperature (T$_B$ = 8 K) and the disappearance of the QTM due to the ferromagnetic coupling of the two Dy spins (Fig. 3).[71,73] However, magnetic studies for the trinuclear Dy$_3$ScN@C$_{80}$-I$_h$ does not show remanence due to a frustrated magnetic ground state, as ferromagnetic coupling in a triangular Dy$_3$N cluster cannot be realized for all three Dy spins at once (Fig. 3). The intriguing properties shown by the Dy$_x$Sc$_{3-x}$N@C$_{80}$-I$_h$ family soon after prompted SMM studies for Dy-nitride,[71,73,74] Dy-sulfide,[70] and Dy-carbide [75,76] clusterfullerenes. Field-induced SMM behavior was also demonstrated for cyano-clusterfullerenes with single metal atoms, TbNC@C$_{82}$ [77] and TbNC@C$_{76}$.[78] Nitride clusterfullerenes give the best SMM, followed by sulfide and carbides clusterfullerenes (containing one and two carbon atoms, in improvement order). The non-metal clusters facilitate the strong single-ion anisotropy needed to trigger a SMM behavior and contribute to the coupling of the magnetic ions in the dinuclear metal-clusterfullerenes, suppressing QTM and thus giving SMM with pronounced remanence. When two or more Dy ions are enclosed within one fullerene cage molecule, the magnetic properties of Dy-containing clusterfullerenes can be tuned by designing exchange interactions between the lanthanide ions.[73] To fully understand SMM properties in EMF, it is important also to consider that fullerenes cages are not just inert containers. They play a relevant role in the relaxation of magnetization, as shown by the changes in the SMM properties of nitride and sulfide clusterfullerenes for different cage sizes and isomers, but it is not yet fully understood.[70,74]

Recently it has been foreseen through ab initio calculations that among all metal clusters the oxide clusterfullerenes, firstly discovered by Stevenson et al.[59,79] may have the largest crystal field splitting with thermal barriers up to 1400 cm^{-1} (Fig. 4), which has arisen their interest as promising platforms to improve SMM in Dy-containing EMF.[70,72] Because their larger values in the energy difference between the ground and the first excited state (ΔE_{1-2}) and the overall crystal field splitting (ΔE_{1-8}), which have been associated to the short Dy—O distances (from 2.041 to 2.029 Å) and rather large Dy−O−Dy angle of 134°, superior SMM properties were predicted for Dy$_2$O@C$_{82}$-C$_{3v}$ than for Dy$_2$ScN@C$_{80}$[70] Even larger crystal field splitting values in mixed-metal Dy—Sc and Dy—Lu oxide clusterfullerenes (DyOM@C$_{2n}$) has been also predicted by Rajaraman et al.[72] Therefore, Dy-oxide clusterfullerenes appear to be a sensible target for designing

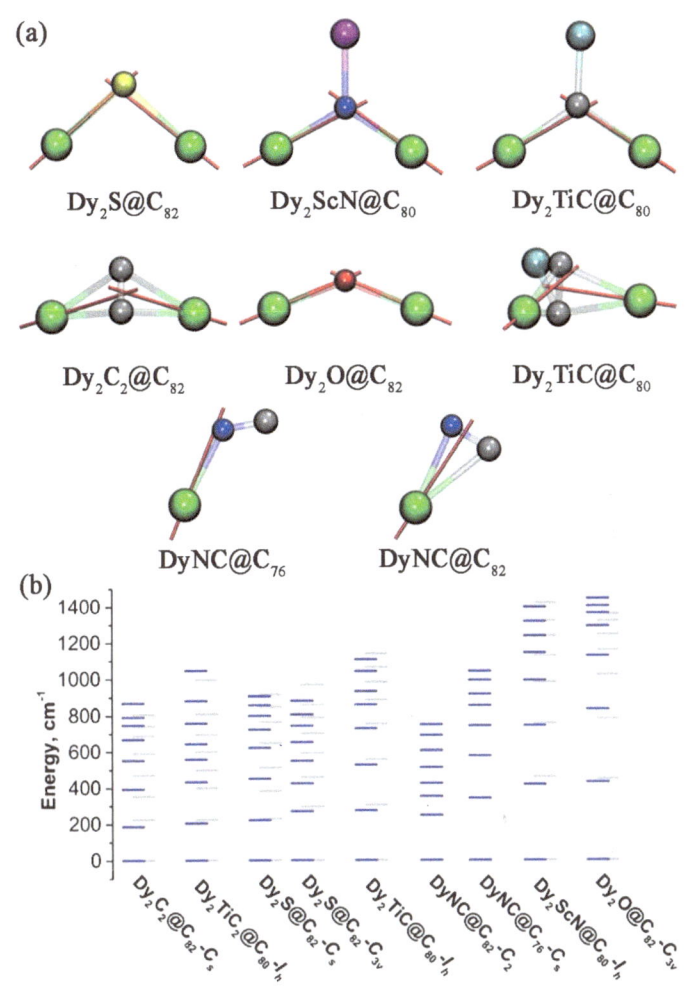

Fig. 4 (A) Structural arrangement of selected endohedral species in selected Dy-containing clusterfullerenes. The magnetic anisotropy axes are shown as red lines for each Dy center according to ab initio calculations in ref. 70. Color code: Dy – green, Sc – magenta, Ti – cyan, S – yellow, C – gray, N – blue, and O – red; carbon cages are omitted for clarity. Note: The compounds not studied experimentally are marked in gray. (B) Calculated energies (cm^{-1}) of crystal field states in different clusterfullerenes (when the molecule has two Dy ions, the energies for each center are given in blue and gray). *Reproduced from Chen CH, Krylov D, Avdoshenko SM, Liu F, Spree L, Yadav R, Alvertis A, Hozoi L., Nenkov K, Kostanyan A, Greber T, Wolter A, Popov A. Selective arc-discharge synthesis of Dy2S-clusterfullerenes and their isomer-dependent single molecule magnetism. Chem Sci. 2017;8:6451–6465.*

SMM–EMF. In 2019, the synthesis, structural characterization and magnetic studies for three isomers of Dy$_2$O@C$_{82}$ (C_s (6), C_{3v} (8) and C_{2v} (9)) were reported for the first time.[72] A modified Krätschmer-Huffman direct current arc-discharge method[80] under a He/CO$_2$ atmosphere was used for their preparation, and multistage high-performance liquid chromatography techniques for their isolation and purification. Single-crystal X-ray structure of the Dy$_2$O@C$_{82}$ isomers co-crystallized with Ni(OEP) revealed that the Dy$_2$O cluster within the fullerene cage has a bent shape (Dy–(μ_2-O)–Dy) and shows an unusual short Dy—O bond distance (Fig. 5), which leads to both a strong axial ligand field (magnetic anisotropy) and an unprecedentedly strong antiferromagnetic exchange coupling between the Dy ion magnetic moments along the Dy—O bonds. It is worth noting that the Dy···Dy exchange coupling in Dy$_2$O@C$_{82}$ is the strongest among all dinuclear Dy SMM. All three isomers display a SMM behavior with a broad magnetic hysteresis at low temperatures, in which the relaxation of magnetization follows an Arrhenius law being the energy barrier the difference in energy between the antiferromagnetically and the ferromagnetically-coupled states. A strong correlation of the relaxation of magnetization with the fullerene cage isomerism via the spin-phonon coupling and the energy transfer from the spin system to the lattice has been hypothesized for explaining the difference in behavior between the three isomers. The C_s isomer with a less symmetric cage in which the endohedral cluster has a more fixed position exhibit the shortest magnetization relaxation times (relaxing therefore faster), whereas the C_{3v} isomer with a more isotropic symmetric cage in which the cluster move freely shows the longest relaxation times. A restricted motion of the endohedral cluster points therefore to the stronger coupling to the cage vibrations, which facilitates the spin-lattice relaxation. On the contrary, for the free motion a weakly coupling with the cage vibrations can be anticipated restricting the spin-lattice energy transfer. Thus, the cage protects the endohedral species not only chemically from the environment, but also from the lattice phonons. Apart from the dynamical properties, the fullerene cage isomer affects structural parameters of the endohedral species (bond lengths and angles), and therefore values of the exchange coupling constants. However, the strong dependence observed for the exchange interactions in Dy$_2$O@C$_{82}$ and the sign of the coupling (antiferromagnetic) is still an open question. Future investigations in the Dy$_2$O@C$_{2n}$ clusterfullerene family may reveal better SMM performances.

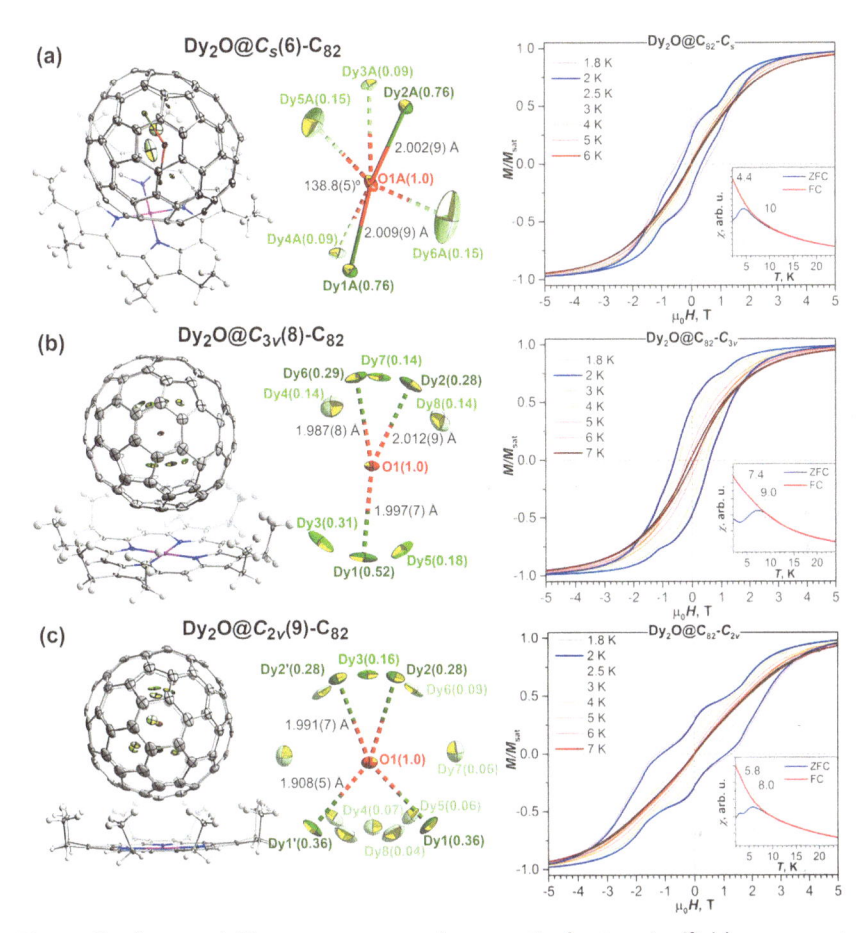

Fig. 5 Single-crystal X-ray structure and magnetic hysteresis (field sweep rate 2.9 mT s^{-1}) together with an inset showing the determination of the blocking temperature T_B (temperature sweep rate 5 K min^{-1}) for the isomers (A) Dy$_2$O@C$_s$(6)-C$_{82}$; (B) Dy$_2$O @C$_{3v}$(8)-C$_{82}$; (C) Dy$_2$O @C$_{2v}$(9)-C$_{82}$ co-crystallized with Ni(OEP). For each isomer, the crystal structure is shown on the left (structures are oriented so that molecular symmetry plane is parallel to the paper), and enlargement of the endohedral Dy$_2$O unit with disordered Dy sites and selected structural parameters are shown in the center. Solvent molecules are omitted for clarity. *Adapted from Yang W, Velkos G, Liu F, Sudarkova S, Wang Y, Zhuang J, Zhang H, Li X, Zhang X, Büchner B, Avdoshenko S, Popov A, Chen N. Single molecule magnetism with strong magnetic anisotropy and enhanced Dy•••Dy coupling in three isomers of dy-oxide clusterfullerene Dy2O@C82. Adv Sci. 2019;6:1901352.*

2.2 Dimetallofullerenes molecule magnets: Single-electron lanthanide-lanthanide bonds

Although metal-metal bonds are well-known in the chemistry of transition metals, lanthanide-lanthanide bonds have been only isolated within fullerene cages and can be stabilized as single-electron Ln–Ln bond in di-EMF, offering a unique opportunity for engineering strong exchange couplings in combination with single-ion magnetic anisotropy.[63] Thus, a characteristic feature in di-EMF is the transfer of metal valence electrons to the carbon cage acting as electron acceptors. The Ln–Ln bonding molecular orbital is one of the frontier orbitals in di-EMFs, and its population depends on its energy in relation to the cage molecular orbitals.[81] As shown in Fig. 5, for the early lanthanides such as the case of La, because the valence molecular orbitals of La—La dimer show a relatively high energy, a complete transfer of six valence electrons takes place, yielding La ions with a formal charge of +3. For the second half of the lanthanide row (from Gd onwards) the relatively low energy of the Ln–Ln bonding orbital restricts the electron transfer to only five electrons leaving each ion with a formal charge of +2.5, and therefore a single electron bond between them. An associated problem is the formation of radicals ($Ln_2@C_{80}$) that are unstable and very difficult to extract due to the rapidly polymerization in presence of some solvents.[82] Stabilization of this unique electronic configuration can be achieved by either the substitution of just one carbon atom in the cage by a nitrogen atom, giving azafullerenes $Ln_2@C_{79}N$,[83] or by the extraction in a polar solvent (such as dimethylformamide) and the quenching through the functionalization of the cage with a chemical such as benzyl bromide to form monoadducts $Ln_2@C_{80}(CH_2Ph)$,[84] transforming one sp^2 carbon of the cage into an sp^3 carbon. By using these two strategies and the aid of computational studies,[81,82] a number of $Ln_2@C_{79}N$ and $Ln_2@C_{80}(CH_2Ph)$ EMF featuring a single-electron Ln – Ln bond have been reported in the last few years. Azafullerenes $Ln_2@C_{79}N$ (Ln=Y, Tb, Gd) were the first di-EMF discovered in 2008, while the first stable $La_2@C_{80}$ monoadducts with single-electron La—La bonds were isolate years later, in 2015 and 2016.[85,86] Computational studies by Rajaraman et al. predicted for $Gd_2@C_{79}N$ and $Dy_2@C_{79}N$ giant exchange coupling and a large magnetization barrier, respectively.[87] The coupling constant between Gd and the unpaired electron on the Gd—Gd bond for $Gd_2@C_{79}N$ were experimentally confirmed 3 years later in 2018 by the groups of Popov and Gao.[88,89] A high blocking temperature of magnetization and giant coercivity were also reported for $Tb_2@C_{79}N$ in 2019.[90] Between 2017 and 2019, $Ln_2@C_{80}(CH_2Ph)$ monoadducts for Dy and Tb showing excellent SMM performance was demonstrated by Popov and co-workers.[84,91] The record set earlier by $Dy_2ScN@C_{80}$ was broken by the $Dy_2@C_{80}(CH_2Ph)$ monoadduct (a three spin system $\{Dy^{3+}-e-Dy^{3+}\}$ with the Dy ions coupling ferromagnetically) showing a truly remarkable blocking temperature of magnetization of 21.9 K and opening of the hysteresis loop between 1.8 and 21 K that marked a breakthrough for the field (Fig. 6).

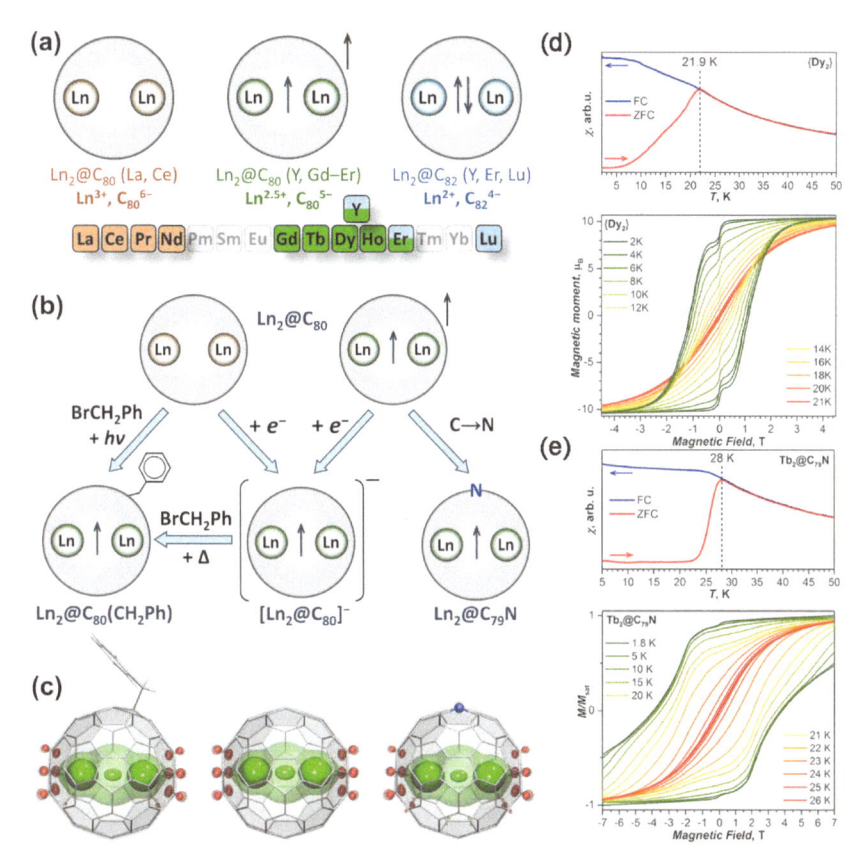

Fig. 6 (A) Oxidation states of lanthanides in dimetallofullerenes. Color code: brown for Ln^{3+}, green for $Ln^{2.5+}$, and blue for Ln^{2+}. Stabilization (B) and spin density distributions (C) of di-EMFs with a single-electron Ln—Ln bond in the form of $Ln_2@C_{80}(CH_2Ph)$, $[Ln_2@C_{80}]^-$, and $Ln_2@C_{79}N$. (D and E) Magnetization blocking temperatures and magnetic hysteresis of $\{Dy_2\}$ and $Tb_2@C_{79}N$. The magnetic field was 0.2–0.3 T, and the temperature sweep rate was 5 K min^{-1}. The magnetic field sweep rate was 3 mT s^{-1} for $\{Dy_2\}$ and 9.5 mT s^{-1} for $Tb_2@C_{79}N$. *Adapted from Liu F, Spree L, Krylov DS, Velkos G, Avdoshenko S, Popov A. Single-electron lanthanide-lanthanide bonds inside fullerenes toward robust redox-active molecular magnets. Acc Chem Res. 2019;52:2981–2993.*

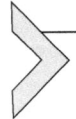

3. Carbon nanotubes: One-dimensional host nanocontainers

Since the discovery more than two decades ago that molecules can be transported and inserted into the confinements of the 1D channels of nanotubes, the ability of carbon nanotubes to act as nanocontainers for a wide variety of species continues invigorating the present research.[92] Carbon nanotubes (CNT) with one macroscopic and two nanoscopic dimensions provide excellent means to achieve the coupling of bistable species with the macroscopic world for harnessing their properties, without attendant problems of deterioration of their functionality. As CNT are excellent electrical conductors, the obtained hybrids materials can be in principle electrically controlled. Such control is essential for addressability in data storage, spintronics or quantum information processing applications.[93,94] Besides, CNT are chemically, mechanically and thermally very robust that is clearly very advantageous for device fabrication. CNT also acts as host containers protecting the guest species from the surroundings and contributing, for example in SMM, to preclude decoherence that is highly detrimental for quantum information processing applications.[95] When species are encapsulated in CNT, they become arranged in geometrically regular 1D arrays (the so-called peapod structures), some of which do not exist outside nanotubes.[96] Thus, as a result of confinement, structural and dynamic properties of the encapsulated species change drastically giving rise to synergistic effects.[97] There are also recent evidences showing that their chemical reactivity can be also greatly affected controlling selectivity of products in heterogeneous catalysis and energy-related application.[97–100] On the contrary, the effects of the encapsulated species on the physicochemical properties of nanotubes are less well understood, which in principle may be due to the intrinsic complexity of CNT properties or the polydispersity of nanotube samples. In this review, we will focus on examples of filled nanotubes involving single-walled (SWCNT) and multi-walled (MWCNT) carbon nanotubes showing internal diameters below 10 nm, in which nanoscale confinement takes place. The scope of this chapter has been also limited to examples only where the nanotubes were filled after their formation (i.e., cases where carbon nanotubes are co-formed or formed around the non-carbon material are not included), and where the presence of guest-species inside nanotubes is unambiguously confirmed by transmission electron microscopy (TEM), which still remains the only direct method of verification for nanotube filling. Before discussing recent examples of CNT containing bistable magnetic species, such as EMF and molecular clusters, some general methodology for filling CNT avoiding the deterioration of guest species will be revised next.

3.1 Strategies for filling CNT

Considering the nature of the guest bistable magnetic species encapsulated in CNT, one of the next two general methods for filling nanotubes have been used, which involve either sublimation techniques in gas phase or capillarity forces in solution. When the magnetic species to be encapsulated can be sublimed (as in the case of fullerenes), a mixture of guest specie and carbon nanotubes with open termini is typically mixed in a sealed tube and heated at temperatures near or above the sublimation point of the guest specie. This mechanism has been studied in detailed for fullerene molecules.[101,102] For confinement of C_{60}, initially adsorbed on the CNT exterior surfaces, a small activation barrier (about 0.3 eV) is required prior to insertion into CNT, that is greatly compensate upon encapsulation due to the significantly enhancement of van der Waals interactions in the CNT interiors, where the surface of the contact is maximized. Thus, the filling in this case strongly depends on the CNT diameter and the size of the guest species to achieve a snug fit, so the van der Waals gap of ~0.3 nm must be carefully considered when choosing the container. When the magnetic species to be encapsulated cannot be sublimed (as in the case of most molecular clusters exhibiting limited thermal stability), encapsulation of the guest species can be achieved by immersion of nanotubes with open termini in solutions containing the guest species after sonication. After immersion in a solvent with surface tension below 200 mN/m, spontaneously filling of CNT with the solvent molecules can occur according to the Young–Laplace law[103] that is referred as capillary filling. However, it is important to note that the efficiency of this method is lower compared to the filling in gas phase, as the guest species in this case are surrounded by a large number of solvent molecules, which interfere with the filling process. Although the solvent selected enables the transport of guest species from the bulk solution phase into carbon nanotubes, the filling process under these conditions is not yet fully understood. Aiming at improving filling using capillarity forces for molecular clusters, supercritical fluids such as *sc*-CO_2 have been used on its own or mixed with conventional solvents to tune surface tensions.[104]

3.2 EMF-SMM as guest molecules in carbon nanotubes: One-dimensional peapod structures

Endohedral metallofullerenes involving metallo and clusterfullerenes (such as $M@C_n$, $M_2@C_n$ and $M_3N@C_n$ with M = Gd, Sc, La, Ti, Tb, Er …) [105–107] have been inserted into SWNT using sublimation methods involving gas phase techniques (10^{-6} Torr) at high temperatures (450–500 °C) identical to those used for the encapsulation of C_{60}. The advancement on high-resolution TEM techniques has led to increasingly detailed studies of the structure of EMF peapods (one-dimensional arrays formed inside CNT). Thus, using aberration-corrected transmission electron microscopes orientational changes of EMF fullerenes inside SWNT can be demonstrated at the real atomic level [108], as precise determination of both the orientation and position of each fullerene cage with respect to the SWNT and also the intra-cage orientation of the endohedral cluster can be obtained.[109]

Among all the known EMF@SWNT structures, no examples of EMF-SMM inserted in SWNT were reported until very recently.[110,111] Within the SWCNT interiors, EMF-SMM can pack forming regular one-dimensional spin arrays,[112] in which neighboring intermolecular dipole–dipole interactions can be suppressed (that is a known strategy for enhancing SMM properties).[113] Peapods structures in SWCNT with metallic or semiconducting properties are of great promise for applications in spintronic devices and quantum computation. In 2018 Avdoshenko et al. reported the first EMF-SMM/SWNT hybrid involving the insertion of $Dy_2ScN@C_{80}$ SMM.[110] Magnetic data of the hybrid obtained at low temperature from X-ray magnetic circular dichroism at the Dy M_5-edge and the SQUID-magnetization shows the disappearance of magnetic hysteresis, and therefore the decrease on the magnetic bistability, in the peapod structure, as a consequence of the encapsulation of $Dy_2ScN@C_{80}$ in SWNT. Ab initio calculations further demonstrate that the encapsulation give rise to a preferential orientation of the endohedral units triggering an alignment of the molecular anisotropy axis and the magnetic ordering. Shortly after, $DySc_2N@C_{80}$ was encapsulated in SWCNT by Nakanishi et al.[111] (Fig. 7). In contrast to the previous report, the magnetic hysteresis in this case was retained upon encapsulation and an increase in the coercivity field values and longer relaxation times compared to those for the unencapsulated SMM were observed. The authors attributed this fact to a dilution effect due to the encapsulation that could partially suppress the QTM relaxation of $DySc_2N@C_{80}$.

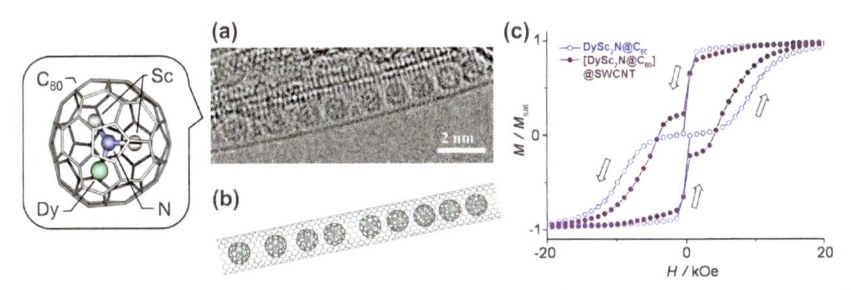

Fig. 7 (A) Image of DySc$_2$N@C$_{80}$ and a transmission electron microscopy image of [DySc$_2$N@C$_{80}$]@SWCNT. (B) Scheme of (a). (C) M vs H plots for DySc$_2$N@C$_{80}$ (filled circles) and [DySc$_2$N@C$_{80}$]@SWCNT (open circles) at 1.8 K. Arrows indicate the direction of the measurements. *Adapted from Nakanishi R, Satoh J, Katoh K, Zhang H, Breedlove BK, Nishijima M, Nakanishi Y, Omachi H, Shinohara H, Yamashita M. DySc2N@C80 single-molecule magnetic metallofullerene encapsulated in a single-walled carbon nanotube. J Am Chem Soc. 2018;140:10955–10959.*

3.3 Other molecule magnets confined in carbon nanotubes

The first SMM encapsulated in CNT was reported by Giménez-Lopez et al.[39], demonstrating a successful methodology for encapsulation of intact Mn$_{12}$Ac SMM in graphitized MWCNT (Fig. 8). Earlier attempts for obtaining SMM-CNT hybrids yielded composites with SMM adsorbed on the external surface of SWCNT, and thus exposed to the detrimental effects of the environment.[114–116] The non-covalent interactions, which are responsible for efficient transport and encapsulation of the guest molecules into nanotubes using supercritical CO_2, allow the Mn$_{12}$Ac molecules to stay mobile within the nanotube and align along the applied magnetic field. The confinement was confirmed by transmission electron microscopy showing the presence of discrete, freestanding entities identified as Mn$_{12}$Ac clusters inside the Mn$_{12}$Ac@CNT hybrid. Additionally, the SMM decomposition was ruled out by the absence of the characteristic crystal lattice structure of Mn$_3$O$_4$ (that is the expected decomposition product). Magnetic measurements on Mn$_{12}$Ac@CNT indicated strong frequency dependence of both χ' and χ'' signals resulting from the slow relaxation of magnetization. Although the encapsulated Mn$_{12}$Ac molecules retain their SMM properties, the host nanotubes provide an alternative pathway for relaxation of SMM magnetization. In turn, the magnetic anisotropy of the guest molecules translates into the anisotropic magnetoresistance behavior of the host nanotubes changing their electrical resistance response. This breakthrough in the field represents a new class of hybrid structures SMM@CNT combining the functional properties of nanotubes and magnetic molecules.

Fig. 8 (A) Transmission electron micrograph of graphitized multiwall CNT filled with Mn$_{12}$Ac (scale bar: 5 nm) and a scheme of the packing of Mn$_{12}$Ac molecules in a nanotube. (B) Schematic representation of the device used for magnetoresistance measurements. Inset: representation of the preferential alignment of Mn$_{12}$Ac molecules in the internal cavity of nanotube with respect to the magnetic field. In-phase (C) and out-of-phase (D) components of the AC susceptibility of Mn$_{12}$Ac @GMWNT measured with an AC field of 3.5 Oe at different frequencies, showing slow relaxation of the magnetization of the encapsulated Mn$_{12}$Ac molecules. (E) Hysteresis curves of magnetization recorded for Mn$_{12}$Ac @GMWNT at 1.8 K and for the control sample Mn$_{12}$Ac (**2**) (black). (F) Derivatives of the hysteresis curves shown in (E). Magnetoresistance measurements of nanotubes hosting Mn$_{12}$Ac SMM. *Adapted from Giménez-López MC, Moro F, La Torre A, Gómez-García C, Brown ., van Slageren J, Khlobystov A. Encapsulation of single-molecule magnets in carbon nanotubes. Nat Commun. 2011;2:407.*

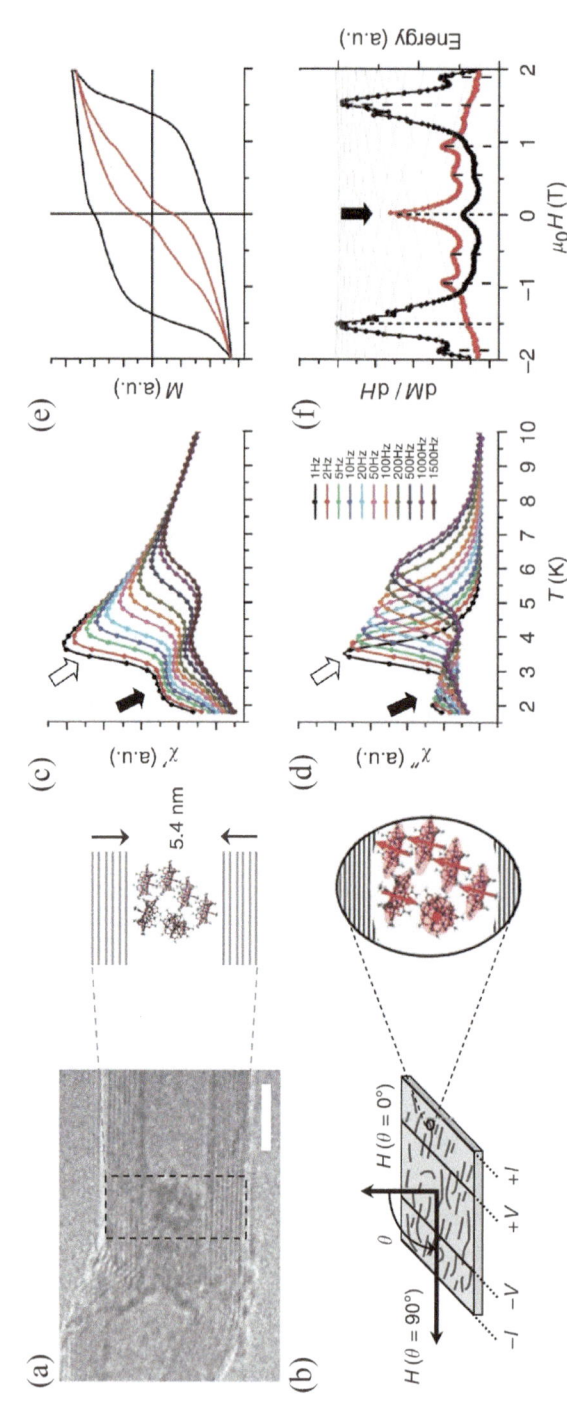

This work was preceded 6 years later by the encapsulation of a Dy acetylacetonato SMM through capillarity forces into MWCNT by Nakanishi et al.[117] As in the previous study, encapsulation was confirmed by transmission electron microscopy and energy-dispersive X-ray spectroscopy analysis. Alternating current magnetic susceptibility measurements, both the in-phase and out-of-phase signals, were clearly frequency dependent, indicating that $Dy(acac)_3(OH_2)_2$ complexes still exhibited SMM like properties. Nevertheless, the absence of hysteresis was attributed to a dilution effect upon confinement. Very recently a trinuclear nickel(II) acetylacetonate SMM containing frustrated spins has been encapsulated in SWNT by Domanov et al.[118] As a consequence of nanoscale confinement, frustration is reverse ressulting in an anisotropic reduction of the superexchange interactions. It is hypothesized that the ferromagnetic-to-antiferromagnetic transition observed after confinement can be attributed to a change in the angle between the three oxo ligands (M–O–M) that it is known to be a crucial parameter for the Goodenough–Kanamori–Anderson rules.

From these works it can be concluded that both relaxation times and exchange interaction in SMM can be tuned by changing coordination environments upon confinement in CNT.

4. Metal-organic frameworks: As three-dimensional platforms

Metal-organic frameworks (MOF) are porous crystalline materials composed by organic linkers connected to metallic centres. Due to their exceptional structural properties, such as their porosity and large internal surface areas, as well as their tunable properties, MOF have been technologically exploited as storage of fuel, capture of gases, for drug delivery, catalysis or sensing.

Magnetic, electrical or optical properties can be modulated by either changing organic linkers or functional centres and by encapsulating functional active guest species in the MOF pores. Two types of magnetic MOF can be found: those with cooperative properties (including long range magnetic order and spin crossover) and those exhibiting molecular properties (such as single molecule magnetism). This kind of three-dimensional structures offer the possibility of taking advantage of their porosity, confining functional molecules inside, giving rise to unique multifunctional materials with unique properties arisen as result of synergistic interactions or the combination of different properties. Moreover, MOF can also be considered

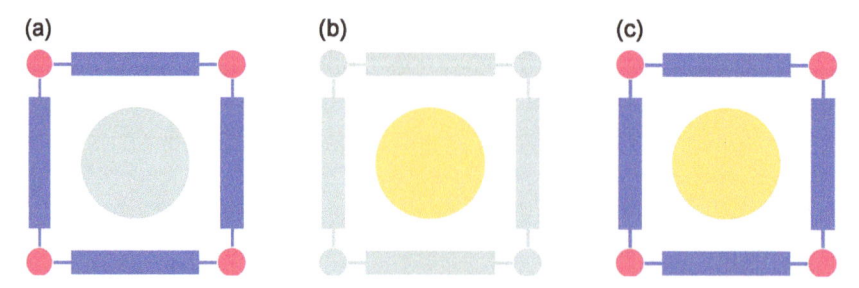

Fig. 9 Schematic representation of (A) a magnetic MOF with a non-functional guest, (B) a non-magnetic MOF with a functional guest, and (C) a magnetic-MOF with a functional guest. From Mínguez Espallargas G, Coronado E. Magnetic functionalities in MOFs: from the framework to the pore. Chem Soc Rev. 2018;47:533–557.

as a versatile platform to create well-organized structures of magnetic centres; such control is very advantageous in quantum computing applications.[119]

Depending on the structure and the filling of their pores, MOF can be divided in three groups: (i) magnetic MOF; (ii) magnetic MOF with non-magnetic guest; and (iii) nonmagnetic MOF with magnetic guest (Fig. 9). Here we will focus on recent advances of magnetic MOF and magnetic MOF including guests.

4.1 Spin crossover into three-dimensional structures

4.1.1 Host-guest chemistry in SCO MOF

Pardo et al.[120] used a three-dimensional magnetic manganese(II)–copper(II) MOF as a host for a SCO complex, [Fe^{III}(sal$_2$-trien)]NO$_3$H$_2$O to yield the Fe(sal$_2$-trien)@MOF by in a single-crystal-to-single-crystal process. The new produced material displayed an enhanced structural stability, strengthening the ferromagnetic behavior, and thus its magnetic ordering temperature. Fe(sal$_2$-trien) hosted in the MOF pores shows spin transition behavior, in contrast to the precursor salt. The authors demonstrated for the first time the post-synthetic encapsulation of a SCO complex within a magnetic three-dimensional MOF.

Due to the switching properties of these systems their sensing properties have been studied recently by Serre et al.[121] Al(III) tricarboxylate (MOF MIL-100(Al)) was employed, which displays a remarkable chemical stability and two sets (24 and 27 Å) accessible only through 5.5 Å pentagonal and 8.5 Å hexagonal microporous windows. The diffusion of metal and ligands is only possible through hexagonal windows, but once the complex is formed it cannot leach out of the pores. Iron complex Fe^{III}(sal$_2$trien)$^+$ was used due to its good stability and the thermal switching close to room temperature.

The resulting $Fe^{III}(sal_2trien)^+$@MOF retains the crystallinity and partial porosity, while the spin state of the inclusion complex can be modulated at room temperature by water uptake in the MOF pores. The effect of the adsorption of solvent molecules on the magnetic properties could be useful for sensing applications.

Effect of pressure on SCO properties was reported by Real et al.[122] using two-dimensional coordination polymers (CP) {Fe(MeOH)$_2$[Pd(SCN)$_4$]}· pz and porous three-dimensional-CP {Fe-(pz)[Pd(SCN)$_4$]} obtained by desolvation of the two-dimensional CP. Despite the fact that these compounds are high-spin due to the weak ligand field from NCS groups, SCO behavior was observed at pressure above 0.3 GPa. The same trend with the pressure is observed for Pt-based compounds and the substitution of pz by other ligands do not affect to the behavior. Only a few examples of complexes present SCO above 400 K and their stability is low, losing their magnetic properties after heating. Tong's group reported two MOF with high stability.[123] The variable-temperature powder X-ray diffraction analysis showed the framework is stable up to 623 K only limited by the experimental conditions, which makes these MOF highly interesting for sensing application at extreme conditions. Moreover, the MOF are stable after exposure to dimethylformamide, morpholine or pyridine.

Hofmann-like SCO porous coordination polymers represent a very especial type of compounds. They exhibit a cooperative thermal bistability at room temperature with large hysteresis loops and their spin state can be switched by irradiating the sample with a pulsed laser. Furthermore, a reversible switching between HS and LS can be obtained with the uptake of guest molecules.

Bousseksou et al.[124] investigated a series of Hofmann-like clathrate MOF structures with spin crossover at room temperature. The study of MOF {Fe (bpac)[M(CN)$_4$]}·H$_2$O·x(bpac) (M=Pt, Pd, and Ni, bpac = bis(4-pyridyl)acetylene) allowed the authors to demonstrate that spin transition varies with the stoichiometry. When low proportion of bpac molecules is included, different iron centres appear yielding incomplete and less-cooperative spin transitions. On the other hand, when a large number of bpac molecules are included, complete and cooperative spin transitions are observed due to the π-π stacking. Vapor-adsorption experiments showed that the MOF is capable of accept guests other than water or bpac molecules. The physical studies of the modified networks showed the synergy between host–guest chemistry and spin-crossover phenomena. Indeed, the clathration of various aromatic molecules revealed a strong correlation with the change in the transition temperature observed. This work shows how MOF with SCO may be used as sensors by absorption of guests into the framework. Ruiz et al.[125] confirmed this by studying the SCO properties of three clathrates where five guest molecules were used to understand the effect of the guests. Both experimental and theoretical results confirmed the importance of guest size. Whereas small molecules presented similar interaction energies for both spin states, the large guests were stabilized by the HS clathrate,

Further research has been focused in the confinement effect of guests in the pores of MOF. In 2014, Sato's group[126] showed the effect of up-taking and releasing water in the SCO behavior of a series of gold-cluster supramolecular systems. The complexes showed SCO behavior in the dehydrated phase. This work revealed an interesting mechanism different from distortion of the geometry and electronic environment of SCO centres, and takes advantage of the interaction between water and MOF. This water-based switch can be easily tuned by changing the coordination linkages.

Hysteretic tristability (three step switching occurring both upon cooling and upon warming with thermal hysteresis) is rare in the area of multistep SCO materials and until 2017 there was only one example reported for the case of three-dimensional Hofmann-type porous coordination polymers.[127] Tangoulis et al.[128] presented preliminary magnetic measurements of the hybrid microporous Hofmann-like spin crossover polymers $\{Fe(C_4H_4N_2)[M(CN)_4]\}\cdot(C_4H_4N_2)\cdot 1.2H_2O$ where M=Pt or Pd showing a hysteretic three-step and two-step thermal induced spin transition for Pt and Pd, respectively. These results confirm that a multistep thermal hysteresis is possible when the A and B sites of the guest-free $[Fe(C_4H_4N_2)\{Pt(CN)_4\}]$ network are occupied by guest molecules.

Also in 2017 Tong et al.[129] reported multistable systems with hysteretic four-step SCO in a three-dimensional Hofmann-type MOF $[Fe(4-abpt)\{Ag(CN)_2\}_2]\cdot 2DMF\cdot EtOH$ (4-abpt=4-amino-3,5-bis(4-pyridyl)-1,2,4 triazole). The spin transition character of these systems is very sensitive because of the presence of hydrogen bonding interactions. Interestingly, by changing the confined guest, such four-step SCO behavior can be triggered from two-step and one-step SCO systems through guest exchange. Hysteretic two- and one-step SCO behaviors were observed with 2DMF·MeCN and with pure solvents (EtOH or MeCN), respectively. By controlling the guest content, a reversible modulation of four-, two-, and one-stepped SCO behaviors was reported for the first time (Fig. 10).

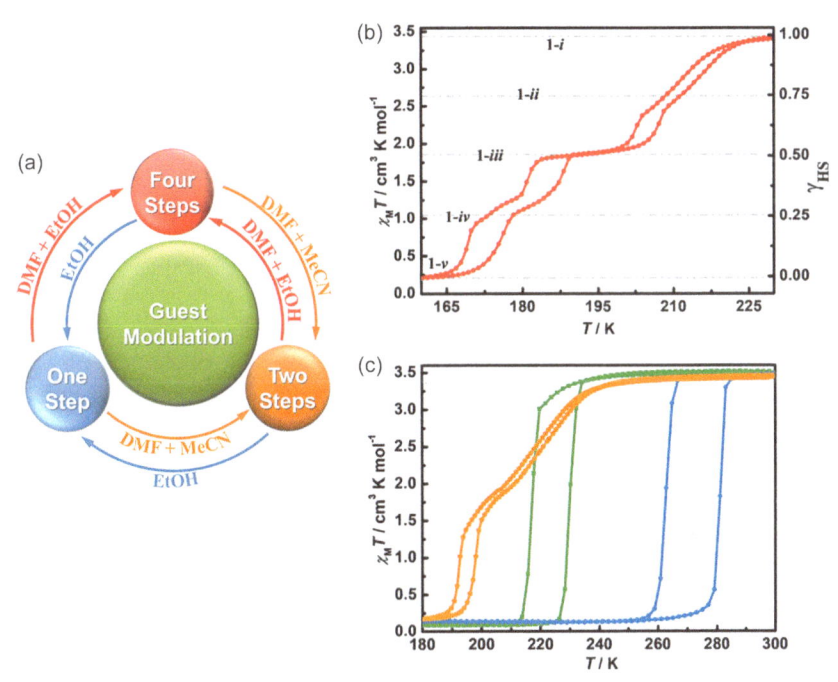

Fig. 10 (A) Graphical representation of reversible modulation of four-, two-, and one-stepped spin-crossover behaviors through guest exchange for MOF [Fe(4-abpt){Ag(CN)$_2$}$_2$]·2DMF·EtOH. (B) Temperature dependence of $\chi_M T$ at 2 K min^{-1} for the solvated compound containg 2 DMF·EtOH. Dashed lines are provided for different HS fractions. (C) Temperature dependence of $\chi_M T$ at 2 K min^{-1} for the next solvated compounds: 2 DMF·MeCN (orange), χEtOH (blue), and γ MeCN (green). *Adapted from Liu W, Peng Y-Y, Wu S-G, Chen Y-C, Hoque MN, Ni Z-P, Chen X-M, Tong M-L. Guest-switchable multi-step spin transitions in an amine-functionalized metal–organic framework. Angew Chem Int Ed. 2017;56:14982–14986.*

In 2017 Real et al. studied the synthesis and properties of a series of new Hofmann-type SCO MOF.[130] For the non-porous material {Fe(bpb)[Ag^I(CN)$_2$]$_2$} a very incomplete SCO was found, due to its intricate structure. On the other hand, the equivalent Ni and Pd systems that yielded two Hofmann-type MOF {Fe(bpb)[M^{II}(CN)$_4$]}·nGuest (M^{II} = Ni, Pd) presented high porosity with either a complete one-step or two-step cooperative SCO behaviors when nitrobenzene or naphthalene are included in the pores interacting by π-π stacking with pyridyl moieties of the framework. However, the application of hydrostatic pressure induces one- and two-step SCO.

New multifunctional materials can be designed by taking advantage of the SCO MOF versatile properties. Zuo's group [131] investigated the effect of the in situ generation of cation radicals of the TTF moiety in TTF(py)$_4$ present in Fe-based MOF on the magnetic properties via iodine doping after immersion in iodine solution. As a result, not only the conductive properties are improved by 2–3 orders of magnitude, but also the control of the magnetic properties via redox state switching. This is a rare example of photo- and electronically switchable spin crossover material.

Calculations can help not only to design better systems with great properties, but also to understand the guest effect in SCO MOF enabling the development of materials with interesting multifunctional properties. Paesani's group used hybrid Monte Carlo (MC) and Molecular Dynamics (MD) simulations to study the variation of the SCO behavior of {Fe(pz)[Pt(CN)$_4$]} upon adsorption of CO_2, SO_2, and CS_2.[132] The presence of CO_2 in the MOF pores was found not to greatly affect the temperature of the transition ($T_{1/2}$). In the case of SO_2 and CS_2, the confinement in the pores helps to stabilize the LS state. These results showed a direct relationship between host–guest interactions and SCO behavior, and demonstrated the ability of these systems to act as sensors. Later, the same group reported a study of the mechanism of modulation of SCO behavior in a Fe MOF upon adsorption of pyrazine.[133] The SCO properties of {Fe(pz)[Pt(CN)$_4$]} vary significantly upon adsorption of guest molecules stabilizing the HS state (10–400 K). The MC/MD simulations showed that the pyrazine molecules adsorbed into pores lie nearly parallel to the pyrazine ligands of the framework. The mobility of both guest pyrazine molecules and pyrazine ligands of the framework is highly suppressed generating high stability in the HS state.

4.1.2 Cooperativity enhancement

Cooperativity is a key parameter in order to achieve technologically useful SCO materials. To improve cooperativity in the solid lattice, the most common and effective strategy is the generation of intermolecular interactions between the individual SCO metal centres, especially through $\pi \cdots \pi$ stacking or hydrogen bonding. Furthermore, the SCO centres can be bridged by the covalent bonding to form polynuclear compounds or polymer species in order to enhance cooperativity. By combining the two strategies the cooperativity in MOFs can be largely enhanced. Upon designing and tuning the host–guest interactions in these systems, new functional MOFs can be created.

The effect of geometry in the cooperativity was studied by Li et al.[134] using a three nuclear Fe(II) MOF. The Fe(II)-tetrazole units that form the nodes are linked to give a rigid structure with different topologies. The authors hypothesized that the geometry in this MOF promoted a cooperativity reduction, as the geometrically favorable structure had a high activation barrier because of the structural arrangement required for the spin transition.

Tong et al.[135] also studied the effect of guests in the MOF properties, including the cooperativity. The authors used the hydrogen bonding strategy to generate strong host–guest interactions. By including sec-butanol in the pores, an asymmetric two-step SCO behavior was observed. To further evaluate the effect of the guest in SCO cooperativity the authors used a series of aromatic guest to exploit the $\pi\cdots\pi$ stacking strategy in the [Fe(dpb){Au(CN)$_2$}$_2$] MOF. Interesting results were obtained for naphthalene, where a hysteresis loop of 73 K was observed. The widths of the thermal hysteresis loops are dramatically changed depending on the nature of the guests. The same trend was observed by Real et al.[136] Although [Fe(tvp)$_2$(NCS)$_2$]·MeOH was the first reported porous SCO MOF of this series, the effect of confinement on the iron centre was not initially studied. Later on, Real et al. investigated the confinement of aromatic molecules and aprotic solvents in the [Fe(tvp)$_2$(NCS)$_2$] MOF and SCO behavior was confirmed by magnetic, calorimetric and X-ray measurements. It is demonstrated that SCO properties can be efficiently tuned by changing the adsorbed solvents. Finally, a 20-fold enhancement effect on the SCO cooperativity was demonstrated by Kepenekian et al.[137] after depositing a Fe(II)-based MOF on Au (111) surface, opening a new avenue for improving SCO MOF properties.

MOF can be then considered perfect three-dimensional platforms to fully exploit SCO properties, enabling the tuning and enhancement of their properties by either modifying the ligands or the pores via absorption of guest molecules.

4.2 Single molecule magnets in three-dimensional structures

SMM need to be protected from the environment to maintain their properties. Due to their well-ordered porous cavities, MOF provide a perfect scenario to achieve long-range ordered of SMM in a controlled environment, maintaining their properties or even improving them.

Some MOF can have flexible or dynamic structures, related to the flexibility of the metal-ligand coordination, the organic ligands, and/or noncovalent interactions. This is particularly useful when aiming to obtain functional materials with tuneable or switching properties, as in the case of magnetic materials that are very sensitive to structural or external stimulus changes. The synthesis of functional SMM MOF is challenging. Subtle modifications in the synthesis of SMM MOF can in principle influence enormously, and in unpredictably way, the dimensionality of the network and generate a quenching in the magnet-like properties. Here, the recent advances of SMM MOF and their challenging synthetic strategies and properties are presented.

4.2.1 MOF exhibiting slow magnetic relaxation
A great number of lanthanide metal–organic frameworks with interesting magnetic properties have been reported, including lanthanides and metal transition-based MOF.

4.2.1.1 Lanthanide MOF
In recent years the research in lanthanide MOF has been mainly focused on the use of Dy, and specifically in the reversible switching of the slow magnetic relaxation. Feng et al.[138] reported the tuneable slow magnetic relaxation behaviors in a Dy-based MOF driven by the absorption/desorption of terminal coordinated water molecules and other guest molecules. Qiu and co-workers reported the synthesis of $Dy(BTC)(H_2O)DMF$ in 2006 from which an anhydrous phase was obtained by calcination under vacuum at 240 °C for 12 h.[139] A single-crystal-to-single-crystal transformation concomitant with the removal of terminal coordinated water molecules and guest molecules was observed. The absorption–desorption of water molecules led to a distinct coordination geometry for the Dy(III) ions responsible for the adjustable slow magnetic relaxation behaviors.

Fine-tuning the Ln ion coordination geometry is not easy, due to their large radius and high coordination number. As mentioned above, the geometry around the lanthanide ions plays an important role and can be chemically constrained during synthesis,[140–143] but this possibility remains highly unexplored. It is also worth mentioning that Ln-based MOF have been also receiving enormous interest due to their excellent optical properties through an "antenna effect," such as large Stokes shifts and high color purity for the development of supported recyclable sensors.[144,145] Nonetheless, advances in this area are out of the scope of this chapter.

4.2.1.2 3d and 4d-based MOF

The synthesis of 3d- and 4d- SMM is gaining attention in the last 5 years due to the ability of the transition metals to create strongly coupled systems.[146,147] The first Co(II)-based magnet reported by Tomás et al.[148] exhibiting a blocking temperature of 2.7 K, in which the slow magnetic relaxation was fine-tune through structural changes, was produced after guest absorption and anion exchange. Murugesu et al.[149] reported also a Co(II)-based crystalline MOF displaying SMM-like behavior under applied static fields, where each metallic node acts as a nanomagnet, and the behavior controlled via manipulation of the coordination environment of the nodes. In 2016, Cornia et al. prepared a tetrairon(III) complex-based MOF by the reaction of tetrairon salt with silver(I) perchlorate.[150] The magnetic properties were preserved in the MOF, which displayed slow magnetic relaxation.

Zhou and co-workers reported on the hydrothermal synthesis of three different single crystalline solvates in a single reaction. The dehydrate [Cu(tzc)(dpp)]n·2H$_2$O transformed into an anhydrous phase [Cu(tzc)(dpp)]$_n$ through [Cu(tzc)(dpp)]$_n$·H$_2$O.[151] Upon heating, the anhydrous phase transformed into a new anhydrous polymorph, which upon cooling gave a new phase. The de/resolution produced a drastic change in their magnetic properties: the dihydrate shows antiferromagnetic exchange interactions, whereas trimorphic anhydrate showed ferromagnetic properties. The high structural diversity of eight different MOF materials with identical metal/ligand composition is unprecedented.

4.2.2 MOF encapsulating bistable magnetic guests

The magnetic properties in these systems can be added or modulated by the removal, reabsorption, or exchange of guest molecules, connected to the metal centre or absorbed on their pores. In this respect, works have been focused mainly on the confinement of Mn$_{12}$ molecules, endohedral metallofullerenes (EMF) or polyoxomethalates (POM).

4.2.2.1 Mn$_{12}$ SMM encapsulation in MOF

In recent years, the confinement of magnetic species in three-dimensional structures have been reported. Molecular magnets were firstly inserted in MOF cavities by Wriedt et al. in 2015.[44] One of the most studied SMM, Mn$_{12}$Ac, was used as the guest molecule and the mesoporous aluminium-based MOF [Al(OH)(SDC)]$_n$ (CYCU-3) as the host framework. The MOF CYCU-3 consisted of hexagonal one-dimensional channel pores of ~3 nm diameter (Fig. 11A) with high thermal and solvent stability,

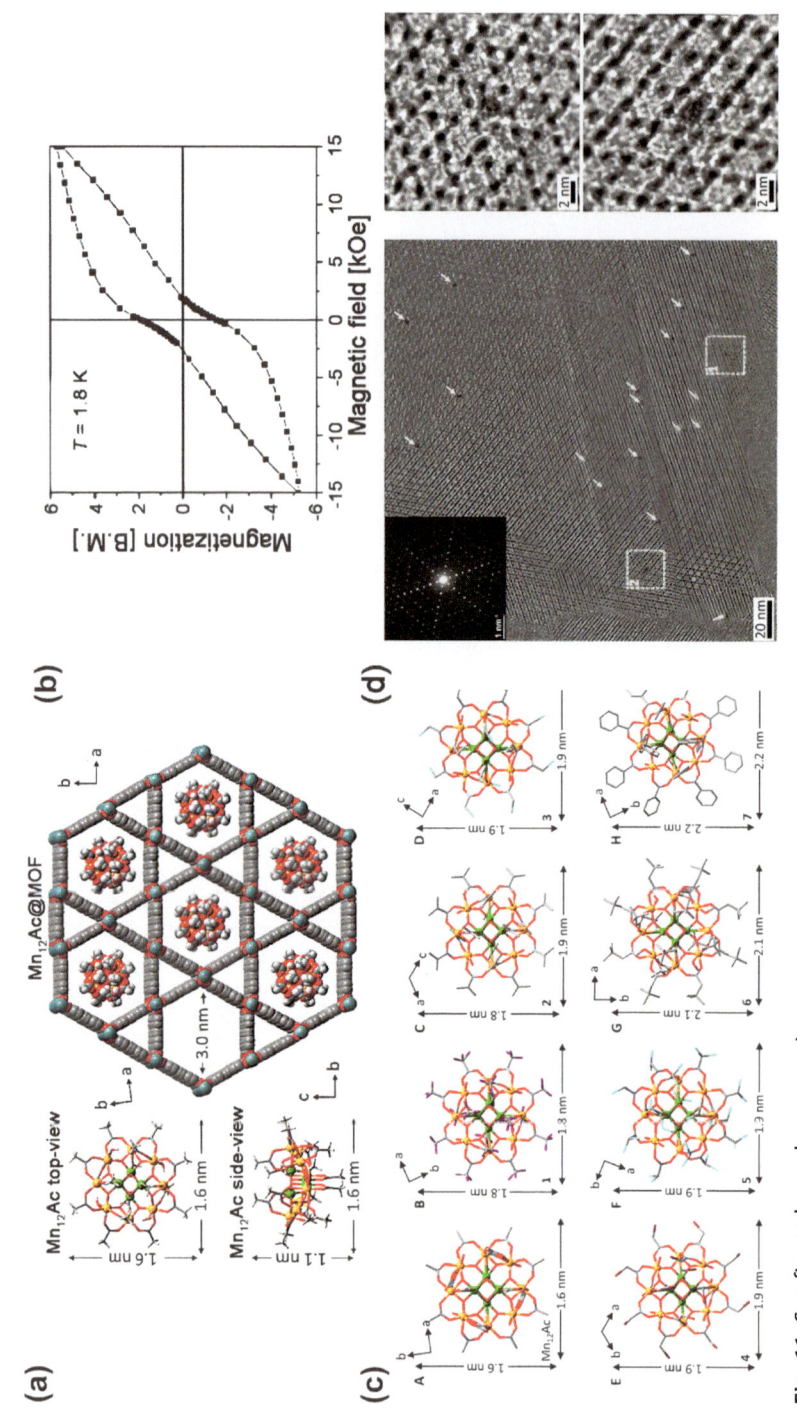

Fig. 11 See figure legend on opposite page.

while the diamagnetic aluminium centres did not influence the SMM properties of Mn$_{12}$Ac. The loading was achieved by soaking the MOF in a Mn$_{12}$Ac solution which led to the composite Mn$_{12}$Ac@MOF containing one molecule per pore. The out-of-phase alternating current susceptibility confirmed that the Mn$_{12}$Ac molecules have been preserved during their incorporation into the MOF cavities. Variable-temperature direct current magnetic measurements of the composite at 1000 Oe also exhibited the typical behavior of Mn$_{12}$Ac derivatives. The narrowing of the hysteresis loop at low fields is consistent with the presence of quantum tunneling of the magnetization and the presence of the faster relaxation isomer of Mn$_{12}$Ac (Fig. 11B). The authors demonstrated that it is possible to incorporate SMM to a MOF maintaining the magnetic properties. The SMM loading occurs at the periphery of the MOF crystals with a single SMM in the transverse direction of the pores to yield a long-range ordered crystalline that is useful for potential SMM-based ultrahigh-density data storage systems.

In 2016, Pardo et al.[47] reported the solid-state incorporation of a Mn(III) porphyrin of formula [Mn^{III}(TPP)(H$_2$O)$_2$]ClO$_4$ into a manganese(II)–copper(II) MOF leading to a MOF containing the MnTPP (MnTPP @MOF) that exhibited a long-range magnetic order and a slow magnetic relaxation.

Fig. 11 (A) Left side: Wireframe structure of [Mn$_{12}$O$_{12}$(O$_2$CCH$_3$)$_{16}$ (OH$_2$)$_4$] and its dimensions; Right side: Schematic representation of CYCU-3 with guest molecules. Structures are shown in a space filling style with realistic size relationship between the atoms. Color scheme: manganese(IV), green; manganese(III), orange; aluminium, turquoise; carbon, gray; oxygen, red; and hydrogen, white. (B) Field-dependent hysteresis of the magnetization for Mn$_{12}$Ac@MOF (C) Crystal structures of selected SMM in the Mn$_{12}$ family [Mn$_{12}$O$_{12}$(O$_2$CR)$_{16}$(OH$_2$)$_4$], with R=CH$_3$ (A), CF$_3$ (B), (CH$_3$)CCH$_2$ (C), CH$_2$Cl (D), CH$_2$Br (E), CHCl$_2$ (F), CH$_2$But (G), and C$_6$H$_5$ (H) Hydrogen atoms and solvent molecules are omitted for the sake of clarity. Color code: Mn^{4+} (green), Mn^{3+} (orange), Br (maroon), Cl (turquoise), F (purple), O (red), and C (gray). (D) Left side: HRTEM image and electron diffraction pattern (inset) of Mn$_{12}$Ac@NU-1000 acquired along the [001] zone axis of NU-1000. Arrows indicate the observed particles that correspond to Mn$_{12}$Ac; Right side: Enlarged images of the highlighted areas in areas 1 and 2, respectively, showing that the clusters of Mn$_{12}$Ac are precisely encapsulated and perfectly fitted in the hexagonal channels of NU-1000. *Adapted from Aulakh D, Pyser JB, Zhang X, Yakovenko A, Dunbar K, Wriedt M. Metal-organic frameworks as platforms for the controlled nanostructuring of single-molecule magnets. J Am Chem Soc, 2015;137:9254–9257; Aulakh D, Xie H, Shen Z, Harley A, Zhang X, Yakovenko AA, Dunbar KR, Wriedt M. Systematic investigation of controlled nanostructuring of mn12 single-molecule magnets templated by metal–organic frameworks.* Inorg Chem. *2017;56:6965–6972. and Aulakh D, Liu L, Varghese JR, Xie H, Islamoglu T, Duell K, Kung C-W, Hsiung C-E, Zhang Y, Drout RJ, Farha OK, Dunbar KR, Han Y, Wriedt M. Direct imaging of isolated single-molecule magnets in metal–organic frameworks. J Am Chem Soc. 2019;141:2997–3005.*

Despite of the disordered position of MnTPP into the MOF, the crystalline structure is preserved upon encapsulation. The authors reported for the first time the single-crystal-to-single-crystal encapsulation of a single-ion magnet within a magnetic MOF and observed the internal field-induced slow magnetic relaxation behavior, as a consequence of the attenuation of the quantum tunneling relaxation of the magnetization.

Further studies using Mn_{12} derivatives as guests were carried out by Wriedt et al. proving that MOF are versatile platforms for the incorporation of any SMM and its nanostructuration.[152] They reported the synthesis and structural characterization of seven different nanocomposites (SMM@MOF) with Mn_{12} derivatives with different chain sizes (Fig. 11C). Extensive characterization confirmed the preservation of the host structure upon SMM confinement, while magnetic measurements revealed slow relaxation of the magnetization.[44] The SMM behavior of the Mn_{12} core is fully retained during the inclusion in the MOF, while thermal stability is enhanced. Recently, for the first time, the same group reported the direct visualization of guest molecules encapsulated in the pores of a MOF using HRTEM.[153] A low-dose HRTEM technique was employed to probe the locations of $Mn_{12}Ac$ in MOF. HRTEM image of $Mn_{12}Ac$@MOF acquired along the [001] direction is shown in Fig. 11D. The enlarged images show that particles are placed in the primary hexagonal channels, showing the successful encapsulation of $Mn_{12}Ac$ (Fig. 11D).

Finally, Zheng et al. reported the first synthesis of a chiral mesoporous La-based MOF with $Mn_{12}Ac$ encapsulated, reaching a loading capacity of 40.15 mol%, a record-breaking loading capacity for $Mn_{12}Ac$ reported in porous materials.[154] By N_2 adsorption isotherms studies, the authors confirmed that $Mn_{12}Ac$@MOF shows a clear hysteresis loop due to the interaction between host and guest, which produced a change in size. The temperature-dependent magnetic susceptibility showed similar behavior to $Mn_{12}Ac$ and the presence of antiferromagnetic interactions between the metal ions. These results confirm that the encapsulation of SMM in MOF is a perfect approximation to reach hybrid nanostructured incorporating the unique magnetic properties of SMM.

4.2.2.2 EMF encapsulated in MOF

Besides the well-known Mn_{12}, EMF such as $DySc_2N@C_{80}$ represents another type of SMM with extraordinary properties to be encapsulated in MOF. In this type of SMM, the external C_{80} cage can be considered as the local crystal field; by changing this crystal field through encapsulation in MOF, the QTM can be controlled. The incorporation of $DySc_2N@C_{80}$ into the pores of MOF-177 and $DySc_2N@C_{80}$ into DUT-51(Zr) led to the suppression of QTM.[155]

Wang et al. investigated the optically controlled magnetism of EMF ($Sc_3C_2@C_{80}$ and $DySc_2N@C_{80}$) by incorporating them into the pores of photoresponsive azoMOFs (Fig. 12).[156] After irradiation with UV light, the photoisomerization of AzoMOF leads to stronger host-guest interaction between AzoMOF and $Sc_3C_2@C_{80}$ modulating the spin relaxation of $Sc_3C_2@C_{80}$. After photoisomerization, the QTM of $DySc_2N@C_{80}$ is suppressed due to the change in the crystal field of AzoMOF. The strong host-guest interactions between $DySc_2N@C_{80}$ and AzoMOF generated crystal-field splitting, influencing the quantum coherence of $DySc_2N@C_{80}$.

Other type of EMF, dimetallofullerenes based on yttrium, can also be confined in MOF. Wang et al.[45] reported the encapsulation of a spin-active metallofullerene ($Y_2@C_{79}N$) in MOF-177 crystals creating a paramagnetic system taking advantage of π-π interaction between nitrogen containing hexatomic ring in $Y_2@C_{79}N$ and MOF-177. The confinement plays a significant role in steering the electron spin on endohedral fullerene, leading to anisotropic paramagnetic solid system that may have applications in high-density data storage and nanoscale nuclear magnetic resonance.

4.2.2.3 Encapsulation of POM in MOF

Polyoxometalates (POM) are bulky anions showing electronic versatility with interesting physical and chemical properties, which make them suitable for applications in the fields of chemistry, materials, nanotechnology, or biology, among others. By their encapsulation in MOF, interesting hybrid systems with multifunctionality can be isolated. Coronado's group [157] reported the encapsulation of $[Mo_6O_{19}]^{2-}$ polyanions with Lindqvist structure in a Ln-based MOF. The anion-exchange capability of the MOF is proven through POM confinement that leaves SIM behavior of the Ln nodes unaltered upon the incorporation. The hybrids were obtained by immersing the MOF in a POM solution. The incorporation of the POM did not cause changes in the Ln coordination environment and the alternating current out-of-phase signal for POM@MOF is identical to the one observed for the MOF as expected. Recently, Li et al. reported the encapsulation of a Dawson-type POM in a framework of a Metal-Organic Nanotubes (MONT) showing moderate antiferromagnetic Cu—Cu interactions.[158] Applications of MONT are limited because of their low chemical stability. Using these POM as templates, the authors were capable of forming MONT. Both MONT (HUST-200 and HUST-201) exhibited exceptional chemical stabilities as a result of the encapsulation. Remarkably, HUST-200 is a stable and highly efficient electrocatalyst for the hydrogen production in acidic aqueous medium.

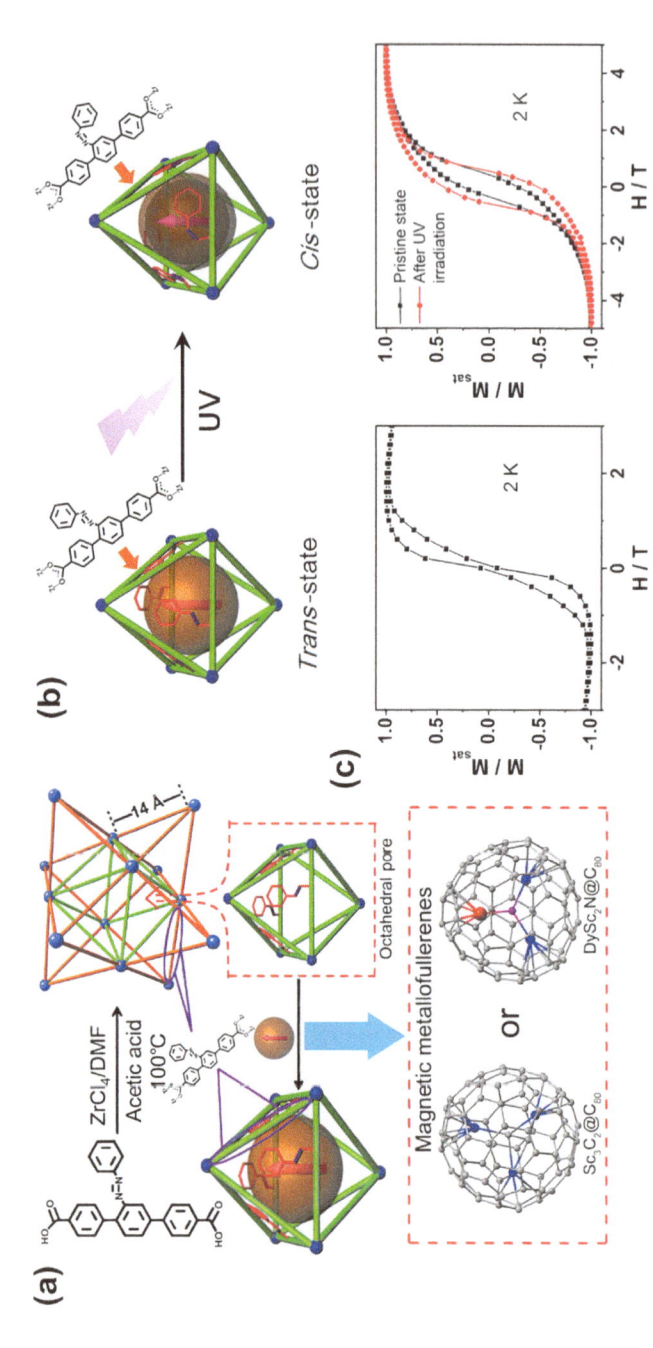

Fig. 12 (A) Schematic illustration of the synthesis of AzoMOF, with azobenzene units as building blocks, and construction of metallofullerene@AzoMOF. Metallofullerene (dark orange sphere) occupies the octahedral pore, which is highlighted in green color. The schematic structures of $Sc_3C_2@C_{80}$ and $DySc_2N@C_{80}$ are shown at the bottom (blue, Zr; gray, C; and orange, Dy). (B) Schematic representation of the effect of UV irradiation on the host−guest interaction between metallofullerene and AzoMOF. c) Hysteresis loops for $DySc_2N@C_{80}$ at 2 K and Hysteresis loops for $DySc_2N@C_{80}$@AzoMOF in the pristine state (black) and after irradiation with UV for 30 min (red) at 2 K. *Adapted from Meng H, Zhao C, Nie M, Wang C, Wang T. Optically controlled molecular metallofullerene magnetism via an azobenzene-functionalized metal–organic framework. ACS Appl Mater Interfaces 2018;10:32607–32612.*

5. Conclusions

Although much effort has been placed during years on learning about magnetic bistability of unconfined functional nanostructures, only very recently the development and understanding of these phenomena under confinement has begun to be mastered. Over the past decade, the limited number of synthetic methodologies, combined with the lack of the control of the nanostructuration processes have been holding back the development of molecule-based magnetic devices. This chapter highlights the strategies and methodologies developed to overcome such challenges and discusses the fundamental knowledge recently gained on the behavior of confined guest nanoscale switches, especially molecular magnets and SCO systems, within fullerene cages, carbon nanotubes and metal-organic frameworks.

Although studies on endohedral metallofullerenes with single-molecule magnet behavior have been only pursued for the last few years and further studies are required to fully understand the observed properties, some important conclusions can be drawn. It is apparent the strong correlation between the size of the exchange constant and the relaxation of magnetization of endohedral species with the isomerism of the fullerene cage. In addition, the size and the shape of the fullerene cage can also affect the structural parameters of the endohedral clusters, such bond lengths and angles. On the other hand, only very few studies on heterometallic lanthanides dimetallofullerenes has been reported to date, so further investigations may improve the control over spin – spin interactions, frontier orbital energies, and single-ion magnetic anisotropy. The influence of the cage functionalization on the spin properties of dimetallofullerenes has not been studied to date. Single-electron Ln–Ln bonds in dimetallofullerenes has been analyzed as three-spin [$Ln^{3+}-e-Ln^{3+}$] systems, however the validity of such approach remains questionable and further theoretical studies are needed. Spins confined within carbon cages are isolated and protected from the environment but still can be manipulated externally, which has huge advantages for quantum information processing. The first experiments in this direction have demonstrated the great potential of single-electron Ln – Ln bonds.

The relative chemical and thermal stability of fullerene cages in endohedral metallofullerenes has allowed the construction of supramolecular peapod nanostructures within single-walled carbon nanotubes. The magnetic hysteresis of endohedral metallofullerenes showing single-molecule behavior was shown to be retained after encapsulation, and an increase in the coercitivity field values and longer relaxation times compared to those for the unencapsulated molecules were also observed. Thus, encapsulation appears to partially suppress quantum tunneling of the magnetization in endohedral fullerenes. Only a few examples of molecular clusters showing magnetic bistability encapsulated in multi-walled carbon nanotubes has been reported to date, since the breakthrough of the encapsulation of the archetypical Mn_{12} single-molecule magnet. In this case, the encapsulated molecule magnets held by non-covalent interactions within the carbon nanotubes were capable to stay mobile within the nanotube and align along the applied magnetic field. Furthermore, from the reported examples become clear that both relaxation times and exchange interaction in molecule magnets can be tuned by changing coordination environments upon confinement in carbon nanotubes. It is worth noting that although single-molecule magnets in the form of either endohedral metallofullerenes or molecule clusters have been encapsulated in the one-dimensional internal channels of nanotubes, control of the inter-molecular distances between individual molecules has not yet been achieved.

In recent years, the long-range ordered multi-dimensional porous cavities in metal-organic frameworks have been employed for the confinement of a variety of molecular magnets including single-ion magnets and paramagnetic endohedral fullerenes. Although a huge number of examples have been reported, significant challenges lie ahead.

The encapsulation of SCO in MOF has proven to be a good strategy to fully exploit SCO properties, enabling the tuning and enhancement of their properties by either modifying the synthesis ligands or the pores via absorption of guest molecules. Cooperativity is a key parameter in order to achieve technology useful SCO based materials. The improvement of the cooperativity in the solid lattice is required and normally is reached by the generation of intermolecular interactions between the individual SCO metal centres, especially $\pi \cdots \pi$ stacking or hydrogen bonding. Through covalent bonds in the framework of MOF the metal centres can be bridged offering a versatile way to obtain polynuclear compounds, in which the cooperativity is greatly improved. Moreover, designing and tuning the host-guest interactions in SCO-MOF systems, new multifunctional functional materials can be created.

Regarding SMM, their organization to allow non-trivial interactions is a necessary step towards a complex molecular quantum device, for which MOF offer a perfect platform. The ability to tune the spatial arrangement of more than one non-identical single-ion magnet is precisely what is currently needed to advance in the global control paradigm of quantum computing. Furthermore, SMM molecules need to be protected from the environment to maintain their properties. Due to their chemical stability and their well-ordered porous cavities, MOF provide a perfect scenario to achieve long-range ordered SMM in a controlled environment maintaining their properties or even improving them while the SMM remain protected from the environment. Finally, by introducing subtle modifications in the synthesis of SMM-MOF it is possible to change drastically their properties, which will lead to a big step forward in the field of SMMs.

The synthesis, study and understanding of the different hybrid nanostructures presented here have been strongly linked in recent years to the improvement on the instrumentation and techniques, such as chromatography, transmission electron microscopy and X-ray diffraction, for the isolation for example of endohedral metallofullerenes magnets and the confirmation of confined guest molecules in nanotubes and metal-organic frameworks. It is not difficult to foresee that in the near future these advances will give a boost to the development of real applications for these nanostructures, ranging from molecular spintronics to quantum computing and sensors.

Acknowledgments

Authors would like to thank financial support from the Ministry of Science of Spain (RYC-2016-20258, RTI2018-101097-A-I00 and EIN2019-103246 for M.G-.L. and FJC2018-0370444-I for A.L-.M.), the European Research Council (ERC) (Starting Grant (NANOCOMP-679124) for M.G-.L.), the Xunta de Galicia (Centro Singular de Investigación de Galicia accreditation 2019-2022, ED431G 2019/03 and ED431B 2018/16) and the European Union (European Regional Development Fund - ERDF).

References

1. Coronado E. Molecular magnetism: from chemical design to spin control in molecules and devices. *Nat Rev Mater.* 2020;5:87–104.
2. (a) Gatteschi D, Sessoli R, Villain J. *Molecular Nanomagnets.* New York: Oxford University Press; 2006. (b) Benelli C, Gatteschi D, eds. *Introduction to Molecular Magnetism: From Transition Metals to Lanthanides.* Weinheim: Wiley-VCH Verlag GmbH & Co. KGaA; 2015.
3. Gatteschi D, Sessoli R. Quantum tunneling of magnetization and related phenomena in molecular materials. *Angew Chem Int Ed.* 2003;42:268–297.
4. Friedman JR, Sarachik MP, Tejada J, Ziolo R. Macroscopic measurements of resonant magnetization tunneling in high spin molecules. *Phys Rev Lett.* 1996;76:3830–3833.
5. Thomas L, Lionti F, Ballou R, Gatteschi D, Sessoli R, Barbara B. Macroscopic quantum tunnelling of magnetization in a single crystal of nanomagnets. *Nature.* 1996;383: 145–147.

6. Ardavan A, Rival O, Morton JL, et al. Will spin-relaxation times in molecular magnets permit quantum information processing. *Phys Rev Lett*. 2007;98, 057201.
7. Schlegel C, van Slageren J, Manoli M, Brechin EK, Dressel M. Direct observation of quantum coherence in single-molecule magnets. *Phys Rev Lett*. 2008;101:147203.
8. Sessoli R, Gatteschi D, Caneschi A, Novak MA. Magnetic bistability in a metal-ion cluster. *Nature*. 1993;365:141.
9. Escalera-Moreno L, Baldovi J, Gaita-Ariño A, Coronado E. Exploring the high-temperature frontier in molecular nanomagnets: from lanthanides to actinides. *Inorg Chem*. 2019;58:11883–11892.
10. Woodruff DN, Winpenny R, Layfield R. Lanthanide single-molecule magnets. *Chem Rev*. 2013;113:5110–5148.
11. Liddle ST, van Slageren J. Improving f-element single molecule magnets. *Chem Soc Rev*. 2015;44:6655–6669.
12. Spree L, Popov A. Recent advances in single molecule magnetism of dysprosium-metallofullerenes. *Dalton Trans*. 2019;48:2861–2871.
13. Goodwin C, Ortu F, Reta D, Chilton N, Mills D. Molecular magnetic hysteresis at 60 kelvin in dysprosocenium. *Nature*. 2017;548:439.
14. Guo F, Day B, Chen Y-C, Tong M-L, Mansikkamäki A, Layfield R. A dysprosium metallocene single-molecule magnet functioning at the axial limit. *Angew Chem Int Ed*. 2017;56:11445–11449.
15. Guo F, Day B, Chen Y-C, Tong M-L, Mansikkamäki A, Layfield R. Magentic hysteresis up to 80 kelvin in a dysprosium metallocene single-molecule magnet. *Science*. 2018;362:1400–1403.
16. McClain K, Gould C, Chakarawet K, et al. High-temperature magnetic blocking and magneto-structural correlations in a series of dysprosium(iii) metallocenium single-molecule magnets. *Chem Sci*. 2018;9:8492–8503.
17. Halcrow MA. *Spin-Crossover Materials: Properties and Applications*. Wiley. A John Wiley & Sons, Ltd; 2013.
18. Kahn O. *Molecular Magnetism*. Weinheim, New York: VCH-Verlag; 1993. ISBN:3-527-89566-3.
19. Coronado E, Giménez-López MC, Levchenko G, et al. Pressure-tuning of magnetism and linkage isomerism in iron (II) hexacyanochromate. *J Am Chem Soc*. 2005;127:4580–4581.
20. Coronado E, Giménez-Lopez MC, Korzeniak T, et al. Pressure-induced magnetic switching and linkage isomerism in K 0.4Fe4[Cr(CN)6]2.8·16H 2O: X-ray absorption and magnetic circular dichroism studies. *J Am Chem Soc*. 2008;130:15519–15532.
21. Giménez-Lopez MC, Clemente-León M, Giménez-Saiz C. Unravelling the spin-state of solvated [Fe(bpp)2]2+ spin-crossover complexes: structure–function relationship. *Dalton Trans*. 2018;47:10453–10462.
22. Gütlich P, Hauser A. Thermal and light-induced spin crossover in iron(II) complexes. *Coord Chem Rev*. 1990;97:1–22.
23. Craig GA, Roubeau O, Aromí G. Spin state switching in 2,6-bis(pyrazol-3-yl)pyridine (3-bpp) based Fe(II) complexes. *Coord Chem Rev*. 2014;269:13–31.
24. Halcrow MA. Structure: function relationships in molecular spin-crossover complexes. *Chem Soc Rev*. 2011;40:4119–4142.
25. Harzmann GD, Frisenda RJ, van der Zant HS, Mayor M. Single-molecule spin switch based on voltage-triggered distortion of the coordination sphere. *Angew Chem Int Ed*. 2015;54:13425–13430.
26. Martinho PN, Rajnak C, Ruben M. Nanoparticles, thin films and surface patterns from spin-crossover materials and electrical spin state control. In: Halcrow MA, ed. *Spin-Crossover Materials: Properties and Applications*. Oxford, UK: John Wiley & Sons Ltd; 2013:375–404.

27. Frisenda R, Harzmann GD, Celis Gil JA, Thijssen JM, Mayor M, van der Zant HSJ. Stretching-induced conductance increase in a spin-crossover molecule. *Nano Lett.* 2016;16:4733–4737.
28. Bernien M, Naggert H, Arruda LM, et al. Highly efficient thermal and light-induced spin-state switching of an Fe(II) complex in direct contact with a solid surface. *ACS Nano*. 2015;9:8960–8966.
29. Dugay J, Aarts M, Giménez-Marqués M, et al. Phase transitions in spin-crossover thin films probed by graphene transport measurements. *Nano Lett.* 2017;17:186–193.
30. Senthil Kumar K, Ruben M. Emerging trends in spin crossover (SCO) based functional materials and devices. *Coord Chem Rev.* 2017;346:176–205.
31. Clemente-León M, Soyer H, Coronado E, Mingotaud C, Gómez-García C, Delhaès P. Langmuir-blodgett films of single-molecule nanomagnets. *Angew Chem Int Ed.* 1998;37:2842–2845.
32. Ghirri A, Corradini V, Cervetti C, et al. Deposition of functionalized Cr_7Ni Molecular rings on graphite from the liquid phase. *Adv Funct Mater.* 2010;20:1552–1560.
33. Saywell A, Magnano G, Satterley C, et al. Self-assembled aggregates formed by single-molecule magnets on a gold surface. *Nat Commun.* 2010;1:75.
34. Gimenez-Lopez MC, Gardener J, Shaw A, et al. Endohedral metallofullerenes in self-assembled monolayers. *Phys Chem Chem Phys.* 2010;12:123–131.
35. Saywell A, Britton A, Taleb A, et al. Single molecule magnets on a gold surface: in situ electrospray deposition, x-ray absorption and photoemission. *Nanotechnology.* 2011;22:075704.
36. Cornia A, Mannini M, Sainctavit P, Sessoli P. Chemical strategies and characterization tools for the organization of single molecule magnets on surfaces. *Chem Soc Rev.* 2011;40:3076–3091.
37. Domingo N, Bellido E, Ruiz-Molina D. Advances on structuring, integration and magnetic characterization of molecular nanomagnets on surfaces and devices. *Chem Soc Rev.* 2012;41:258–302.
38. Coradin T, Larionova J, Smith A, et al. Magnetic nanocomposites built by controlled incorporation of magnetic clusters into mesoporous silicates. *Adv Mater.* 2002;14:896–898.
39. Giménez-López MC, Moro F, La Torre A, et al. Encapsulation of single-molecule magnets in carbon nanotubes. *Nat Commun.* 2011;2:407.
40. Giménez-López MC, Chuvilin A, Kaiser U, Khlobystov A. Functionalised endohedral fullerenes in single-walled carbon nanotubes. *Chem Commun.* 2011;47:2116–2118.
41. Li Y, Gao F, Beves J, Li Y-Z, Zuo J-L. A giant metallo-supramolecular cage encapsulating a single-molecule magnet. *Chem Commun.* 2013;49:3658.
42. Clemente-León M, Coronado E, Gómez-García C, et al. Insertion of a single-molecule magnet inside a ferromagnetic lattice based on a 3D bimetallic oxalate network: towards molecular analogues of permanent magnets. *Chem Eur J.* 2014;20:1669–1676.
43. Clemente-León M, Coronado E, Giménez-López MC, et al. Insertion of a spin crossover Fe[III] complex into an oxalate-based layered material: coexistence of spin canting and spin crossover in a hybrid magnet. *Inorg Chem.* 2008;47:9111–9120.
44. Aulakh D, Pyser JB, Zhang X, Yakovenko A, Dunbar K, Wriedt M. Metal-organic frameworks as platforms for the controlled nanostructuring of single-molecule magnets. *J Am Chem Soc.* 2015;137:9254–9257.
45. Feng Y, Wang T, Li Y, et al. Steering metallofullerene electron spin in porous metal–organic framework. *J Am Chem Soc.* 2015;137:15055–15060.
46. Pardo E, Burguete P, Ruiz-Garcia R, et al. Ordered mesoporous silicas as host for the incorporation and aggregation of octanuclear nickel(ii) single-molecule magnets: a bottom-up approach to new magnetic nanocomposite materials. *J Mater Chem.* 2006;16:2702–2714.

47. Mon M, Pascual-Álvarez A, Grancha T, et al. Solid-state molecular nanomagnet inclusion into a magnetic metal–organic framework: interplay of the magnetic properties. *Chem Eur J*. 2016;22:539–545.
48. Popov A, Yang S, Dunsch L. Endohedral fullerenes. *Chem Rev*. 2013;113:5989–6113.
49. Yang S, Wei T, Jin F. When metal cluster meet carbon cages: endohedral clusterfullerenes. *Chem Soc Rev*. 2017;46:5005–5058.
50. Lu X, Feng L, Akasaka T, Nagase S. Current status and future developments of endohedral matellofullerenes. *Chem Soc Rev*. 2012;41:7723–7760.
51. Kurotobi K, Murata YA. Single molecule of water encapsulated in fullerene C_{60}. *Science*. 2011;333:613–616.
52. Olmstead MM, de Bettencourt-Dias A, Stevenson S, Dorn HC, Balch AL. Crystallographic characterization of the structure of the endohedral fullerene {$Er_2@C_{82}$ Isomer I} with C_s cage symmetry and multiple sites for erbium along a band of ten contiguous hexagons. *J Am Chem Soc*. 2002;124:4172–4173.
53. Xu W, Feng L, Calvaresi M, et al. An experimentally observed trimetallo-fullerene $Sm_3@I_h$-C_{80}: encapsulation of three metal atoms in a cage without a nonmetallic mediator. *J Am Chem Soc*. 2013;135:4187–4190.
54. Krause M, Ziegs F, Popov AA, Dunsch L. Entrapped bonded hydrogen in a fullerene: the five-atom cluster Sc_3CH in C_{80}. *ChemPhysChem*. 2007;8:537–540.
55. Wang CR, Kai T, Tomiyama T, et al. A scandium carbide endohedral metallofullerene: $(Sc_2C_2)@C_{84}$. *Angew Chem Int Ed*. 2001;40:397–399.
56. Wang TS, Feng L, Wu JY, et al. Planar quinary cluster inside a fullerene cage: synthesis and structural characterizations of $Sc_3NC@C_{80-I_h}$. *J Am Chem Soc*. 2010; 132:16362–16364.
57. Stevenson S, Rice G, Glass T, et al. Small-bandgap endohedral metallofullerenes in high yield and purity. *Nature*. 1999;401:55–57.
58. Dunsch L, Yang S, Zhang L, Svitova A, Oswald S, Popov AA. Metal sulfide in a C_{82} fullerene cage: a new form of endohedral clusterfullerenes. *J Am Chem Soc*. 2010;132:5413–5421.
59. Stevenson S, Mackey MA, Stuart MA, et al. A distorted tetrahedral metal oxide cluster inside an icosahedral carbon cage. synthesis, isolation, and structural characterization of $Sc_4(\mu_3$-$O)_2@I_h$-C_{80}. *J Am Chem Soc*. 2008;130:11844–11845.
60. Gan LH, Wu R, Tian JL, Fowler PW. An atlas of endohedral Sc_2S cluster fullerenes. *Phys Chem Chem Phys*. 2017;19:419–425.
61. Westerström R, Dreiser J, Piamonteze C, et al. An endohedral single-molecule magnet with long relaxation times: DySc2N@C80. *J Am Chem Soc*. 2012;134:9840–9843.
62. Feng L, Hao Y, Liu A, Slanina Z. Trapping metallic oxide clusters inside fullerene cages. *Acc Chem Res*. 2019;52:1802–1811.
63. Liu F, Spree L, Krylov DS, Velkos G, Avdoshenko S, Popov A. Single-electron lanthanide-lanthanide bonds inside fullerenes toward robust redox-active molecular magnets. *Acc Chem Res*. 2019;52:2981–2993.
64. Zhang JF, Fatouros PP, Shu CY, et al. High relaxivity trimetallic nitride (Gd_3N) Metallofullerene MRI contrast agents with optimized functionality. *Bioconjug Chem*. 2010;21:610–615.
65. Lu X, Bao L, Akasaka T, Nagase S. Recent progress in the chemistry of endohedral metallofullerenes. *Chem Commun*. 2014;50:14701–14715.
66. Hou J, Yang Z, Li Z, Chai H, Zhao R. Electric-field-induced spin switch of endohedral dodecahedrane heterodimers $H@C_{20}H_n$–$C_{20}H_n@M$ (M= Cu, Ag and Au, n = 15, 18, and 19): a theoretical study. *J Mol Model*. 2017;23:242.
67. Chattopadhyaya M, Alam MM, Sen S, Chakrabarti S. Electrostatic spin crossover and concomitant electrically operated spin switch action in a ti-based endohedral metallofullerene polymer. *Phys Rev Lett*. 2012;109:257204–257205.

68. Konarev D, Khasanov S, Shestakov A, et al. Spin crossover in anionic cobalt-bridged fullerene $(Bu_4N^+)\{Co(Ph_3P)\}_2(\mu_2\text{-Cl}-)(\mu_2\text{-}\eta^2,\eta^2\text{-}C_{60})_2$ dimers. *J Am Chem Soc.* 2016;138:16592–16595.
69. Singh M, Shukla P, Khatua M, Rajaraman G. A design criteria to achieve giant ising-type anisotropy in Co^{II}-encapsulated metallofullerenes. *Chem Eur J.* 2020;26: 464–477.
70. Chen CH, Krylov D, Avdoshenko SM, et al. Selective arc-discharge synthesis of Dy_2S-clusterfullerenes and their isomer-dependent single molecule magnetism. *Chem Sci.* 2017;8:6451–6465.
71. Krylov D, Liu F, Avdoshenko SM, et al. Record-high thermal barrier of the relaxation of magnetization in the nitride clusterfullerene $Dy_2ScN@C_{80}$-Ih. *Chem Commun.* 2017;53:7901–7904.
72. Singh MK, Rajaraman G. Acquiring a record barrier height for magnetization reversal in lanthanide encapsulated fullerene molecules using DFT and *ab initio* calculations. *Chem Commun.* 2016;52:14047–14050.
73. Westerström R, Dreiser J, Piamonteze C, et al. Tunneling, remanence, and frustration in dysprosium-based endohedral single-molecule magnets. *Phys Rev B.* 2014;89: 060406.
74. Schlesier C, Spree L, Kostanyan A, et al. Strong carbon cage influence on the single molecule magnetism in Dy–Sc nitride clusterfullerenes. *Chem Commun.* 2018;54: 9730–9733.
75. Junghans K, Schlesier C, Kostanyan A, et al. Methane as a selectivity booster in the arc-discharge synthesis of endohedral fullerenes: selective synthesis of the single-molecule magnet $Dy_2TiC@C_{80}$ and its congener $Dy_2TiC_2@C_{80}$. *Angew Chem Int Ed.* 2015;54:13411–13415.
76. Brandenburg A, Krylov D, Beger A, Wolter A, Büchner B, Popov A. Carbide clusterfullerene $DyYTiC@C_{80}$ featuring three different metals in the endohedral cluster and its single-ion magnetism. *Chem Commun.* 2018;54:10683–10686.
77. Liu F, Gao C-L, Deng Q, et al. Triangular monometallic cyanide cluster entrapped in carbon cage with geometry-dependent molecular magnetism. *J Am Chem Soc.* 2016;138:14764–14771.
78. Liu F, Wang S, Gao C-L, et al. Mononuclear clusterfullerene single-molecule magnet containing strained fused-pentagons stabilized by a nearly linear metal cyanide cluster. *Angew Chem Int Ed.* 2017;56:1830–1834.
79. Mercado B, Olmstead M, Beavers C, et al. A seven atom cluster in a carbon cage, the crystallographically determined structure of $Sc_4(mu_3\text{-O})_3@Ih\text{-}C_{80}$. *Chem Commun.* 2010;46:279–281.
80. Kratschmer W, Lamb LD, Fostiropoulos K, Huffman DR. Solid C60: a new form of carbon. *Nature.* 1990;347:354–358.
81. Popov A, Avdoshenko S, Pendás A, Dunsch L. Bonding between strongly repulsive metal atoms: an oxymoron made real in a confined space of endohedral metallofullerenes. *Chem Commun.* 2012;48:8031–8050.
82. Wang Z, Kitaura R, Shinohara H. Metal-dependent stability of pristine and functionalized unconventional dimetallofullerene $M_2@Ih\text{-}C_{80}$. *J Phys Chem C.* 2014;118:13953–13958.
83. Fu W, Zhang J, Fuhrer T, et al. $Gd_2@C_{79}N$: isolation, characterization, and monoadduct formation of a very stable heterofullerene with a magnetic Spin State Of $S = 15/2$. *J Am Chem Soc.* 2011;133:9741–9750.
84. Liu F, Krylov DS, Spree L, et al. Single molecule magnet with an unpaired electron trapped between two lanthanide ions inside a fullerene. *Nat Commun.* 2017;8:16098.
85. Bao L, Chen M, Pan C, et al. Crystallographic evidence for direct metal−metal bonding in a stable open-shell $La_2@I_h\text{-}C_{80}$ derivative. *Angew Chem Int Ed.* 2016;55:4242–4246.

86. Yamada M, Kurihara H, Suzuki M, et al. Hiding and recovering electrons in a dimetallic endohedral fullerene: air-stable products from radical additions. *J Am Chem Soc.* 2015;137:232–238.
87. Rajaraman G, Singh MK, Yadav N. Record high magnetic exchange and magnetization blockade in $Ln_2@C_{79}N$ (Ln = Gd(III) and Dy(III)) molecules: a theoretical perspective. *Chem Commun.* 2015;51:17732–17735.
88. Velkos G, Krylov DS, Kirkpatrick K, et al. Giant exchange coupling and field-induced slow relaxation of magnetization in $Gd_2@C_{79}N$ with a single-electron Gd-Gd bond. *Chem Commun.* 2018;54:2902–2905.
89. Hu Z, Dong B-W, Liu Z, et al. Endohedral metallofullerene as molecular high spin qubit: diverse rabi cycles in $Gd_2@C_{79}N$. *J Am Chem Soc.* 2018;140:1123–1130.
90. Velkos G, Krylov D, Kirkpatrick K, et al. High blocking temperature of magnetization and giant coercivity in the azafullerene $Tb_2@C_{79}N$ with a single-electron Tb−Tb bond. *Angew Chem Int Ed.* 2019;58:5891–5896.
91. Liu F, Velkos G, Krylov DS, et al. Air-stable redox-active nanomagnets with lanthanide spins radical-bridged by a metal−metal bond. *Nat Commun.* 2019;10:571.
92. Smith BW, Monthioux M, Luzzi DE. Encapsulated C_{60} in carbon nanotubes. *Nature.* 1998;396:323–324.
93. Bogani L, Wernsdorfer W. Molecular spintronics using single-molecule magnets. *Nat Mater.* 2008;7:179–186.
94. Trif M, Troiani F, Stepanenko D, Loss D. Spin-electric coupling in molecular magnets. *Phys Rev Lett.* 2008;101:217201.
95. Benjamin S, Ardavan A, Briggs A, et al. Towards a fullerene-based quantum computer. *J Phys Condens Matter.* 2006;18:S867–S883.
96. Chuvilin A, Bichoutskaia E, Giménez-López MC, Kaiser U, Khlobystov AN. Self-assembly of a sulphur-terminated graphene nanoribbon within a single-walled carbon nanotube. *Nat Mater.* 2011;10:687–692.
97. Giménez-López MDC, La Torre A, Fay MW, Brown PD, Khlobystov AN. Assembly and magnetic bistability of Mn3O4 nanoparticles encapsulated in hollow carbon nanofibers. *Angew Chem Int Ed.* 2013;52:2051–2054.
98. Aygün M, Stoppiello C, Lebedeva MA, et al. Comparison of alkene hydrogenation in carbon nanoreactors of different diameters: probing the effects of nanoscale confinement on ruthenium nanoparticle catalysis. *J Mater Chem A.* 2017;5:21467–21477.
99. Aygün M, Chamberlain TW, Giménez-López M, Khlobystov AN. Magnetically recyclable catalytic carbon nanoreactors. *Adv Funct Mater.* 2018;28:1802869.
100. Giménez-López M, Kurtoglu A, Walsh DA, Khlobystov AN. Extremely stable platinum-amorphous carbon electrocatalyst within hollow graphitized carbon nanofibers for the oxygen reduction reaction. *Adv Mater.* 2016;28:9103–9108.
101. Ulbricht H, Moos G, Hertel T. Interaction of C60 with carbon nanotubes and graphite. *Phys Rev Lett.* 2003;90:095501.
102. Girifalco LA, Hodak M. Van der Waals binding energies in graphitic structures. *Phys Rev B.* 2002;65:125404.
103. Mattia D, Gogotsi Y. Static and dynamic behaviour of liquids inside carbon nanotubes. *Microfluid Nanofluid.* 2008;5:289–305.
104. Giménez-López M, Moro F, La Torre A, Van Slageren J, Khlobystov AN. Encapsulation of single-molecule magnets in carbon nanotubes. *Nat Commun.* 2011;2:407.
105. Warner JH, Watt AAR, Ge L, et al. Dynamics of paramagnetic metallofullerenes in carbon nanotube peapods. *Nano Lett.* 2008;8:1005–1010.
106. Guan L, Suenaga K, Okubo S, Okazaki T, Iijima S. Metallic wires of lanthanum atoms inside carbon nanotubes. *J Am Chem Soc.* 2008;130:2162–2163.

107. Kitaura R, Imazu N, Kobayashi K, Shinohara H. Fabrication of metal nanowires in carbon nanotubes via versatile nano-template reaction. *Nano Lett.* 2008;8:693–699.
108. Sato Y, Suenaga K, Okubo S, Okazaki T, Iijima S. Structures of D_{5d}-C_{80} and Ih-$Er_3N@C_{80}$ fullerenes and their rotation inside carbon nanotubes demonstrated by aberration-corrected electron microscopy. *Nano Lett.* 2007;7:3704–3708.
109. Giménez-López M, Chuvilin A, Kaiser U, Khlobystov AN. Functionalised endohedral fullerenes in single-walled carbon nanotubes. *Chem Commun.* 2011;47:2116–2118.
110. Avdoshenko S, Fritz F, Schlesier C, et al. Partial magnetic ordering in one-dimensional arrays of endofullerene single-molecule magnet peapods. *Nanoscale.* 2018;10:18153–18160.
111. Nakanishi R, Satoh J, Katoh K, et al. $DySc_2N@C_{80}$ single-molecule magnetic metallofullerene encapsulated in a single-walled carbon nanotube. *J Am Chem Soc.* 2018;140:10955–10959.
112. Katoh K, Yamashita S, Yasuda N, et al. Control of the spin dynamics of single-molecule magnets by using a quasi one-dimensional arrangement. *Angew Chem Int Ed.* 2018;57:9262–9267.
113. Habib F, Lin P-H, Long J, Korobkov I, Wernsdorfer W, Murugesu M. The use of magnetic dilution to elucidate the slow magnetic relaxation effects of a Dy_2 single-molecule magne. *J Am Chem Soc.* 2011;133:8830–8833.
114. Bogani L, Danieli C, Biavardi E, et al. Single-molecule-magnet carbon-nanotube hybrids. *Angew Chem Int Ed.* 2009;48:746–750.
115. Kyatskaya S, Galán Mascarós JR, Bogani L, et al. Anchoring of rare-earth-based single-molecule magnets on single-walled carbon nanotubes. *J Am Chem Soc.* 2009;131:15143–15151.
116. Giusti A, Charron G, Mazerat S, et al. Magnetic bistability of individual single-molecule magnets grafted on single-wall carbon nanotubes. *Angew Chem Int Ed.* 2009;48:4949–4952.
117. Nakanishi R, Yatoo M, Katoh K, Breedlove B, Yamashita M. Dysprosium acetylacetonato single-molecule magnet encapsulated in carbon nanotubes. *Materials.* 2017;10:7.
118. Domanov O, Weschke E, Saito T, et al. Exchange coupling in a frustrated trimetric molecular magnet reversed by a 1D nano-confinement. *Nanoscale.* 2019;11:10615.
119. Mínguez Espallargas G, Coronado E. Magnetic functionalities in MOFs: from the framework to the pore. *Chem Soc Rev.* 2018;47:533–557.
120. Abhervé A, Grancha T, Ferrando-Soria J, et al. Spin-crossover complex encapsulation within a magnetic metal–organic framework. *Chem Commun.* 2016;52:7360–7363.
121. Tissot A, Kesse X, Giannopoulou S, et al. A spin crossover porous hybrid architecture for potential sensing applications. *Chem Commun.* 2019;55:194–197.
122. Muñoz-Lara FJ, Arcís-Castillo Z, Muñoz MC, Rodríguez-Velamazán JA, Gaspar AB, Real JA. Heterobimetallic MOFs containing tetrathiocyanometallate building blocks: pressure-induced spin crossover in the porous {FeII(pz)[PdII(SCN)4]} 3D coordination polymer. *Inorg Chem.* 2012;51:11126–11132.
123. Bao X, Guo P-H, Liu W, et al. Remarkably high-temperature spin transition exhibited by new 2D metal–organic frameworks. *Chem Sci.* 2012;3:1629–1633.
124. Bartual-Murgui C, Salmon L, Akou A, et al. Synergetic effect of host–guest chemistry and spin crossover in 3d hofmann-like metal–organic frameworks [Fe(bpac)M(CN)4] (M=Pt, Pd, Ni). *Chem Eur J.* 2012;18:507–516.
125. Aravena D, Castillo ZA, Muñoz MC, et al. Guest Modulation of spin-crossover transition temperature in a porous iron(II) metal–organic framework: experimental and periodic DFT studies. *Chem Eur J.* 2014;20:12864–12873.

126. Xu H, Xu Z-L, Sato O. Water-switching of spin crossover in a gold cluster supramolecular system: from metal–organic frameworks to catenane. *Microporous Mesoporous Mater*. 2014;197:72–76.
127. Sciortino NF, Scherl-Gruenwald KR, Chastanet G, et al. Hysteretic three-step spin crossover in a thermo- and photochromic 3D pillared Hofmann-type metal–organic framework. *Angew Chem Int Ed*. 2012;51:10154–10158.
128. Polyzou CD, Lalioti N, Psycharis V, Tangoulis V. Guest induced hysteretic tristability in 3D pillared Hofmann-type microporous metal–organic frameworks. *New J Chem*. 2017;41:12384–12387.
129. Liu W, Peng Y-Y, Wu S-G, et al. Guest-switchable multi-step spin transitions in an amine-functionalized metal–organic framework. *Angew Chem Int Ed*. 2017;56: 14982–14986.
130. Piñeiro-López L, Valverde-Muñoz FJ, Seredyuk M, Muñoz MC, Haukka M, Real JA. Guest induced strong cooperative one- and two-step spin transitions in highly porous iron(II) Hofmann-type metal–organic frameworks. *Inorg Chem*. 2017;56:7038–7047.
131. Wang H-Y, Ge J-Y, Hua C, et al. Photo- and electronically switchable spin-crossover iron(II) metal–organic frameworks based on a tetrathiafulvalene ligand. *Angew Chem Int Ed*. 2017;56:5465–5470.
132. Pham CH, Paesani F. Guest-dependent stabilization of the low-spin state in spin-crossover metal-organic frameworks. *Inorg Chem*. 2018;57:9839–9843.
133. Pham CH, Paesani F. Spin crossover in the {Fe(pz)[Pt(CN)4]} metal–organic framework upon pyrazine adsorption. *J Phys Chem Lett*. 2016;7:4022–4026.
134. Yan Z, Li M, Gao H-L, Huang X-C, Li D. High-spin versus spin-crossover versus low-spin: geometry intervention in cooperativity in a 3D polymorphic iron(ii)–tetrazole MOFs system. *Chem Commun*. 2012;48:3960–3962.
135. Li J-Y, He C-T, Chen Y-C, et al. Tunable cooperativity in a spin-crossover Hoffman-like metal–organic framework material by aromatic guests. *J Mater Chem C*. 2015;3:7830–7835.
136. Romero-Morcillo T, De la Pinta N, Callejo LM, et al. Nanoporosity, inclusion chemistry, and spin crossover in orthogonally interlocked two-dimensional metal–organic frameworks. *Chem Eur J*. 2015;21:12112–12120.
137. Groizard T, Papior N, Le Guennic B, Robert V, Kepenekian M. Enhanced cooperativity in supported spin-crossover metal–organic frameworks. *J Phys Chem Lett*. 2017;8:3415–3420.
138. Zhou Q, Yang F, Xin B, et al. Reversible switching of slow magnetic relaxation in a classic lanthanide metal–organic framework system. *Chem Commun*. 2013;49: 8244–8246.
139. Guo X, Zhu G, Li Z, Sun F, Yang Z, Qiu S. A lanthanide metal–organic framework with high thermal stability and available Lewis-acid metal sites. *Chem Commun*. 2006;3172–3174.
140. Liu K, Li H, Zhang X, Shi W, Cheng P. Constraining and tuning the coordination geometry of a lanthanide ion in metal–organic frameworks: approach toward a single-molecule magnet. *Inorg Chem*. 2015;54:10224–10231.
141. Huang G, Fernandez-Garcia G, Badiane I, et al. Magnetic slow relaxation in a metal–organic framework made of chains of ferromagnetically coupled single-molecule magnets. *Chem Eur J*. 2018;24:6983–6991.
142. She S, Gu X, Yang Y. Field-induced single molecule magnet behaviour of a three-dimensional Dy(III)-based complex. *Inorg Chem Commun*. 2019;110:107584.
143. Wang M, Meng X, Song F, et al. Reversible structural transformation induced switchable single-molecule magnet behaviour in lanthanide metal–organic frameworks. *Chem Commun*. 2018;54:10183–10186.

144. Xu H, Fang M, Cao C-S, Qiao W-Z, Zhao B. Unique (3,4,10)-connected lanthanide–organic framework as a recyclable chemical sensor for detecting Al3+. *Inorg Chem.* 2016;55:4790–4794.
145. Li J-M, Huo R, Li X, Sun H-L. Lanthanide–organic frameworks constructed from 2,7-naphthalenedisulfonate and 1H-imidazo[4,5-f][1,10]-phenanthroline: synthesis, structure, and luminescence with near-visible light excitation and magnetic properties. *Inorg Chem.* 2019;58:9855–9865.
146. Son K, Kim RK, Kim S, Schütz G, Choi KM, Oh H. Metal organic frameworks as tunable linear magnets. *Phys Status Solidi A.* 2020;217:1901000.
147. Lin L-D, Ge R, Zhang J, et al. Construction of two high-nuclear 3d-4d heterometallic cluster organic frameworks by introducing a bifunctional tripodal alcohol as a structure-directing agent. *Chem Asian J.* 2019;14:1985–1991.
148. Campo J, Falvello LR, Forcén-Vázquez E, Sáenz de Pipaón C, Palacio F, Tomás M. A symmetric, triply interlaced 3-D anionic MOF that exhibits both magnetic order and SMM behaviour. *Dalton Trans.* 2016;45:16764–16768.
149. Brunet G, Safin DA, Jover J, Ruiz E, Murugesu M. Single-molecule magnetism arising from cobalt(ii) nodes of a crystalline sponge. *J Mater Chem C.* 2017;5:835–841.
150. Rigamonti L, Cotton C, Nava A, et al. Diamondoid Structure in a metal–organic framework of Fe4 single-molecule magnets. *Chem Eur J.* 2016;22:13705–13714.
151. Wriedt M, Yakovenko AA, Halder GJ, Prosvirin AV, Dunbar KR, Zhou H-C. Reversible switching from antiferro- to ferromagnetic behaviour by solvent-mediated, thermally-induced phase transitions in a trimorphic MOF-based magnetic sponge system. *J Am Chem Soc.* 2013;135:4040–4050.
152. Aulakh D, Xie H, Shen Z, et al. Systematic investigation of controlled nanostructuring of Mn12 single-molecule magnets templated by metal–organic frameworks. *Inorg Chem.* 2017;56:6965–6972.
153. Aulakh D, Liu L, Varghese JR, et al. Direct imaging of isolated single-molecule magnets in metal–organic frameworks. *J Am Chem Soc.* 2019;141:2997–3005.
154. Chen H-J, Zheng X-Y, Zhao Y-R, et al. A record-breaking loading capacity for single-molecule magnet Mn12 clusters achieved in a mesoporous Ln-MOF. *ACS Appl Electron Mater.* 2019;1:804–809.
155. Li Y, Wang T, Meng H, et al. Controlling the magnetic properties of dysprosium metallofullerene within metal–organic frameworks. *Dalton Trans.* 2016;45:19226–19229.
156. Meng H, Zhao C, Nie M, Wang C, Wang T. Optically controlled molecular metallofullerene magnetism via an azobenzene-functionalized metal–organic framework. *ACS Appl Mater Interfaces.* 2018;10:32607–32612.
157. Baldoví JJ, Coronado E, Gaita-Ariño A, Gamer C, Giménez-Marqués M, Mínguez EG. A SIM-MOF: three-dimensional organisation of single-ion magnets with anion-exchange capabilities. *Chem Eur J.* 2014;20:10695–10702.
158. Zhang L, Li S, Gómez-García CJ, et al. Two novel polyoxometalate-encapsulated metal–organic nanotube frameworks as stable and highly efficient electrocatalysts for hydrogen evolution reaction. *ACS Appl Mater Interfaces.* 2018;10:31498–31504.

CHAPTER FIVE

Recent advances in the synthesis and application of tris(pyridyl) ligands containing metallic and semimetallic p-block bridgeheads

Andrew J. Peel[a,*], Jessica E. Waters[a], Alex J. Plajer[b], Raúl García-Rodríguez[c], and Dominic S. Wright[a,*]
[a]Department of Chemistry, Lensfield Road, Cambridge, United Kingdom
[b]Inorganic Chemistry Research Laboratory, South Parks Road, Oxford, United Kingdom
[c]GIR MIOMeT-IU, Cinquima, Química Inorgánica, Facultad de Ciencias, Universidad de Valladolid, Valladolid, Spain
*Corresponding authors: e-mail address: ajp208@cam.ac.uk; dsw1000@cam.ac.uk

Contents

1. Introduction	193
2. Group 13 tris(pyridyl) ligands	194
2.1 Boron	194
2.2 Aluminium	195
2.3 Gallium and indium	213
3. Group 14 tris(pyridyl) ligands	217
3.1 Silicon and germanium ligands	218
3.2 Tin and lead	223
4. Group 15 tris(pyridyl) ligands	234
4.1 Phosphorus	234
4.2 Arsenic, antimony and bismuth	234
5. Outlook	240
Acknowledgments	241
References	241

1. Introduction

Tripodal ligands are a cornerstone for building new inorganic and organometallic structures,[1] their applications range from single site homogeneous catalysts,[2] to scaffolds in bioinorganic chemistry,[3–6] and the construction of supramolecular architectures.[7,8] In particular, C_3-symmetric

Fig. 1 (A) Tris(2-pyridyl)- and (B) tris(pyrazolyl) tripodal ligand containing non-metal bridgehead atoms or groups.

ligands consisting of heterocycles connected by a bridgehead group have come to the fore because of their strong affinity for metal ions.[9] While Trofimenko's eponymous anionic tris(pyrazolyl)borate (so-called 'scorpionate') ligands are famous examples, a diverse range of tris(2-pyridyl) ligands Y(2-py)$_3$ (Y=bridgehead group such as (BR)$^-$,[10] CR,[5,11–18] COR'/NH$_2$,[19–23] N,[24–31] P or PO[32–40] and As[41,42]; 2-py=2-pyridyl) have also been investigated extensively over several decades (Fig. 1).

Much of the versatility among tripodal ligands stems from the ability to tune their properties by modifying the substitution pattern of the heterocyclic ring. This has the potential to change not only the ligand bite angle but also the overall donor strength and steric domain of the ligand. A less well investigated means of varying the electronic and steric characteristics of such ligands is by changing the bridgehead atom to a semi-metallic or metallic element. This has previously been explored in a small number of tris(pyrazolyl)borate analogues that incorporate heavier p-block element bridgehead atoms,[43–47] and we highlighted the chemistry of tris(2-pyridyl) ligands of this type in a short review in 2012.[48] Apart from increasing the ligand bite angle, incorporating larger bridgehead atoms is expected to introduce further possibilities, such as variability in the oxidation state of the bridgehead atom and potentially redox interplay between the bridgehead and the N-coordinated metal center. In this article, the evolution of tris(pyridyl) ligands containing the metallic and semi-metallic p-block elements in the last decade is reviewed and new-found applications of these systems in diverse settings are highlighted.

2. Group 13 tris(pyridyl) ligands
2.1 Boron

Though not the first Group 13 pyridyl 'ate systems to be reported, the neutral species 4-^{t}BuC$_6$H$_4$B(2-py)$_2$(1-H-2-py) (**1H**), a tris(2-pyridyl)borate precursor, was first synthesized in 2012 by Jäkle and co-workers who reacted an arylboron bromide with pre-isolated 2-pyridyl Grignard reagent [Mg$_2$(2-py)$_2$Cl$_2$(THF)$_3$] (Scheme 1). The suitability of tris(2-pyridyl)borates

Scheme 1 Synthesis of **1**.

for benchtop use is a great practical convenience. Access to **1H** enabled coordination studies with transition metal ions, with $^tBuC_6H_4B(2\text{-py})_3^-$ (**1**) acting as a facially coordinating tridentate ligand in sandwich complexes of the type [M(**1**)$_2$] (M = Mg^{2+}, Mn^{2+}, Co^{2+}, Cu^{2+}, Ru^{2+}; Fig. 2).[49] For the case of [Fe(**1**)$_2$], bond length analysis and spectroscopic measurements indicated Fe(II) to be low spin, suggesting that borate anion **1** possesses strong σ-donor characteristics. Subsequent work has explored the nitroxide-mediated polymerization of (styryl)B(2-py)$_3$(1-H-2-py) (**2H**),[50] and the formation of a ROMP-derived block co-polymer incorporating the anion **2** that can react to form a complex with Cu(II) (the metal cation proving exchangeable with Fe^{2+}).[51] Other functionalised borate precursors have also been explored.[10]

2.2 Aluminium

Valence isoelectronic with tris(pyrazolyl)borates, anionic tris(pyridyl) aluminium(III) species have been the subject of intensive study since their inception in 2004.[52] The lower electronegativity of aluminium compared

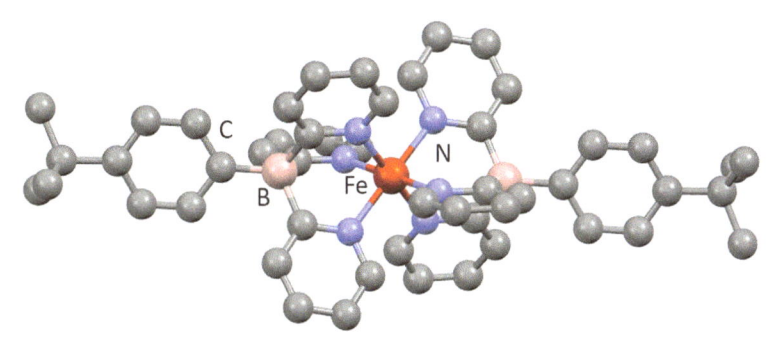

Fig. 2 Molecular structure of [Fe(**1**)$_2$].

to boron imparts greater ionic character to Al—C bonds in organometallic tripodal ligands, which in turn has a marked impact on the reactivity of these aluminates. The first complex in this ligand class to be reported was [Li(THF)MeAl(2-py)₃] [Li(THF)(**3**)], synthesized by direct combination of *in situ* generated 2-Li-py and commercially available MeAl₂Cl in THF (Scheme 2). This complex displays a tridentate coordination mode common to lighter, neutral tris(2-pyridyl) species (Fig. 3).[53] Removal of the THF from [Li(THF)(**3**)] was accomplished by heating the solid complex under vacuum, followed by recrystallisation from toluene (Scheme 3). X-ray diffraction revealed the product to be a dimer, [Li(**3**)]₂, which incorporates a central Li₂N₂ core with each Li centre 4-coordinate (Fig. 4). NMR spectroscopy studies in d_8-toluene indicated an equilibrium involving the dissociation of THF from [Li(THF)(**3**)] at low concentration, while the possibility of dimerization at high concentration was suggested to be unlikely based on comparison of the spectrum of a concentrated sample with that of isolated [**3**]₂. Lability of the Al—C bonds in (**3**)⁻, which gives rise to its ability to act as a soft carbanion source, was established by reaction of CuCl with [Li(THF)(**3**)]. In contrast to the combination of 2-Li-py with CuCl, which results in decomposition, using [Li(THF)(**3**)] in place of the lithio-pyridyl reagent gave an isolable, trimeric organocopper species [Cu(2-py)]₃ (**4**).

$$\text{MeAlCl}_2 + 3(2\text{-Li-py}) \xrightarrow[-2\text{ LiCl}]{\text{THF}} [\text{Li(THF)MeAl(2-py)}_3]$$
$$[\text{Li(THF)}(\mathbf{3})]$$

Scheme 2 Synthesis of [Li(THF)(**3**)].

In contrast to previously observed ligand transfer behavior (*e.g.*, giving rise to trimeric copper complex **4**), combination of [Li(THF)(**3**)] with either FeCl₂ or Cp₂Mn resulted in intact transfer of **3**⁻ to give sandwich complexes [M(**3**)₂] (M = Fe²⁺, Mn²⁺) (Scheme 4 and Fig. 5).[54] ¹H NMR spectroscopy and magnetic measurements indicated that [Fe(**3**)₂] is paramagnetic and analysis of the bond lengths found in the crystal structures of [Fe(**3**)₂] and [Mn(**3**)₂] suggested that these are high-spin complexes (*cf.* [Fe(**1**)₂], which is low-spin). The catalytic activity of [Fe(**3**)₂] was then investigated in the epoxidation of styrene. [Fe(**3**)₂] proved to be a very selective epoxidation catalyst at modest temperatures, producing 96.5% epoxide, no benzaldehyde nor polymerization products and minimal diol formation after 1 h at 65 °C (45.7% conversion). Longer reaction times lead to increased conversion,

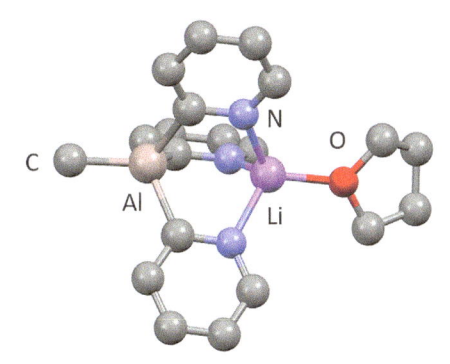

Fig. 3 Molecular structure of [Li(THF)(**3**)].

$$2\ [\text{Li(THF)MeAl(2-py)}_3] \xrightarrow[-2\ \text{THF}]{60°\text{C}/0.1\ \text{mm Hg}} [\text{LiMeAl(2-py)}_3]_2$$

[Li(THF)(**3**)]   [Li(**3**)]$_2$

Scheme 3 De-solvation and dimerisation of [Li(THF)(**3**)].

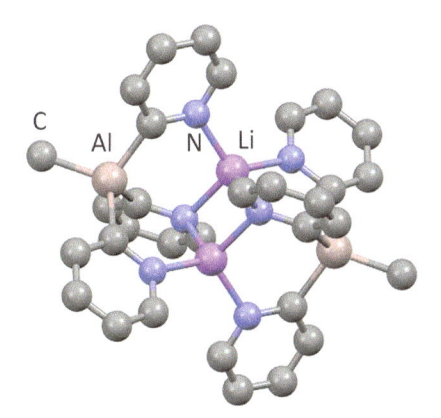

Fig. 4 Molecular structure of [Li(**3**)]$_2$.

$$2\ [\text{Li(THF)MeAl(2-py)}_3] \xrightarrow[\substack{M = \text{Fe, X = Cl} \\ M = \text{Mn, X = Cp}}]{MX_2} [M(\mathbf{3})_2]$$

[Li(THF)(**3**)]

Scheme 4 Synthesis of [M(**3**)$_2$] (M = Fe^{2+}, Mn^{2+}).

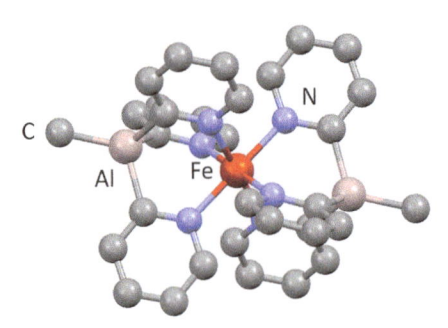

Fig. 5 Molecular structure of [Fe(**3**)$_2$].

as expected, but at the cost of selectivity; a similar trend is observed with increasing temperature. At 100 °C, 39% epoxide was detectable after 0.5 h but none at 4 h. Significantly, while other established catalytic systems required (organo)peroxide oxidants, dry air can be used as the source of oxygen in this system—a major cost-saving and environmental advantage.

The reaction of other sources of s-block and d-block metals with [Li(THF)(**3**)] gave mixed outcomes.[55] Sandwich and half-sandwich complexes were obtained when this reagent was combined with CaI$_2$ or ZnCl$_2$, respectively, to give [(**3**)$_2$Ca] and [(**3**)ZnCl], the latter being obtained in low crystalline yield ostensibly due to its high solubility (Fig. 6A and B). On the other hand, reaction of [Li(THF)(**3**)] with Mo(C$_7$H$_8$)(CO)$_3$ gives a trimetallic complex [(THF)Li(μ-CO)Mo(CO)$_2$MeAl(2-py)$_3$] (**5**) with a Mo(CO)$_3$ moiety capped by **3** and terminated by Li(THF)$_3$ (Fig. 6C).

Attempts to produce higher homologues of Group 13 met with variable success. Comproportionation of InCl$_3$ and Me$_3$In, to MeInCl$_2$ followed by reaction with 2-Li-py resulted in decomposition, whereas one-pot reaction of ^{n}BuLi and InCl$_3$, followed by treatment with 2-Li-2-py gave indate [Li(THF)^{n}BuIn(2-py)$_3$] [Li(THF)(**6**)] in low yield. X-ray crystallography reveals the same structure-type as seen with [Li(THF)(**3**)]. However, a similar route does not produce the Ga equivalent due to reductive decomposition releasing elemental gallium (gallium complexes have subsequently been accessed by other routes, see Section 2.3). Similarly, decomposition occurred when [Li(THF)(**6**)] was combined with FeBr$_2$ or Cp$_2$Mn. On the other hand, reaction of the indate with (C$_7$H$_8$)Mo(CO)$_3$ successfully produced a similar complex to **5** (*viz*. Fig. 6C). However, as a likely consequence of the larger bridgehead substituent (^{n}Bu *vs* Me), the product [(THF)$_2$Li(μ-CO)Mo(CO)$_2^n$BuIn(2-py)$_3$]$_\infty$ (**7**) differs in forming a polymer in the solid state (Fig. 7). Analysis of the IR spectra of the aluminate and indate complexes of Mo (**5** and **7**) indicates that the bridgehead atom has

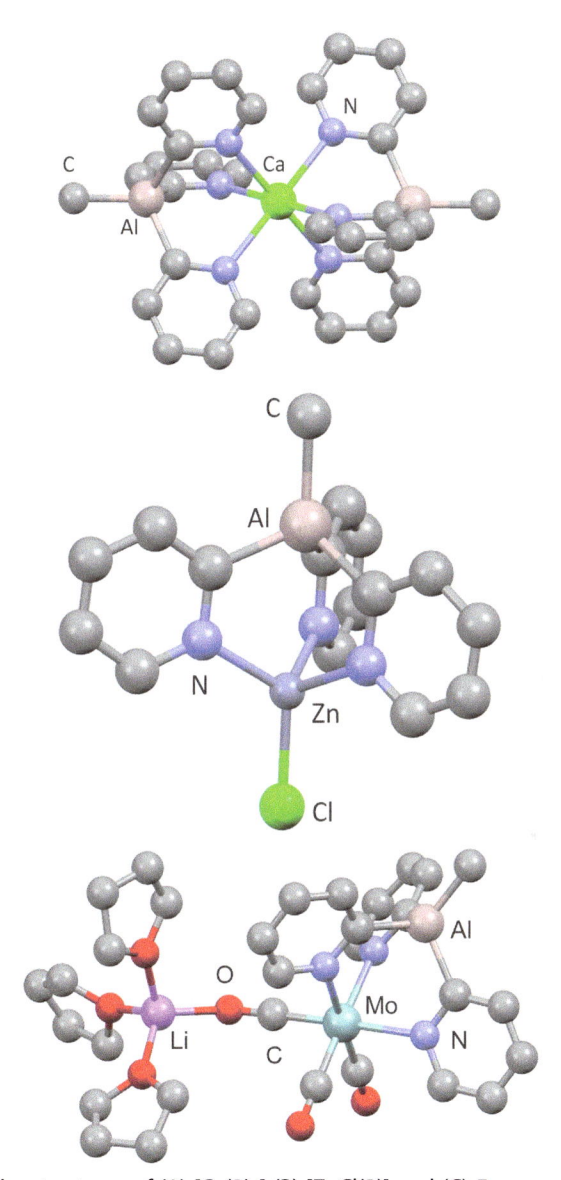

Fig. 6 Molecular structures of (A) [Ca(**3**)₂] (B) [ZnCl(**3**)] and (C) **5**.

little effect on the donor/acceptor characteristics of the pyridyl ligands. Furthermore, comparison of metric parameters in [Li(THF)(**3**)] and [Li(THF)(**6**)] suggests that the increased bite angle seen in the latter is primarily a result of a difference of *ca.* 0.2 Å between the length of the respective Al—C and In—C bonds.

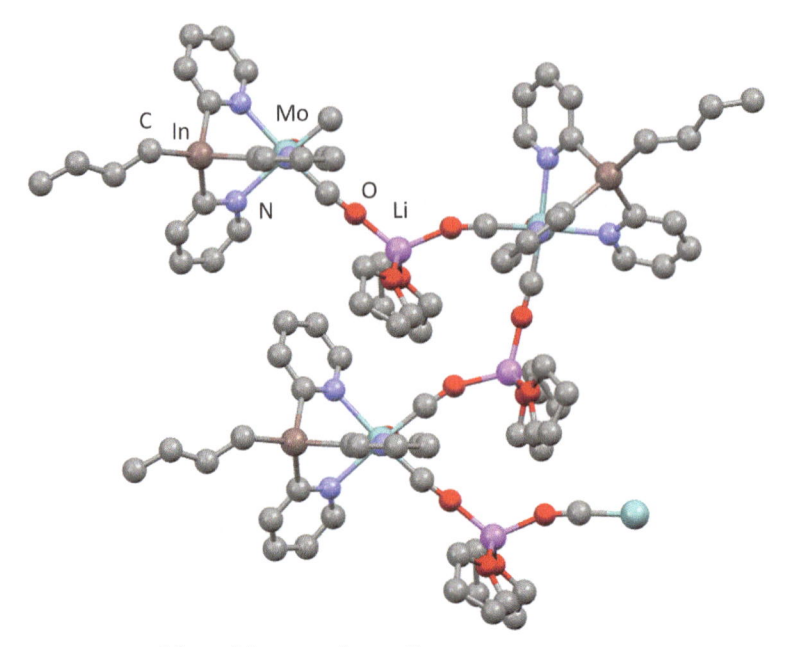

Fig. 7 A section of the solid-state polymer **7**.

In order to obtain ligands of the type [RAl(2-py′)]⁻ (R=organyl, 2-py′=substituted 2-pyridyl) that allow for variation in the R-group and the substitution pattern of the 2-pyridyl moiety, a more general route to pyridyl aluminates was adopted since many of the precursors RAlCl$_2$ are not commercially available (Scheme 5).[56] This enabled R=Et, ^{n}Bu, ^{s}Bu and ^{t}Bu bridgehead substituents to be introduced into pyridyl aluminates, which seemingly do not cause major distortions to the overall structure of the complexes. On the other hand, with R′=Me and py′=3-Me-2-py (**8**), 5-Me-2-py (**9**) and 6-Me-2-py (**10**) much greater structural variations are observed in [(THF)$_3$Li(μ-X)Li(**8**)] (X=Cl, Br; [these are crystallographically disordered in a 1:1 ratio]), [Li(THF)(**9**)] and [Li(THF)(**10**)]. Notably, while [Li(THF)(**9**)] shows largely the same structure-type as seen for [Li(THF)(**3**)], Me-substitution at the 3-position in [(THF)$_3$Li(μ-X)Li(**8**)] results in a reduced bite angle (due to steric interaction with the Al*Me* group). This can account for the complexation of the less sterically demanding halide bridged [Li–X–Li(THF)$_3$]$^+$ ion (Fig. 8). Meanwhile, 6-Me substitution of the 2-py moiety in [Li(THF)(**10**)] gives a distorted tetrahedral coordination environment for the Li-centre that is associated with twisting of one of the pyridyl groups by *ca.* 24° (Fig. 9).

Scheme 5 A general route to lithium tris(pyridyl)aluminates.

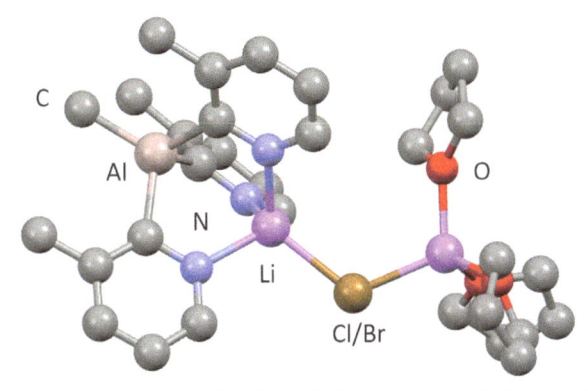

Fig. 8 Molecular structure of [(THF)$_3$Li(μ-X)Li(**8**)] (X=Cl$_{0.5}$Br$_{0.5}$). The [Li(THF)]$^+$ counter ion usually found with aluminates has been replaced by [Li–X–Li(THF)$_3$]$^+$ in this structure.

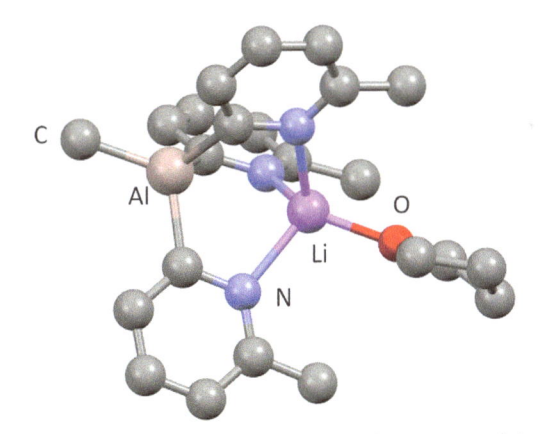

Fig. 9 Molecular structure of [Li(THF)(**10**)], showing the canting of the pyridyl groups.

Subsequently, a more extensive range of 2-pyridyl aluminate complexes of Ca^{2+}, Fe^{2+} and Mn^{2+} has been explored, where the bridgehead and substitution of the 2-pyridyl substituents were varied. *Re*-evaluation of previously determined Fe—N bond lengths, and those reported in this work led to the suggestion of spin equilibria (spin-crossover) operating between 180 and 298 K for Fe(II) sandwich complexes with these ligands.[57]

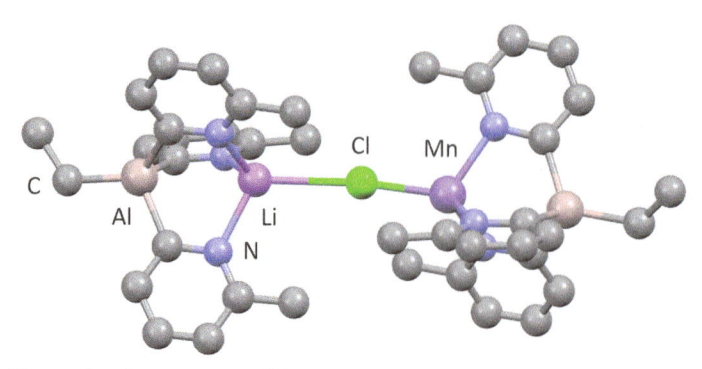

Fig. 10 The molecular structure of **12**.

Reaction of [Li(THF){EtAl(6-Me-2-py)₃}] [Li(THF)(**11**)] with MnCl₂ gives [{EtAl(6-Me-2py)₃}Mn(μ-Cl)Li{(6-Me-2py)₃AlEt}] (**12**), rather than a sandwich complex, presumably resulting from the desire to avoid unfavorable interactions between the pyridyl Me substituents of **11**⁻ (Fig. 10). Interestingly this steric congestion is tolerated in the sandwich complex [Ca(**11**)₂], but not for Mn(II) with a smaller ionic radius (Ca²⁺ 1.14 Å, Mn²⁺ 0.97 Å). Furthermore, the isolation of **12** (containing a M–X–Li fragment) hints that halide-bridged bimetallic complexes of this type may be transmetallation intermediates.

Noting earlier that [Li(THF)(**3**)] can act as soft pyridyl transfer reagent,[52] the ability of tris(2-pyridyl) aluminate reagents [Li(THF)(**11**)] and [Li(THF){EtAl(2-py)₃}] [Li(THF)(**13**)] to transfer groups intact to Sn(II) has been investigated.[58] While 2-pyridyl transfer was observed in all reactions conducted (Scheme 6A), the most important outcome noted was the one-pot reaction forming Janus head ligand [(6-Me-2-py)₃Sn]₂ (**14**), featuring a Sn—Sn bond, (produced from [Li(THF)(**11**)], Fig. 11). While X-ray diffraction confirmed the structure, *in situ* NMR spectroscopy studies

Scheme 6 (A) Synthesis of **14** and (B) likely disproportionation pathway leading to its formation.

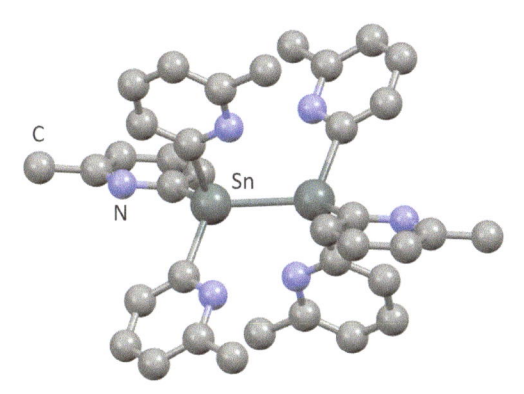

Fig. 11 Molecular structure of **14**, which can be described as Janus head shaped (opposing connected faces).

provided more information on a potential mechanism of formation. NMR-scale reaction of [Li(THF)(**11**)] with SnCl$_2$ at room temperature revealed rapid formation of **14** (without decomposition after 24 h) and concomitant production of [AlCl$_4$]$^-$ (alongside other species, but *not* including the bipyridine [(6-Me-2-py]$_2$). Taken together with the observation from a larger scale reaction of a black metallic precipitate forming during this reaction, disproportionation of Sn(II) to Sn(0) and Sn(IV) is the most likely mechanism of formation of the Janus head ligand (Scheme 6B). Similar *in situ* studies using [Li(THF)(**3**)] and SnCl$_2$ suggest less selective and slower reaction. However, this enabled the isolation of a potential intermediate [{Cl$_2$(2-py)$_2$Sn}Al{Sn(2-py)$_3$}] (**15**) containing an Al(III) centre coordinated by two Sn(II) centred anions, [SnCl$_2$(2-py)]$^{2-}$ and [Sn(2-py)$_3$]$^-$ (Fig. 12).

The reaction of tripodal aluminates with samarium(II) has also been studied.[59] Strikingly different outcomes were observed for unsubstituted tris(2-pyridyl)aluminate and the 6-Me-substituted ligands (Scheme 7).[59] In the case of the former, the reaction mixture proved so sensitive that the only isolable product is the dinuclear Sm(III) complex [Sm{EtAl(2-py)$_3$}{EtAl(2-py)$_2$O}]$_2$ (**16$_2$**) (Fig. 13), which incorporates a new oxy(pyridyl) aluminate ligand. In contrast, the Me groups of **11**$^-$ are capable of stabilizing a Sm(II) sandwich complex [Sm(**11**)$_2$], ostensibly through steric protection of the lanthanoid centre. *In situ* NMR spectroscopy suggested O$_2$ to be the source of oxygen in **16$_2$**, rather than H$_2$O, and this was confirmed by the addition of dry O$_2$ gas to the reaction mixture from which the oxo-complex **16$_2$** was obtained. The solid-state structure of **16$_2$** provides insight into the potential nature of reaction intermediates in the previously

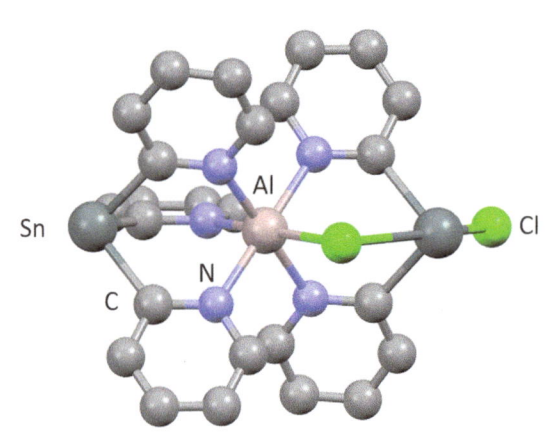

Fig. 12 Molecular structure of **15**.

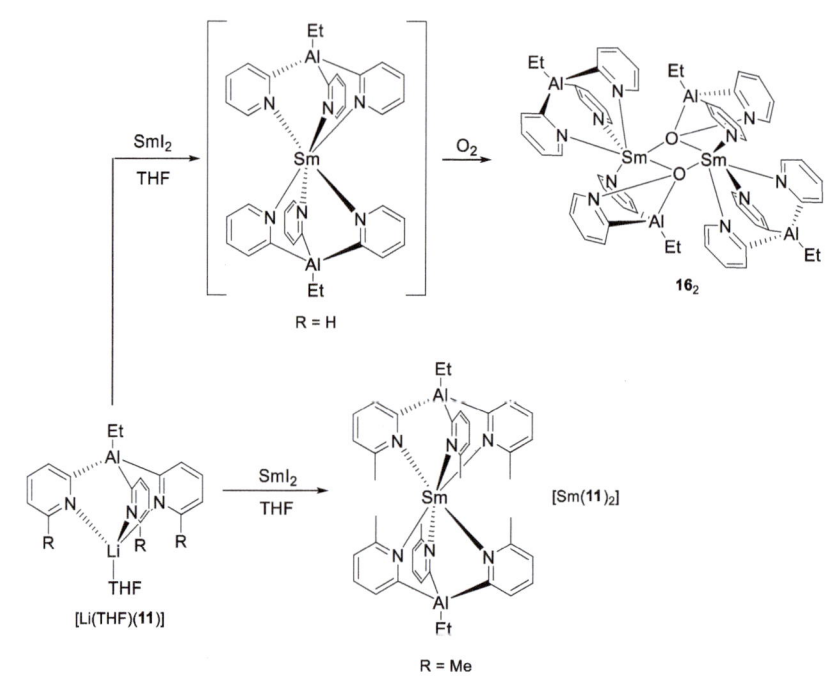

Scheme 7 Proposed reaction leading to **16₂** and [Sm(**11**)₂].

discussed epoxidation of styrene using [(**3**)₂Fe],[54] in which dry air was used as the oxidant. On the other hand, exposure of [(**11**)₂Sm] to O_2 led to the observation of a bipyridine among other products, suggesting a radical-based coupling reaction in this case.

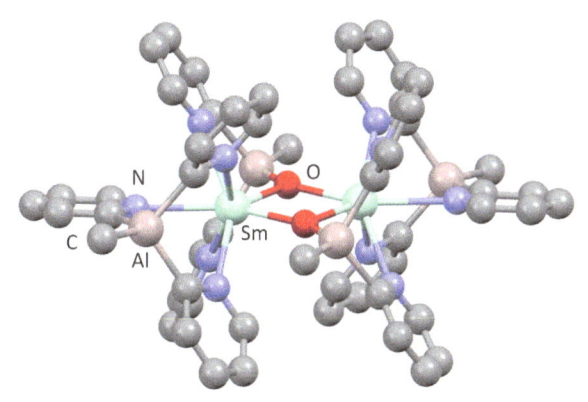

Fig. 13 The molecular structure of **16₂**. The coordination environment of Sm is best described as capped trigonal prismatic.

Further evidence for the strong influence on the donor properties of aluminate ligands by substitution at the 6-position of the pyridyl rings in tris(2-pyridyl)aluminates comes from the structures and reactivity of [EtAl(6-R-2-py)$_3$]$^-$ complexes (R = Me **11**$^-$, Br **17**$^-$, CF$_3$ **18**$^-$).[60] In contrast to [Li(THF)(**3**)], [Li(THF)(**11**)] loses THF without dimerization when heated under vacuum. Additionally, analogues [Li(**17**)] and [Li(**18**)], crystallized as unsolvated monomers even when etherate solvent was used. NMR spectroscopy and cryoscopy indicated retention of the solid-state monomers for all three aluminates upon dissolution in aromatic solvent. In the solid state, these species reveal three-coordinate, pyramidal Li and features that suggest interactions with the 6-R groups (the Li•••R group separation being below the sum of the relevant Van der Waals radii) (Fig. 14). As for previously reported **11**$^-$, whose ability to coordinate Fe(II) was questioned,[57] **17**$^-$ and **18**$^-$ did not form sandwich complexes with Fe(II). Nonetheless, these ligands are valuable in controlled hydrolysis (or alcoholysis) reactions, in which novel functionalised aluminates were produced (Scheme 8). Hydrolysis of [Li(**17**)] resulted in a hydroxyl(pyridyl) aluminate and allowed for the isolation of a crystalline dimer [EtAl(6-Br-2-py)$_2$OH]$_2$ (**19₂**) (Fig. 15). Meanwhile, elimination of EtH is apparently disfavoured, and though counterintuitive on thermodynamic grounds, this parallels the previous observation of the greater reactivity of the pyridyl moieties (*e.g.*, in selective transmetallation to Cu). Attempts to further deprotonate the hydroxy group of **19₂** using Et$_2$Zn gave only a 2-pyridyl transfer product, the zincate [LiZn(6-Br-2-py)$_3$]$_2$ (**20₂**).

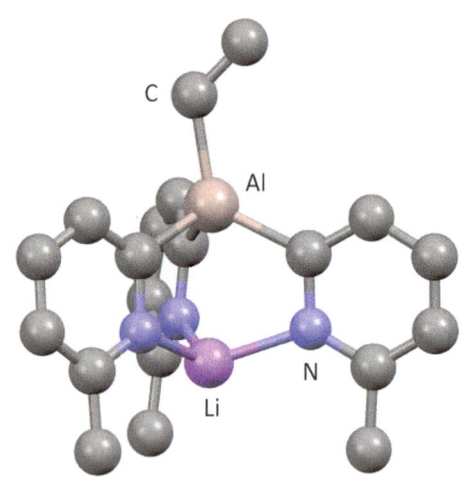

Fig. 14 The molecular structure [Li(**11**)].

Scheme 8 The hydrolysis of [Li (**17**)] (R=Br, R′=H) and alcoholysis of [Li(**13**)] and [Li (**17**)] (R=H, Br; R′=Me).

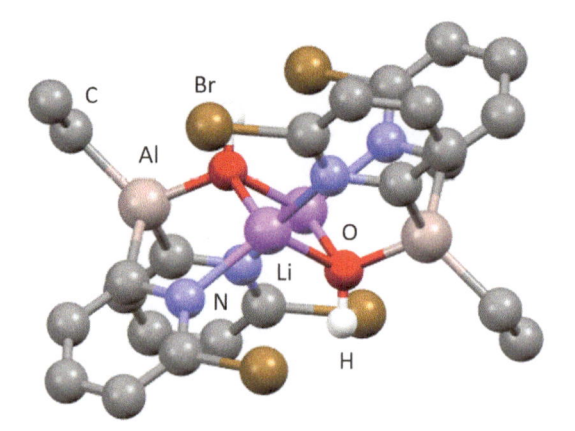

Fig. 15 Molecular structure of **19**$_2$.

Taking advantage of the selective alcoholysis discussed above, selective incorporation of chiral alcohols into the aluminate framework (in the form of an alkoxide) was possible.[61] [Li(**11**)] proved to be an ideal reagent for determination of enantiomeric excess, since the Me groups of **11**⁻ can serve as so-called "reporter" groups in ^{1}H NMR spectroscopy. When achiral MeOH was reacted with [Li(**11**)], only one Me resonance was observed by ^{1}H NMR spectroscopy for the 6-Me group in [LiEtAl(6-Me-2-Py)$_2$(OMe)]. However, when enantiomerically pure 2-butanol, 2-octanol or 1-phenylethanol was used, splitting of the resonance for the 6-Me substituents of the pyridyl rings into two signals (a pair of singlets) was observed in the ^{1}H NMR spectra (Fig. 16A). The observation of two signals for the 6-Me group is due to the formation of homochiral dimers (like that shown in Fig. 16B) in solution (later verified in the solid state by X-ray crystallography (Fig. 16C)) containing two magnetically inequivalent 6-Me groups in the ^{1}H NMR spectrum. If the alcohol is not enantiomerically pure or racemic, three possible dimers can be formed (named according to the chirality of the alcohol): *RR*, *SS*, and *RS* (or *SR*), as now the two enantiomers of the chiral alcohol are present. Hence, the homochiral *RR* and *SS* dimers (which are enantiomers) give rise to a pair of singlets; while the heterochiral (*meso*) *RS* (or *SR*) pair of dimers give rise to two singlets at different chemical shifts (the homo and heterochiral dimers are diastereomers). The proportion of the latter heterochiral dimer present in solution depends on the ratio of the two enantiomers. In other words, as the amount of the other enantiomer in solution increases, a greater amount of the heterochiral dimer can be produced, leading to a 50% homochiral:heterochiral ratio for racemic mixtures in the absence of diastereoselectivity. This provides a simple method to determine *ee* quantitatively by integration of the NMR resonances in mixtures. Notably, simpler-to-interpret ^{7}Li NMR spectroscopy could be used in addition to ^{1}H NMR spectroscopy to determine the *ee*.

Application of the same 2-pyridyl substituted ligand sets to the coordination chemistry of redox stable divalent lanthanoid complexes (Eu(II) and Yb(II)) of tris(2-pyridyl)aluminates focussed on investigating the influence of the electronic demands of the 2-pyridyl ring substituent.[62] Crystals of sandwich complexes with **11**⁻ and **17**⁻ could be obtained from the reactions of the lithium complexes with EuI$_2$ and YbI$_2$, whereas for **18**⁻ no apparent coordination occurred. Though the products obtained were found to be solvent dependent, co-crystallization of LiI could be avoided by using acetonitrile/toluene as the solvent. The steric bulk and electron withdrawing

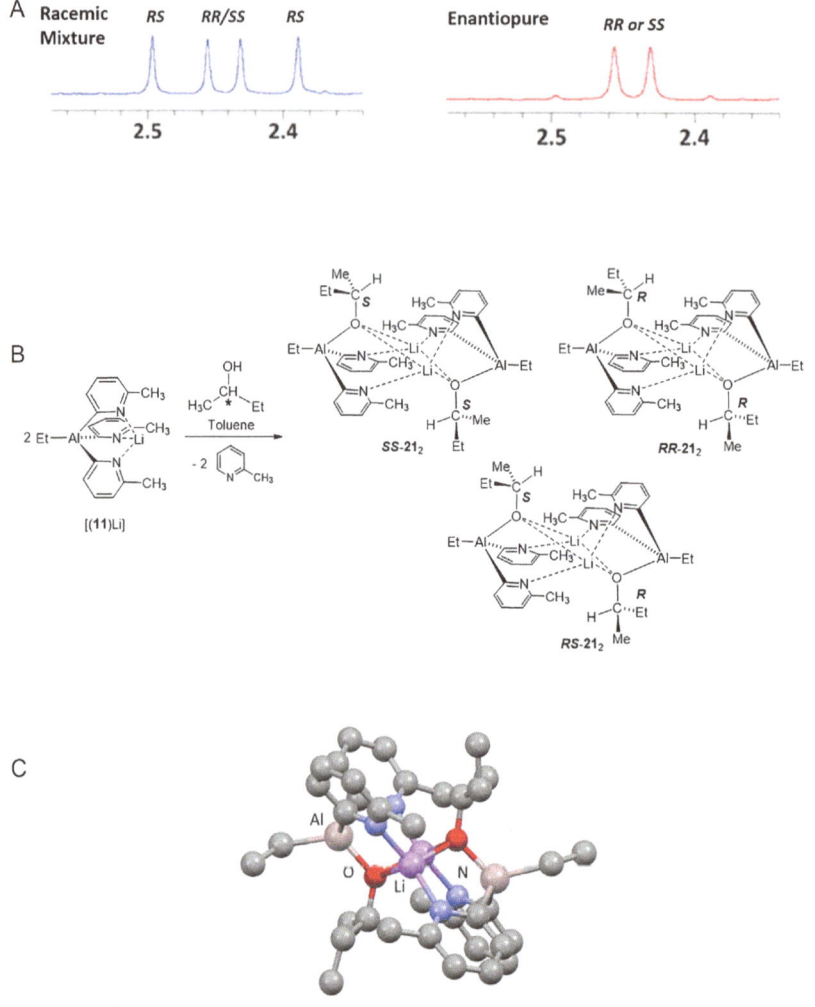

Fig. 16 (A) ^{1}H NMR spectra showing resonances from chiral/achiral aggregates (B) generation of chiral and achiral aggregates *via* deprotonation and (C) the molecular structure of the chiral *RR* dimer [EtAl(6-Me-2-py)$_2${(R)-O(CH)(Me)CH$_2$Me}]$_2$ (**21**$_2$) formed by reaction of [Li(**11**)] with enantiopure (R)-2-butanol.

properties of Br and CF$_3$ substituents are likely responsible for the weaker coordinative properties of **17**$^-$ and **18**$^-$ towards lanthanoid ions. Spectroscopic data suggest that formation of [Yb(**11**)$_2$] involves the intermediate complex [YbI(**11**)], whose structure was determined by X-ray diffraction on a stoichiometrically prepared sample (Fig. 17).

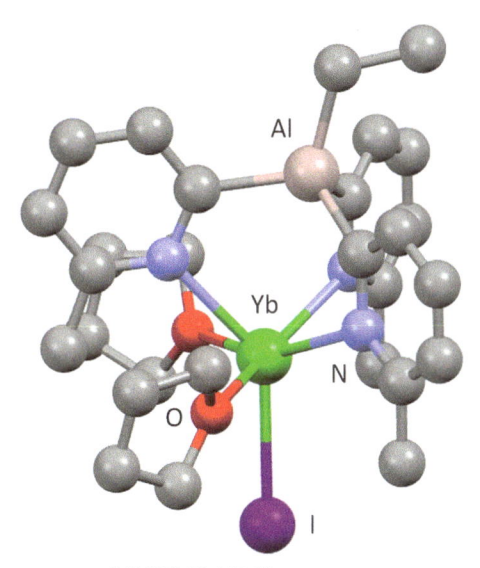

Fig. 17 Molecular structure of [(THF)₂YbI (**11**)].

The main thrust of this work was to uncover dynamic behavior of these lanthanoid systems in solution. This was studied in detail by *in situ* multidimensional NMR spectroscopy (taking advantage of the diamagnetism of Yb(II)). Hence, [Yb(**11**)₂] in THF was unaffected by the addition of LiI, while bromo-substituted [Yb(**17**)₂] showed dynamic behavior even when small amounts of lithium salts were added. This indicated progressive comproportionation to give putative half-sandwich complex [(THF)₂YbI (**17**)] and [Li(**17**)] upon complete addition of one equivalent of LiI (Scheme 9). Treatment of [Yb(**11**)₂] with [Li(**11**)] did not lead to detectable ligand exchange by 2D NMR spectroscopy. On the other hand a mixture of [Yb(**17**)₂] and [Li(**17**)] was shown to exchange components, and when a small amount of LiI was added to the mixture, [YbI(**17**)] was also identified. Lastly, cross-exchange of **11**⁻ and **17**⁻ in Yb sandwich complexes was studied. In this case, when [Yb(**17**)₂] was combined with [Li (**11**)], ligand exchange occurred *via* the putative mixed ligand complex [Yb(**11**)(**17**)], whose existence is supported by NMR spectroscopy.

The potential for different modes of coordination brought about by 3-pyridyl and 4-pyridyl isomers (due to the significant change in bite and bonding vectors of the ligands compared to 2-pyridyl ligands) is an area of increasing interest and this topic is reviewed more extensively for other tripodal ligands with other bridgehead atoms later. García-Rodríguez and

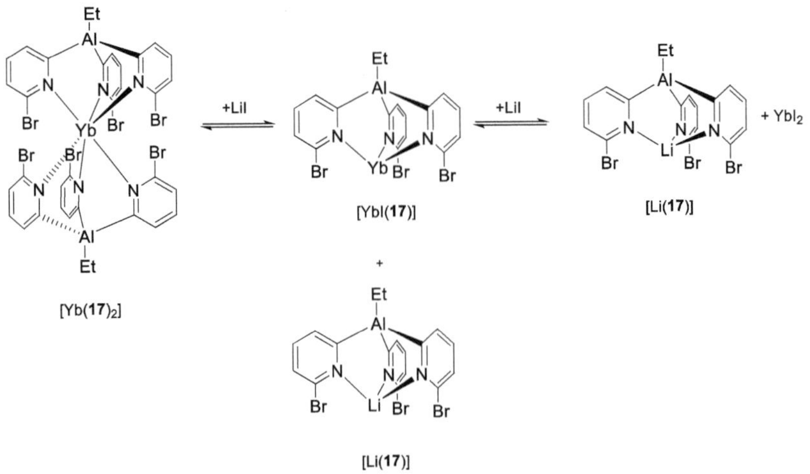

Scheme 9 Series of equilibria involved in Yb/Li exchange.

co-workers introduced 2-, 3- or 4-(CH$_2$O)-py groups into the aluminate framework by reactions of [Li(**11**)] as the base with the corresponding alcohols, giving dimers **22**$_2$, **23**$_2$ and **24**$_2$, respectively (Scheme 10).[63] No secondary interactions with the pyridyl-N atoms and metal centres are seen in the structures of **23**$_2$ and **24**$_2$, however, an Al•••N interaction is evident in **22**$_2$, giving rise to a distorted trigonal bipyramidal coordination environment at the Al(III) centre (Fig. 18).

Interestingly, 2:1 reaction of HOCH$_2$-4-py with [(**11**)Li] gave the polymer [EtAl(6-Me-2-py)(py-2-CH$_2$O)$_2$Li]$_\infty$ (**25**$_\infty$) (Fig. 19). This demonstration of 'double basicity' of the aluminate was explored further by reaction of [(**11**)Li] with chiral (−)-menthol, which produced a solid-state dimer [EtAl(6-Me-2-py)(metholate)$_2$Li]$_2$ (metholate = (1R,2S,5R)-5-methyl-2-propan-2-ylcyclohexan-1-oxide) (**26**$_2$) rather than a polymeric structure (Fig. 20). This reactivity has the potential to introduce four different groups to aluminium and raised the possibility of tetrahedral chiral-at-aluminium 'ates. This was realized by sequential reaction of [Li(**11**)] with MeOH and ^{t}BuOH, releasing two equivalents of 6-Me-2-H-py (Scheme 11) to give [LiEtAl(6-Me-2-py)(^{t}BuO)(MeO)]$_2$ (**27**$_2$) (in which the Al centre has four different substituents, and is therefore chiral). Though the only isolable crystalline product was an achiral *RS* dimer in the solid state (Fig. 21), NMR spectroscopy (including ^{1}H–^{7}Li-HOESY) provides strong evidence for the existence of the other possible *RR* and *SS* dimeric aggregates in solution.

Scheme 10 Deprotonation of (py')CH$_2$OH (py' = 2-, 3- or 4-py), giving **22$_2$**–**24$_2$**.

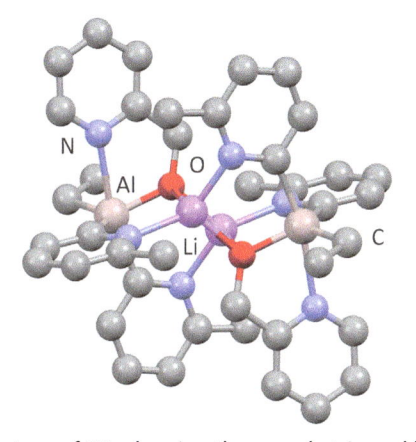

Fig. 18 Molecular structure of **22$_2$** showing the pseudo trigonal bipyramidal coordination geometry of Al.

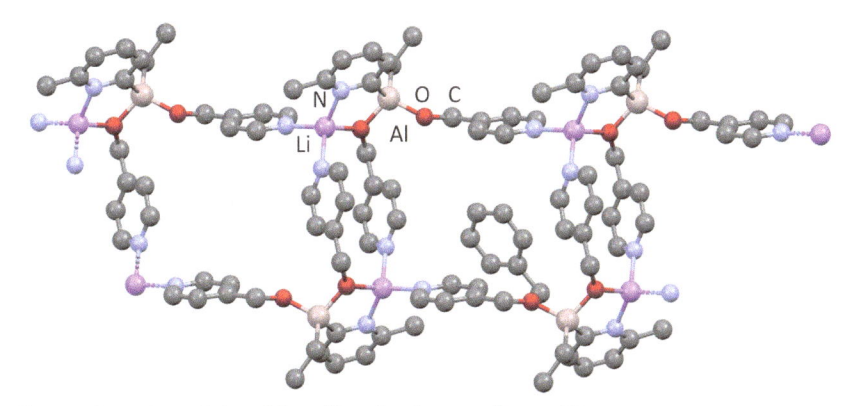

Fig. 19 A section of the ribbon-like aluminate polymer **25$_∞$**.

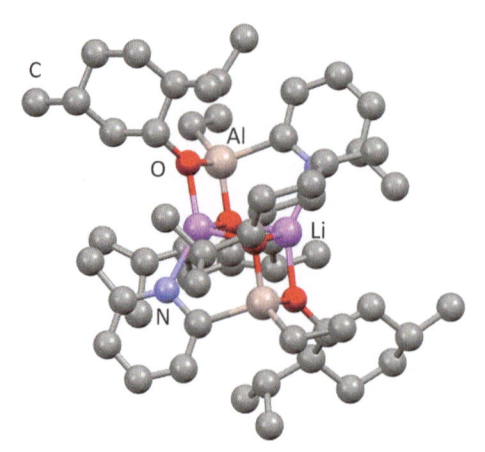

Fig. 20 Molecular structure of **26**₂.

Scheme 11 Stereochemically distinct aluminates formed from chiral-at-Al monomers **27** associating to give chiral or achiral aggregates.

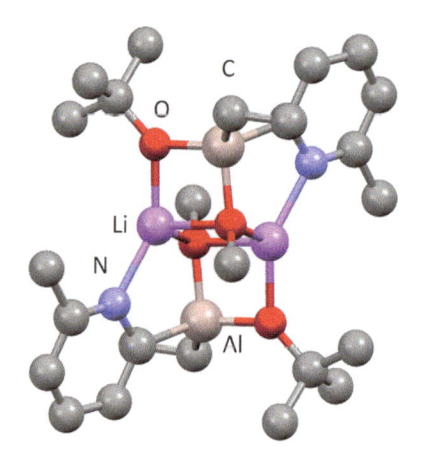

Fig. 21 Molecular structure of centrosymmetric *RS*-**27**₂.

The scope of reactivity of tris(2-pyridyl)aluminates was extended to aldehydes and carboxylic acid groups.[64] In the first instance, benzaldehyde was reacted with [Li(**11**)], resulting in selective insertion into an Al-2-pyridyl bond, giving a structure similar to other dimeric alkoxy(2-pyridyl)

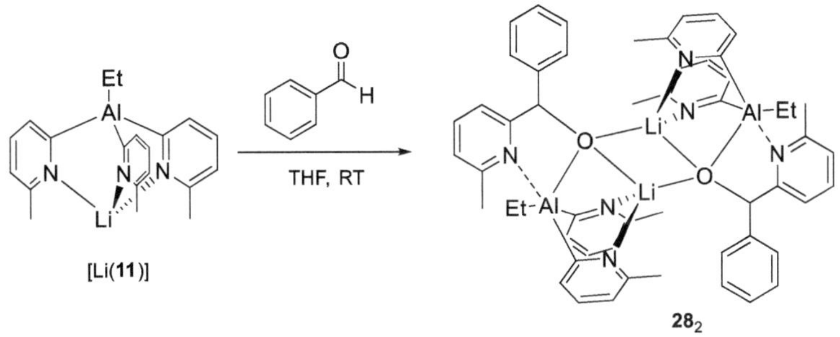

Scheme 12 Nucleophilic addition of 6-Me-2-py⁻ to benzaldehyde.

aluminates (Scheme 12). However, a secondary interaction of the pendant 2-pyridyl function with the Al-bridgehead atom is present in the product [LiEtAl(6-Me-2-py)$_2$(PhCH(6-Me-2-py)O)]$_2$ (**28$_2$**) (Fig. 22, *viz.* Fig. 18). While reaction with aldehydes and ketones was demonstrated, less electrophilic functional groups, such as nitriles and imines are not susceptible to insertion. The reactions of tris(2-pyridyl)aluminates [Li(**11**)], [Li(**17**)] and [Li(**18**)] with benzoic acid were also probed. Reaction of [Li(**11**)] with benzoic acid at room temperature was shown to be non-selective by NMR spectroscopy, while the same reaction at low temperature in toluene formed the anions [EtAl(6-Me-2-py)$_2$(PhCO$_2$)]⁻ (**29⁻**) and [EtAl(6-Me-2-py)(PhCO$_2$)$_2$]⁻ (**30⁻**). Their lithium salts could be isolated in scaled-up reactions: for [Li(**29**)]$_2$, crystallography reveals a dimer, which, while still boasting a Li$_2$O$_2$ core, demonstrates side-on association of the aluminate with respect to the central Li$_2$O$_2$ ring unit (Fig. 23). In line with the expected reduced basicity of the electron withdrawing 2-pyridyl substituents in **17⁻** and **18⁻**, bearing 6-Br and 6-CF$_3$ functionalities, reaction with PhCO$_2$H proved unselective and only very small amounts of one of the products of reaction, [LiEtAl(6-Br-2-py)$_2$(PhCO$_2$)]$_2$ (**31$_2$**), were obtained. Crystallography revealed a previously unobserved bridging mode for a pyridyl-aluminate anion, where each Li centre is coordinated by a 2-pyridyl function from each aluminate ion within the dimer (Fig. 24).

2.3 Gallium and indium

The difficulties in preparing tripodal ligands from gallium and indium have been noted previously.[48] This is due to the greater instability of trivalent heavier Group 13 metal compounds in the presence of strongly reducing organolithium reagents, such as 2-lithio-pyridine reagents. However, just as transfer of 2-pyridyl moieties from an aluminate to SnCl$_2$ is known,[58] stannates and plumbates (which will be discussed in detail in Section 3.2)

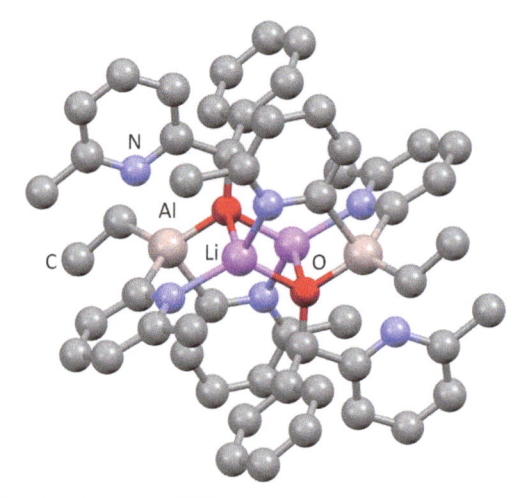

Fig. 22 The molecular structure of **28**₂.

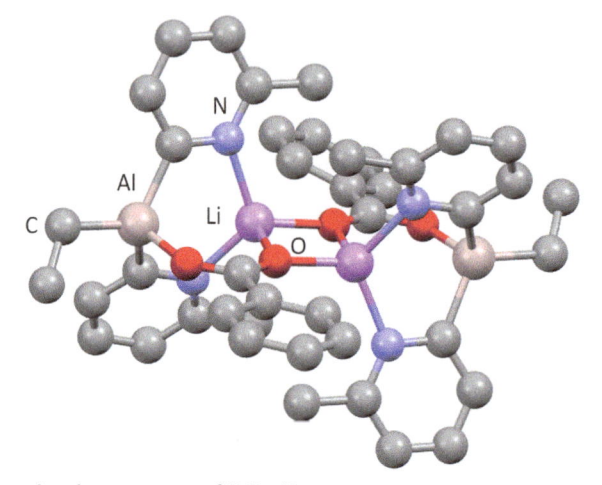

Fig. 23 The molecular structure of [Li(**29**)]₂.

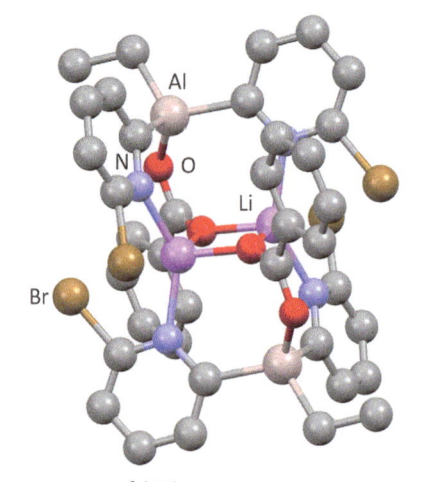

Fig. 24 The molecular structure of (**31**)₂.

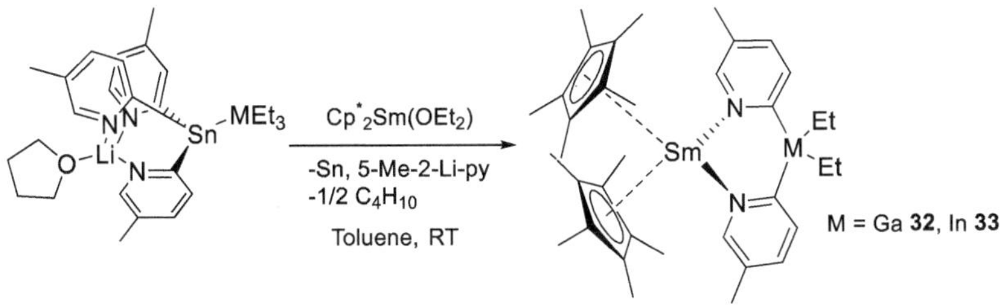

Scheme 13 Production of **32** and **33** accompanied by elimination of Sn.

have proved to be suitable for transmetallation involving heavier Group 13 metals. The first indication of this came from Zeckert who showed that decomposition of tris(2-pyridyl)stannate–MEt$_3$ adducts (M=Ga, In) when treated with a samarocene gave [Cp*$_2$Sm(5-Me-2-py)$_2$MEt$_2$], (M=Ga **32**, M=In **33**) formally containing the bidentate [(5-Me-2-py)$_2$MEt$_2$]$^-$ ion (Scheme 13).[65]

It is not until very recently, however, that this Group 14 to Group 13 2-pyridyl transfer process has been extended to generate a wider range of Ga and In species.[66] This relies on the very facile degradation of a lithium plumbate [LiPb(6-^{t}BuO-2-py)$_3$] [Li(**34**)]. The products of the reactions of [Li(**34**)] with GaR$_3$, dimeric [LiR$_2$Ga(6-^{t}BuO-2-py)$_2$]$_2$ (R=Et, **35**$_2$) and monomeric [Li(THF)R$_2$Ga(6-^{t}BuO-2-py)$_2$] (R=^{i}Pr **36**, ^{t}Bu **37**), are solvent dependent (Scheme 14 and Fig. 25). Interestingly, when ^{t}Bu$_3$Ga was reacted with [Li(**34**)] an adduct, [Li(**34**)(^{t}Bu$_3$Ga)$_3$], was isolated on one occasion. However, this adduct decomposed upon re-dissolution in THF or benzene (decomposition could be followed by NMR spectroscopy in both C$_6$D$_6$ and d_8-THF solution and a crystalline species similar to **35**$_2$ was isolated from C$_6$D$_6$). When less sterically demanding Me$_3$Ga was treated with [Li(**34**)] in THF, the first example of a tris(2-pyridyl)gallate [LiMeGa(6-^{t}BuO-2-py)$_3$] (**38**) was isolated in high crystalline yield (Fig. 26). Similar reactions with R$_3$In (R=Me, Et) reagents gave structurally analogous lithium tris(2-pyridyl)indates (R=Me **39**, Et **40**) (Scheme 15). Overall, these reactions highlight the importance of both the steric properties of the bridgehead substituent in directing the outcome of 2-pyridyl transfer to gallium or indium and the involvement of redox processes during their formation. The last point is shown by the reaction of InI with Li(**34**) in which metallic lead is deposited and a lithium tris(2-pyridyl)indate [Li(THF)(I)In(6-^{t}BuO-2-py)$_3$] (**41**) is formed (containing an In(III) bridgehead, Scheme 16).

Scheme 14 Synthesis of gallates **35–37**.

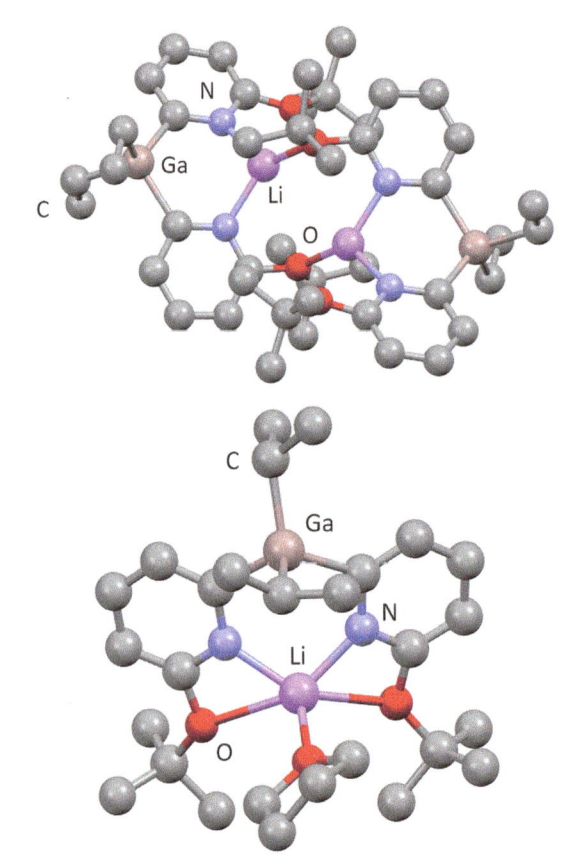

Fig. 25 Molecular structures of **35₂** and **36**.

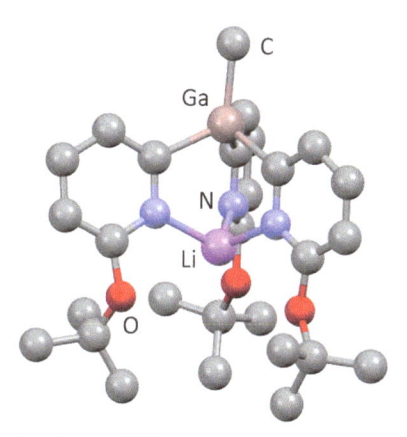

Fig. 26 The molecular structure of **38**.

Scheme 15 Formation of tris(pyridyl)gallates and indates **38–40**.

Scheme 16 Formation of **41** by oxidation of indium(I) iodide.

3. Group 14 tris(pyridyl) ligands

The chemistry of tris(2-pyridyl) ligands bearing a (functionalised) carbon bridgehead atom of the type [RC(2-py)₃] is extensive and has been reviewed elsewhere.[53] In contrast, the chemistry of tripodal ligands containing heavier congeners as bridgeheads is less well explored. Unlike the ligands discussed in Sections 1 and 2, different positional isomers of the pyridyl donor function (3-pyridyl and 4-pyridyl) have been explored in some tris(pyridyl) Group 14 species and this introduces greater versatility.

3.1 Silicon and germanium ligands

The synthesis of RSi(2-py)$_3$ ligands postdates that of heavier Sn-based counterparts (see Section 3.2). Wright and Hopkins reported the synthesis of the first structurally characterized tris(2-pyridyl)silane by reaction of 2-Li-py (generated by Br/Li exchange) with MeSiCl$_3$ at low temperature (Scheme 17).[67] NMR spectroscopy confirmed the silane to be MeSi(2-py)$_3$ (**42**), however, crystallography revealed that the ligand complexes the lithium salt by-products (LiCl/LiBr, Fig. 27). Compared to Sn analogues (to be discussed later), the silane provides a more compact ligand, primarily on account of the shorter C—Si bonds. Another route to ligands of this type is a one-pot reaction of SiCl$_4$ or GeCl$_4$ sequentially with ^{n}BuLi and 2-Li-py.[48]

Scheme 17 Synthesis of silane ligand **42**.

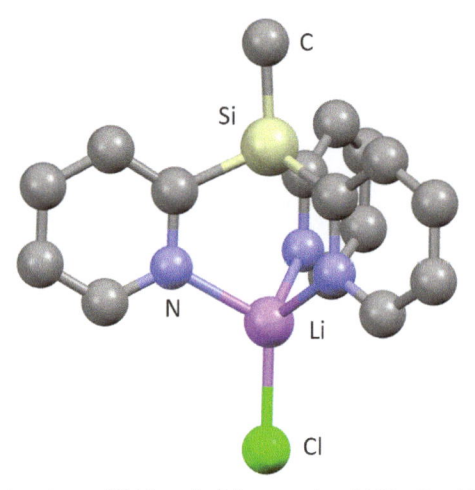

Fig. 27 Molecular structure of lithium halide complex [Li(Cl$_{0.8}$Br$_{0.2}$)(**42**)]. Only Cl is displayed for clarity.

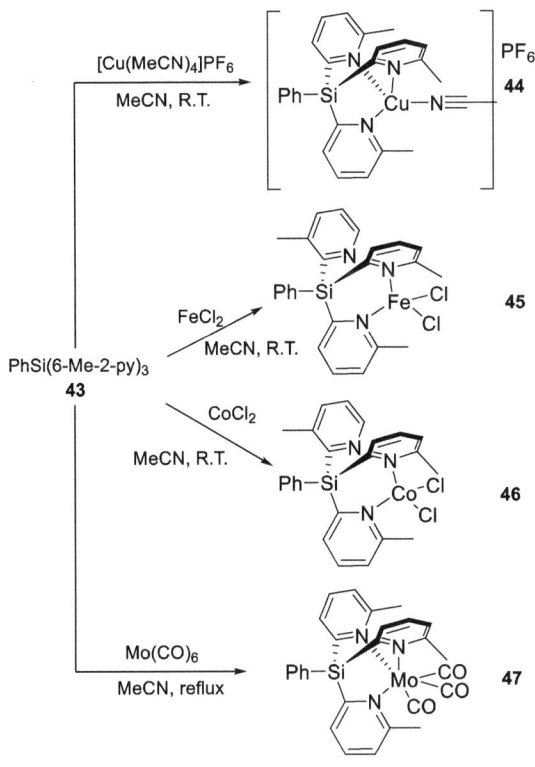

Scheme 18 Syntheses of tris(pyridyl)silane-metal complexes **44–47**.

A precedent for the stabilizing influence of the 6-Br-2-py group from previous reports[68] suggested use of the 6-Me-2-py moiety as a stabilizing ligand in this area (as seen in Section 2). PhSi(6-Me-2-py)$_3$ (**43**) was synthesized by substitution of PhSiCl$_3$ and its coordination chemistry towards metals (Cu(II), Fe(II), Co(II) and Mo; complexes **44–47**) was studied (Scheme 18).[69] Advantages of **43** over the unsubstituted ligand **42** include the higher yield, air stability and the absence of lithium halide coordination when the ligand is extracted in toluene. While the free ligand shows a propeller like pattern in the solid state (Fig. 28), this orientation changes upon coordination to metals (Scheme 18).

The variation in metal coordination geometry that results from bis-coordination of the pyridyl donor in **45** and **46** can be partially ascribed to the steric pressure of the 6-Me ring substituents and the small cation sizes in the case of Fe^{2+} and Co^{2+}, which disfavours the formation of ion separated sandwich complexes but maintains the strong M—Cl bond.

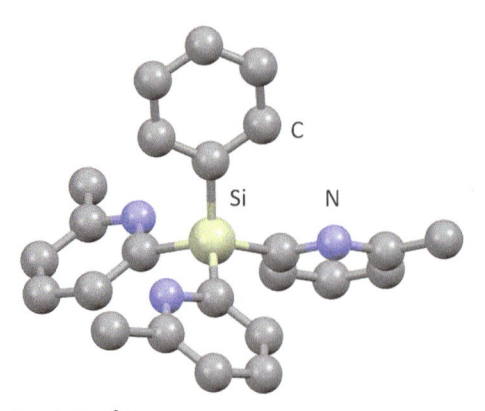

Fig. 28 Molecular structure of **43**.

Variable temperature NMR experiments on **45** and **46** indicate an intramolecular fluxional process, involving the precession of the pyridyl ligands around the metal centre. Difficulties in assigning the low temperature NMR spectra attributed to the paramagnetic ions were overcome by DFT calculations. These results gave a value of the magnetic anisotropy that is consistent with tetrahedral metal centers.

Tris(3-pyridyl)silane ligands have gained attention in recent years in the field of supramolecular chemistry, with applications found in the production of luminescent materials and construction of supramolecular clusters. In contrast to tris(2-pyridyl)silanes, the expanded ligand bite favors the coordination of the 3-pyridyl groups to different metal centres over chelation to a single metal. Boomishankar and co-workers reported the synthesis of MeSi(3-py)$_3$ (**48**) from 3-Li-py and MeSiCl$_3$ at low temperature, followed by purification using column chromatography.[8] The related quinolyl ligand MeSi(3-qy)$_3$ (qy=quinolyl) (**49**) was prepared in a similar manner. When combined with three equivalents of CuI under solvothermal conditions, MOFs [(Cu$_6$I$_6$)(**48**)$_6$]$_n$ and [(Cu$_6$I$_6$)(**49**)$_6$]$_n$ were obtained (Fig. 29). Crystallography reveals 2D topologies best described as hexagonal sheets containing Cu$_6$I$_6$ clusters linked by the silane ligands and showing intersheet π-interactions. Interestingly, other silane:CuI ratios gave 1D MOFs, containing Cu$_2$I$_2$ or CuI motifs and conversion between the 2D and 1D-type MOFs could be achieved by addition of ligand or CuI under appropriate conditions.

The thermochromic properties of MOFs [(Cu$_6$I$_6$)(**48**)$_6$]$_n$ and [(Cu$_6$I$_6$)(**49**)$_6$]$_n$ are quite different. The former exhibits thermochromism typically associated with Cu$_4$I$_4$ clusters, while the latter exhibits ^{3}XLCT/^{3}MLCT

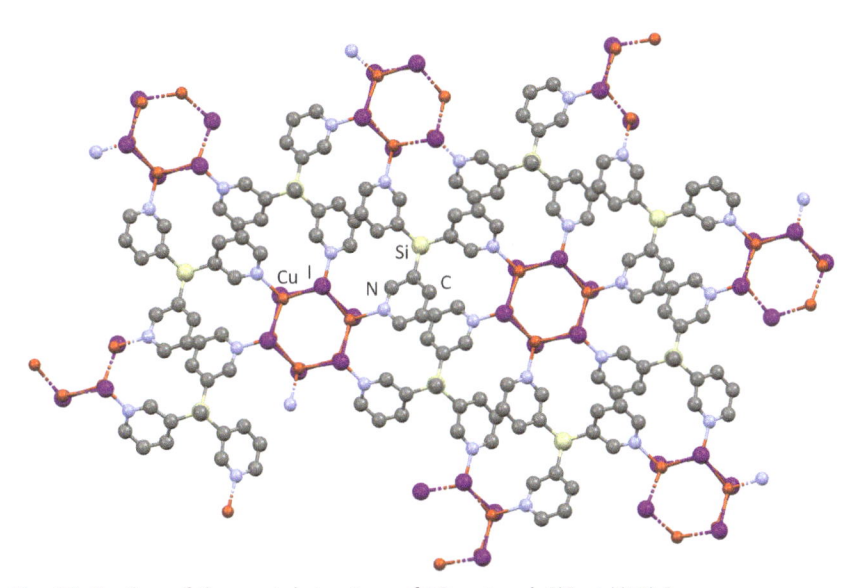

Fig. 29 Section of the crystal structure of 2D network [(Cu$_6$I$_6$)(**48**)$_6$]$_n$.

(iodide-to-ligand and metal-to-ligand charge transfer, respectively) emission with marginal variation in emission wavelength at room and low temperature. On the other hand, mechanochromic luminescence showed a reversal in behavior at 298 K: [(Cu$_6$I$_6$) (**48**)$_6$]$_n$ showed a blue-shifted emission and [(Cu$_6$I$_6$)(**49**)$_6$]$_n$ exhibited a red-shifted emission upon grinding. This was attributed to a difference in the changes in Cu•••Cu interactions in the two samples upon grinding (becoming shorter in [(Cu$_6$I$_6$)(**49**)$_6$]$_n$). On the other hand, the behavior of [(Cu$_6$I$_6$)(**48**)$_6$]$_n$ could be explained by a rise in the LUMO energy due to packing effects.

Boomishankar has also reported the silane bridged Co and Ni-based supramolecular species [Co$_6${MeSi(3-py)}$_8$Cl$_6$(H$_2$O)$_6$]$^{6+}$ (**50**$^{6+}$) and [Ni$_6${MeSi(3-py)}$_8$Cl$_9$(H$_2$O)$_2$]$^{3+}$ (**51**$^{3+}$), both of which have been investigated for their ability to promote photochemical hydrogen evolution using H$_2$O (the synthesis of the Co(II) complex is shown in Scheme 19).[70]

Scheme 19 Synthesis of M$_6$L$_8$ cage **50**Cl$_6$.

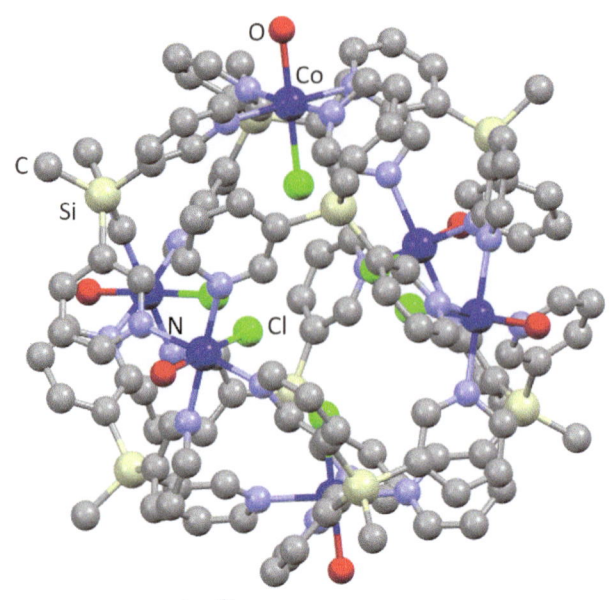

Fig. 30 Molecular structure of **50**[6+].

A crystallographic study of the **50**Cl$_6$ revealed a cage of six Co(II) ions arranged in an octahedron. Each of these ions is coordinated by four N-donor atoms of tris(3-pyridyl) ligands and by internally and externally coordinating by Cl and H$_2$O ligands (Fig. 30).

The rigidity of the tris(3-pyridyl)silane linkers plays a critical role in maintaining a tight nanocage of metal ions, making it impervious to guest molecules and ensuring the cage's integrity in aqueous media. Cyclic voltammetry in buffered aqueous solution between pH 2 and 7.5 revealed a region of pH dependence for the Co^{II}/Co^{I} redox couple, but pH independence for the Co^{III}/Co^{II} redox couple, which suggests a proton-coupled electron transfer process, postulated to be important in the electrochemical proton reduction pathway. Furthermore, electrocatalytic measurements revealed a turnover frequency for hydrogen production of 16 h^{-1}. Photocatalytic activity in aqueous solution was investigated with [Ru(bipy)$_3$]Cl$_2$ as a photosensitiser and ascorbic acid as a sacrificial electron donor. When irradiated at $\lambda = 469$ nm for 2 h, a turnover number of 43 was recorded at a catalyst concentration of 50 μM, compared to negligible activity of CoCl$_2$ under the same conditions, even at much higher concentrations. However, decomposition of the photosensitiser and air-sensitivity pose limiting barriers to the system's utility.

Replacing Co by Ni gave a similar Ni$_6$L$_8$ cage (L = **48**).[71] While the core structure resembles that of **50**$^{6+}$, the overall structure consists of **51**$^{3+}$ cages connected by bridging chloride ligands, forming a 1D coordination network. Cyclic voltammetry revealed two quasi-reversible redox couples which exhibited pH dependant behavior, again suggesting proton-coupled electron transfer. A likely mechanism of H$_2$ production from H$_2$O involves reduction of Ni(II) to Ni(I), followed by protonation to give a Ni(III) hydride species; subsequent reduction to Ni(II) and reaction with another proton completes the formation of hydrogen. Photocatalytic activity was probed in the presence of [Ru(bipy)$_3$]Cl$_2$ and ascorbic acid. This revealed a promising turnover number of 2824 in 69 h.

While the tris(3-pyridyl)silane described above has seen several applications, only one report exists on the use of a tris(4-pyridyl)silane, MeSi(4-py)$_3$ (**52**), where it was observed to bind three equivalents of the strongly Lewis acidic rhodium complex [Rh$_2$(O$_2$CCF$_3$)$_4$] forming an extended pyramid [{Rh$_2$(O$_2$CCF$_3$)$_4$}$_3$(**52**)] terminated by η^2 bonded benzene ligands (Fig. 31).[72]

3.2 Tin and lead

The number of reports on tris(pyridyl)-Sn and -Pb ligands exceeds that of silicon significantly and their chemistry is further enriched by the possibility of having the bridgehead atom in the II or IV oxidation states. In the case of lead, the inert pair effect leads to the overwhelming preference for the II

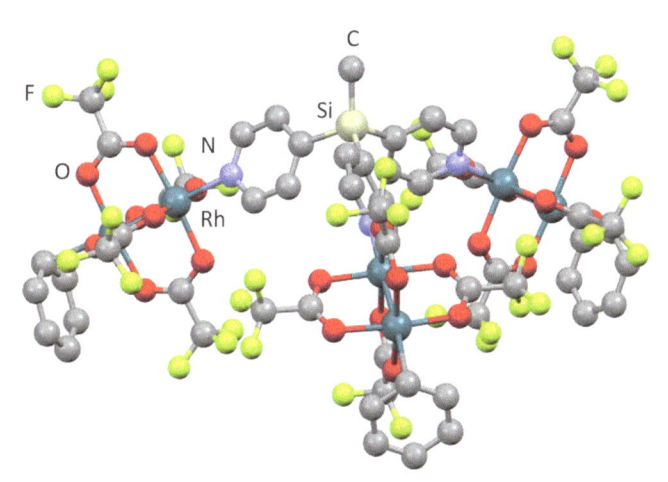

Fig. 31 Molecular structure of tripodal complex [{Rh$_2$(O$_2$CCF$_3$)$_4$}$_3$(**52**)].

oxidation. The groups of Wright and Zeckert have made great strides in understanding the chemistry of tris(pyridyl)-Sn and -Pb compounds.

The synthesis of a Sn(IV) tris(2-pyridyl) ligand was first accomplished by treatment of Cp$_2$Sn^{II} with 2-Li-py.[73] 2-Li-py was generated by Li/Br exchange, however, the unexpected formation of a Sn(IV) bridgehead in the product [^{n}BuSn(2-py)$_3$LiBr]·0.5(THF) [LiBr(**53**)]·0.5(THF) implicated involvement of by-product ^{n}BuBr in its formation (Scheme 20). Two possibilities could account for this, (i) nucleophilic displacement by the stannate intermediate or (ii) oxidative addition of ^{n}BuBr to Sn(II). Reaction of [Li(**53**)]·0.5(THF) with Cu^{II}Cl$_2$ (2:1 ratio) unexpectedly gave [Cu^{I}Br(**53**)]·0.5(THF), containing Cu(I) (Fig. 32). Electrochemical studies confirmed the instability of the Cu(II) complex but indicated that coordination of Cu(I) conferred enhanced stability to the ligand, suggesting electron accepting character. The coordination complex [Mo(CO)$_3$(**53**)] has since been prepared and compared with the closely related complex [Mo(CO)$_3${MeSn(2-py)$_3$}]. IR spectroscopy suggests that there is little influence of the bridgehead group on donor-acceptor characteristics of the ligand.[74]

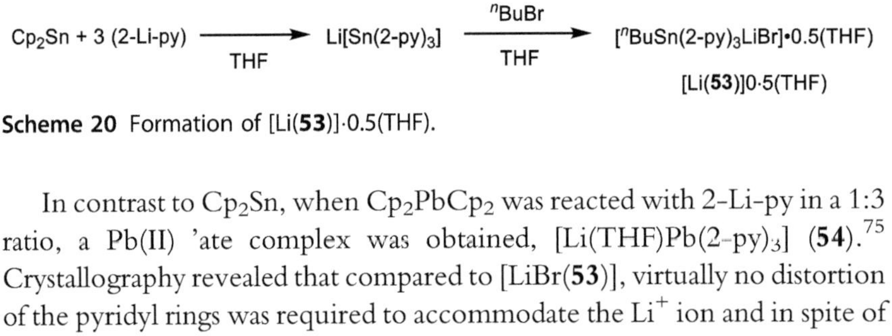

Cp$_2$Sn + 3 (2-Li-py) $\xrightarrow{\text{THF}}$ Li[Sn(2-py)$_3$] $\xrightarrow[\text{THF}]{^n\text{BuBr}}$ [^{n}BuSn(2-py)$_3$LiBr]·0.5(THF) [Li(**53**)]0·5(THF)

Scheme 20 Formation of [Li(**53**)]·0.5(THF).

In contrast to Cp$_2$Sn, when Cp$_2$PbCp$_2$ was reacted with 2-Li-py in a 1:3 ratio, a Pb(II) 'ate complex was obtained, [Li(THF)Pb(2-py)$_3$] (**54**).[75] Crystallography revealed that compared to [LiBr(**53**)], virtually no distortion of the pyridyl rings was required to accommodate the Li$^+$ ion and in spite of the significantly more acute C–M–C angles in the anion **54**, the larger size of the Pb bridgehead results in nearly tetrahedral coordination of the Li centre (Fig. 33).

In 2009, Zeckert and co-workers reported the synthesis and characterization of lanthanoid complexes of a lithium tris(2-pyridyl)stannate [LiSn(5-Me-2-py)$_3$] [Li(**55**)].[76] The 2:1 reaction of the Sn(II) reagent with [Yb(Cp*)$_2$(OEt$_2$)] resulted in sandwich complex [Yb(**55**)$_2$] (Scheme 21 and Fig. 34). This complex exhibits twisting of the 2-pyridyl donor units, a heavily distorted Yb coordination environment and trigonal pyramidal Sn-centres (C–Sn–C *ca.* 96°). On the other hand, the analogous reaction with Eu(II) proceeded much more slowly, and although the same type of sandwich complex was eventually formed, an intermediate trimetallic

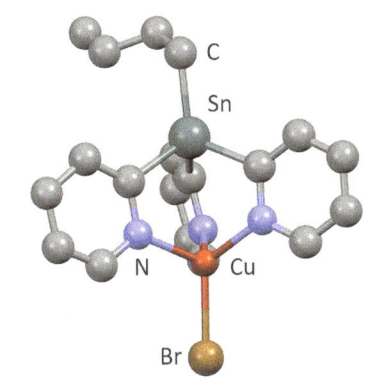

Recent advances in the synthesis and application | 225

Fig. 32 Molecular structure of [CuBr(**53**)].

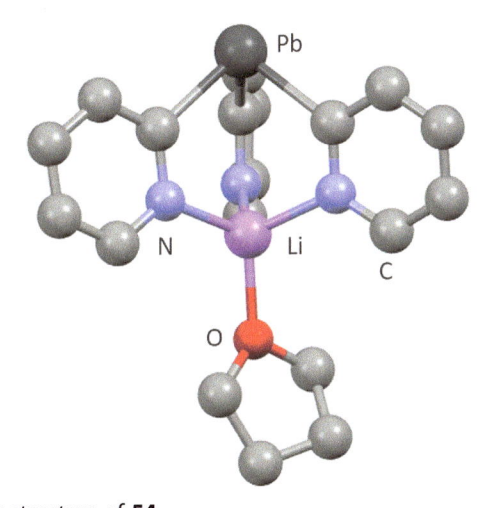

Fig. 33 Molecular structure of **54**.

Scheme 21 Synthesis of lanthanoid stannate complexes [Ln(**55**)$_2$].

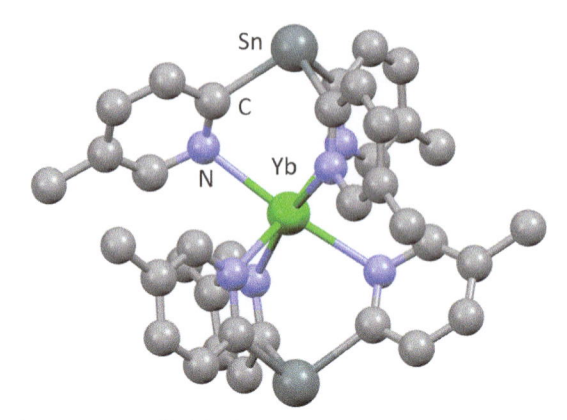

Fig. 34 Molecular structure of [Yb(**55**)$_2$].

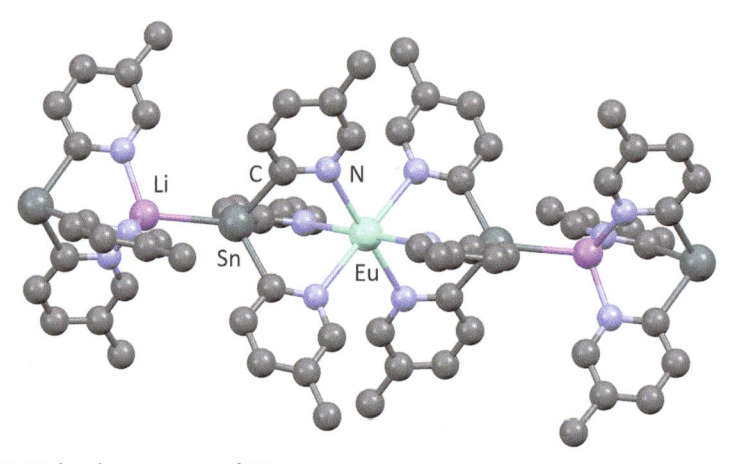

Fig. 35 Molecular structure of **56**.

species was also isolated [Eu{Sn(5-Me-2-py)$_3$-κ^1Sn:$\kappa\ ^3$N}$_2$(Li{Sn(5-Me-2-py)$_3$-κ^3N})$_2$] (**56**) (Fig. 35). Interesting features in **56** include a short Sn—Li bond and the relative 'opening out' of the inner tripodal ligands compared to the outer ones as a result of Li coordination and to accommodate the larger lanthanoid ion.

Subsequent work has explored the ability of [Li(**55**)] to complex trivalent Ln cations (Ln=La(III), Yb(III)) when treated with [Ln(Cp)$_3$] (Scheme 22).[77] The product Sn—Ln bonded complexes display good stability in solution making them amenable to study using multinuclear (^{1}H, ^{7}Li, ^{119}Sn, ^{139}La) NMR spectroscopy. The solid-state structures of both complexes were obtained, showing that the [Li(**55**)] unit is preserved,

Scheme 22 Synthesis of Sn–Ln complexes (Ln=La(III), Yb(III)).

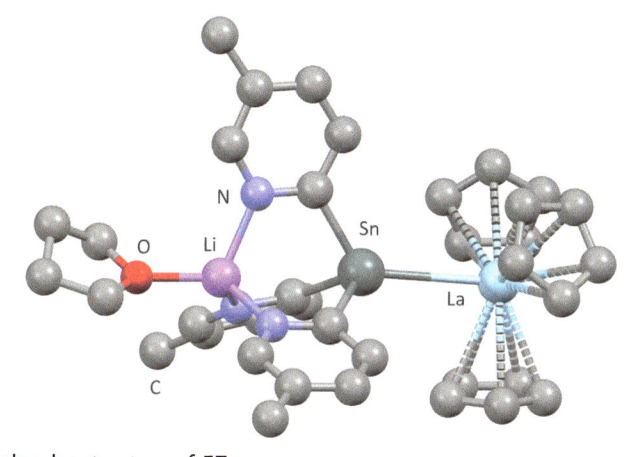

Fig. 36 Molecular structure of **57**.

e.g., in [Cp$_3$LaSn(5-Me-2-py)$_3$Li(THF)] (**57**) (Fig. 36). The Ln–Sn distances are below the sum of the van der Waals radii and DFT calculations indicate that the bond is largely ionic. The coordinative behavior of anion **55** towards trivalent La in **57** is in contrast to divalent species, *e.g.*, Yb(II) in [YbSn(3-Me-2-py)$_3$(Cp*)], which shows a preference for Yb^{II}–N bonding.[78]

Ga—Sn and In—Sn bonded adducts are obtained from reactions of [Li(**55**)] with triorgano Group 13 compounds (Scheme 23).[65] The products of reactions with GaEt$_3$ are solvent dependent, with a polymer involving a (C—H)•••Li interaction (in non-polar solvent) or a simple adduct (in polar solvent) being obtained (Fig. 37). NMR spectroscopy supported the retention of the Sn—M bonds in solution for both Ga and In, but the presence of long Sn—M bonds in the solid-state structures implies that this bonding is weak. Despite the observation of 2-pyridyl ligand transfer in the analogous reactions of organoaluminium reagents, reaction of the sandwich

Scheme 23 Solvent dependant synthesis of Sn–Ga/In bonded adducts. Dashed lines indicate (C—H)•••Li interactions (M = Ga, In).

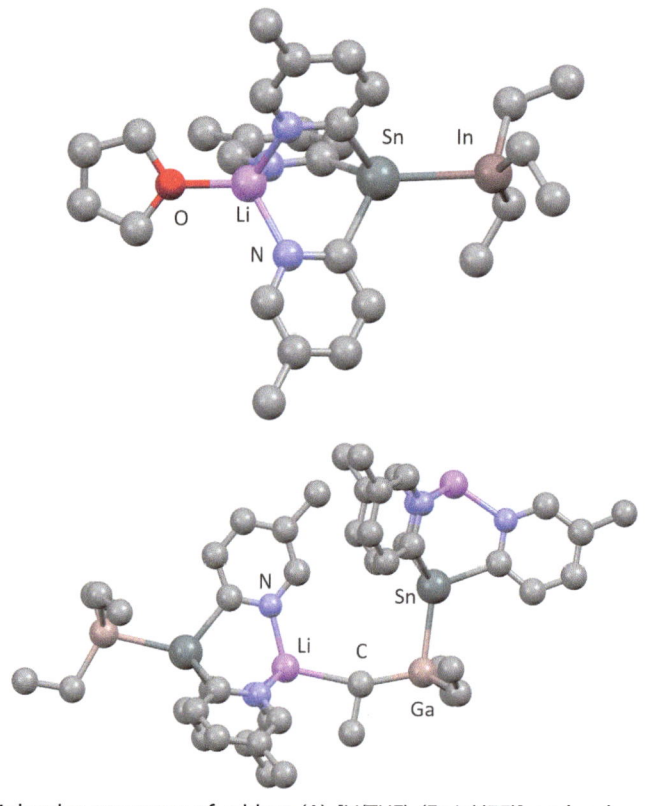

Fig. 37 Molecular structures of adduct (A) [Li(THF) (Et$_3$In)(**55**)] and polymeric (B) [Li (Et$_3$Ga)(**55**)]$_\infty$. (C—H)•••Li interactions (only the carbon centres in this interaction are shown).

compound [Yb{Sn(3-Me-2-py)$_3$}$_2$] with AlMe$_3$ gave the Al—Sn bonded adduct [Yb{Sn(3-Me-2-py)$_3$AlMe$_3$}$_2$] (**58**).

In contrast to the stannate compounds, the tris(2-pyridyl)plumbate ligand **34**$^-$ displays unprecedented redox behavior.[79] Reaction of [Li(**34**)] with [Ln(Cp)$_3$(THF)] (Ln=Sm, Eu) in a 1:1 ratio provided adducts containing Ln—Pb bonds similar to related stannate complexes, [LiCp$_3$Ln(**34**)], but with no external coordination of Li (*viz* Fig. 36).[77] The Eu(III) complex proved unstable towards prolonged storage in solution and decomposed to afford a Janus head ligand [**34**]$_2$ accompanied by reduction of Eu(III) to Eu(II) (Scheme 24). This was postulated to be an intramolecular process involving sterically-induced reduction, which is supported by spin-trapping EPR measurements. A similar result occurs when [Cp*$_2$Eu(OEt$_2$)] is reacted with [Li(**34**)], however, the reaction is slow enough in benzene to allow isolation of a remarkable pentametallic intermediate [(THF)Eu{μ-Pb(6-^{t}BuO-2-py)$_3$Eu(Cp*)$_2$}$_2$] (**59**) (Fig. 38). *Re*-dissolving the solid complex in THF furnished sandwich complex [Eu(**34**)$_2$].

[LiCp$_3$EuPb(6-O^{t}Bu-2-py)$_3$] $\xrightarrow[\text{- LiCp}]{\text{THF}}$ 1/2 [(6-O^{t}Bu-2-py)$_3$Pb)]$_2$ + 1/*n* [Cp$_2$Eu(THF)]$_n$
[**34**]$_2$

Scheme 24 Redox reaction generating Janus head-type diplumbane [**34**]$_2$.

Interestingly, it has also been found that despite being relatively electron-rich, the neutral Sn(IV) pyridyl species Sn(5-Me-2-py)$_4$ and Sn(6-O^{t}Bu-2-py)$_4$ can form hypercoordinate Sn(IV) complexes upon reaction with two further equivalents of the appropriate lithio-pyridines.[80] The octahedral Sn centre then effectively acts as two sets of tris(2-pyridyl)ligands towards the Li$^+$ counterion in a manner similar to that found in mono anionic stannate complexes (see **60** in Fig. 39).

It has been found that the tris(2-pyridyl) stannates and plumbates can act as Group 14 centred nucleophiles.[81] Reaction of [LiSn(6-^{t}BuO-2-py)$_3$] [Li(**61**)] with E{N(SiMe$_3$)$_2$}$_2$ (E=Sn, Pb) led to the formation of a stannate or plumbate [E(**61**)$_3$]$^-$ featuring Sn—Sn or Pb—Sn bonds (Scheme 25). The structure of [LiSn(**61**)$_3$] is shown in Fig. 40. Multinuclear (^{119}Sn and ^{207}Pb) NMR spectroscopic studies were used to monitor the slow decomposition of [LiPb(**61**)$_3$] in solution (which gives metallic Pb) and Sn—Sn bonded, Janus head complex **61**$_2$. The Sn counterpart did not exhibit this behavior.

To date, there is only one recent report on the synthesis and application of a heavy Group 14 tris(3-pyridyl) ligand, PhSn(3-py)$_3$ (**62**).[82]

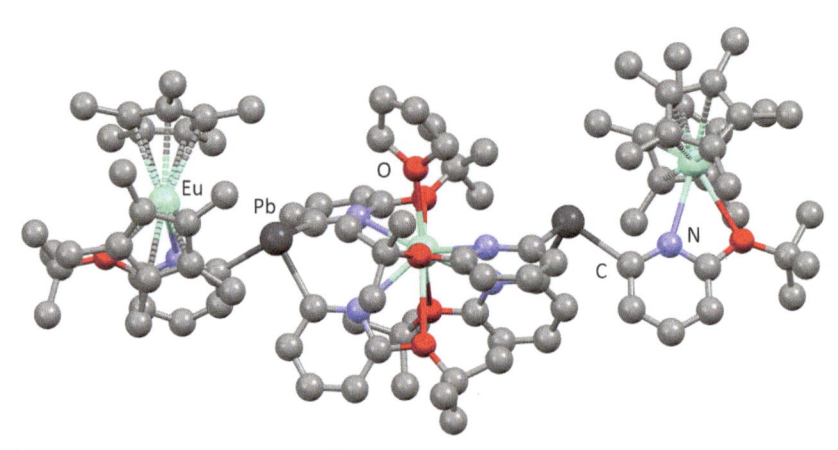

Fig. 38 Molecular structure of Eu(II) complex **59**.

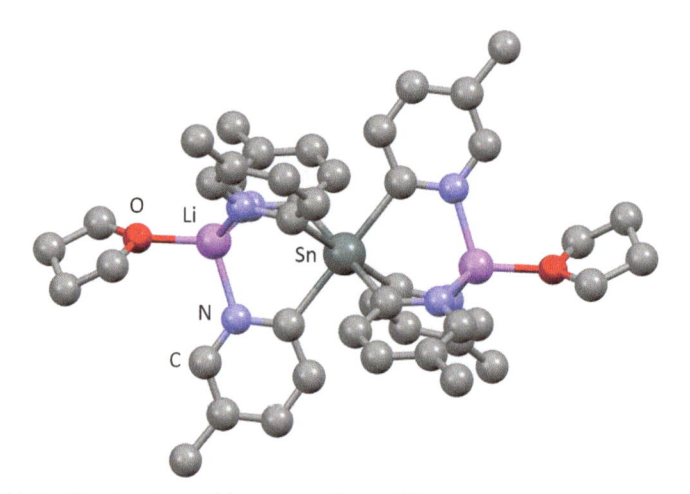

Fig. 39 Molecular structure of hypercoordinate **60**.

Scheme 25 Synthesis of [LiE{**61**}₃]. R = ^{t}BuO, E = Sn, Pb.

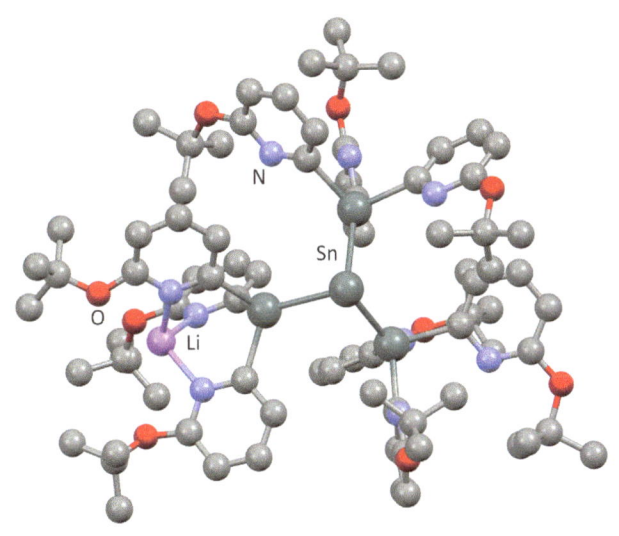

Fig. 40 Molecular structure of [LiSn(**61**)₃].

In the synthesis of this ligand, effective metal halogen exchange of 3-bromopyridine could only be achieved using a turbo-Grignard reagent, with the intermediate 3-magnesiated pyridine being reacted *in situ* with PhSnCl₃ to yield the desired tris(3-pyridyl)stannane (Scheme 26). Advantages of using stannane **62** over the closely related silane **48** (discussed previously) include its air- and moisture-stability and its ease of purification (which does not require column chromatography).

Scheme 26 Synthesis of **62**.

The 1:1 combination of **62** with CuI gave a 1D coordination polymer [(CuI)₂(**62**)₂] containing CuI dimer units (Fig. 41A), which contrasts with the more complex topology obtained using **48**.[8] A radically different outcome was seen when CuI was replaced by [Cu(MeCN)₄]PF₆, whose anion is only weakly coordinating. In this case 1:1 reaction with the ligand produced an anion-encapsulating cage [PhSn(3-py)₃Cu(MeCN)]₄(PF₆)₄ [**63**]₄(PF₆)₄,

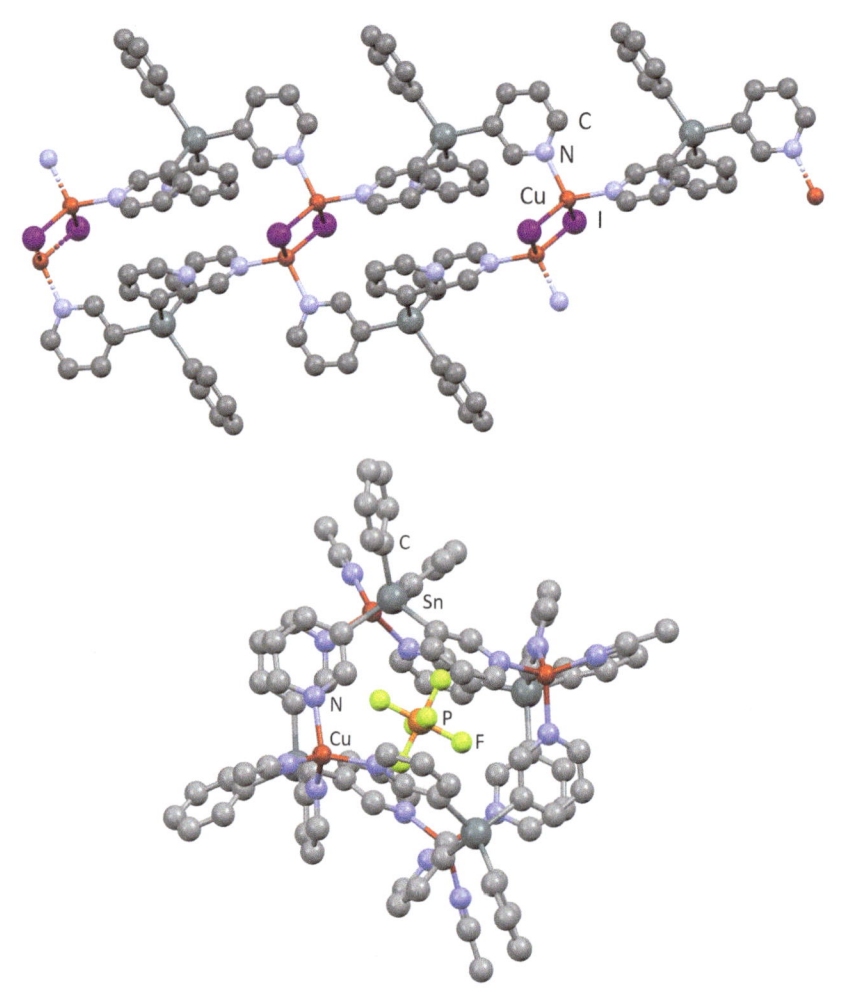

Fig. 41 (A) Section of 1D polymer [(CuI)$_2$(**62**)$_2$], showing that only $\frac{2}{3}$ of the available N-donor atoms of the stannane unit are utilized in coordination, and (B) discrete cage structure [(**63**)$_4$($\square$PF$_6$)]$^{3+}$ with encapsulated PF$_6^-$.

whose structure consists of a cage, [(**63**)$_4$($\square$PF$_6$)]$^{3+}$, with the other three PF$_6^-$ counterions contained in the lattice (Fig. 41B). In depth multinuclear NMR spectroscopic studies in d_3-MeCN suggest dissociation of the intact cage at higher temperature and upon dilution. Furthermore, at room-temperature DOSY analysis indicates different diffusion coefficients for the PF$_6^-$ ions, congruent with the presence of free and encapsulated anions in solution under these conditions. In contrast to the Cu salt, when AgPF$_6$ was used as a metal source, a discrete cage was no longer formed; instead a 2D coordination network {[PhSn(3-py)$_3$Ag(NCMe)][PF$_6$]·Et$_2$O}$_n$ (**64**) was isolated, its structure consisting of layers of connected Ag$_3$Sn$_3$ ring units.

The reaction of **62** with Co(II) was investigated. Initial combination of **62** with methanolic CoCl$_2$ solution eventually yielded pink crystals which converted to a blue solid during isolation, attributed to the loss of a MeOH ligand and consequent change to the coordination geometry at Co. X-ray crystallography on the blue crystals reveals a 1D coordination network [{Co(PhSn(3-py)$_3$)Cl$_2$(MeOH)}$_2$]$_\infty$ [(**65**$_2$)]$_\infty$ in which **62** is tridentate and the Co centres are octahedrally coordinated (Fig. 42A). Reaction of

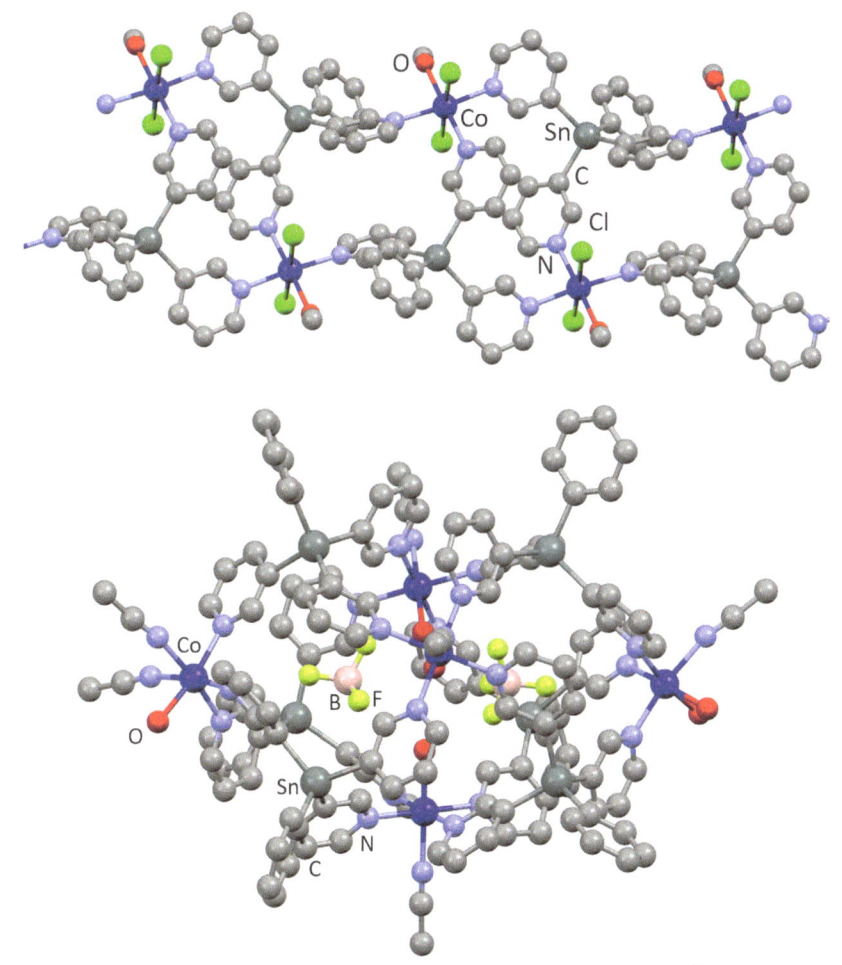

Fig. 42 (A) 1D coordination polymer [**65**$_2$]$_\infty$ and (B) cationic cage **66**$^{8+}$ showing two encapsulated BF$_4^-$ ions.

PhSn(3-py)$_3$ with Co(BF$_4$)$_2$·6H$_2$O, however, once again demonstrates the powerful influence of the *non-coordinating* anion. A remarkable Co$_5$Sn$_6$ based caged structure was obtained, the cation of which is [Co$_5${PhSn(3-py)$_3$}$_6$ (MeCN)$_6$(H$_2$O)$_6$(☐BF$_4$)$_2$]$^{8+}$ (**66**$^{8+}$). The cage framework is best described by the location of the metal atoms, with octahedrally coordinated Co situated at the vertices of a trigonal bipyramid and Sn taking the corners of a trigonal prism (Fig. 42B). Internally-coordinated H$_2$O ligands are F•••H H-bonded to the two encapsulated BF$_4^-$ ions. Although NMR spectroscopy was hampered by paramagnetism, data are consistent with retention of the cage structure in solution.

4. Group 15 tris(pyridyl) ligands

4.1 Phosphorus

Tris(pyridyl)phosphines (and related species) are a major class of tripodal ligands that have found applications in many areas of chemistry and are the most intensively studied members of Group 15 tris(pyridyl) ligands.[83–115] However, ligands featuring (semi-)metallic bridgeheads which are the focus of this review are much rarer.

4.2 Arsenic, antimony and bismuth

Two tripodal arsine ligands have been investigated in the construction of photoluminescent materials. Lithium halogen exchange on 2-bromopyridine, followed by reaction with arsenic trichloride forms As(2-py)$_3$ (**67**) which can subsequently be oxidized by ^{t}BuOOH to give O=As(2-py)$_3$ (**68**) (Scheme 27).[86] This allowed for crystallization of the arsine and a copper complex of the arsine oxide (Fig. 43). The latter has been studied, alongside tris(2-pyridyl)phosphine chalcogenides copper complexes, for its luminescent properties. In the solid state, [CuI(**68**)] shows approximate tetrahedral geometry at Cu with some deviation of the halide from the bridgehead•••copper vector due to packing effects (and compared to the phosphorus analogue, a slightly larger bridgehead bite angle). Overall, the suite of complexes [though dominated by phosphorus-based tris(2-pyridyl)ligands], demonstrate

Scheme 27 Synthesis of **67** and **68**.

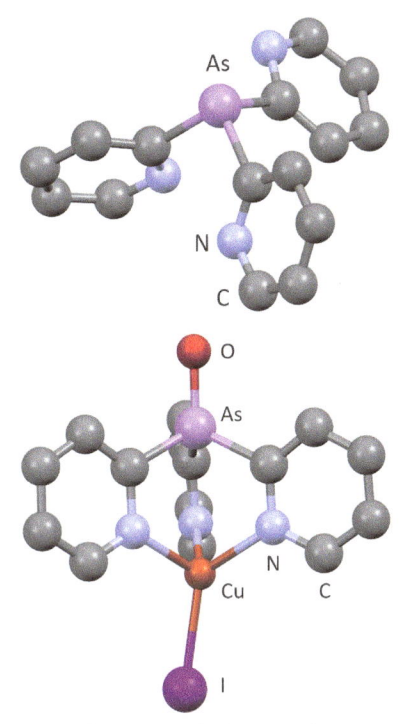

Fig. 43 Molecular structures of (A) **67** and (B) [CuI(**68**)].

desirable properties for OLEDs, which is ascribed to a combination of phosphorescence and TADF (Thermally Activated Delayed Fluorescence) emission paths.

The effects of bridgehead atom modification have scarcely been investigated in tris(2-pyridyl)pnictogenides. However, significant advances have been made by Wright and co-workers, who have structurally characterized (6-Me-2-py)$_3$E (E = As **69a**, Sb **69b**, Bi **69c**) and their complexes.[116] In the solid state, these ligands exhibit typical propeller-blade conformation (Fig. 44A, shown for bismuth **69c**), with the Me-groups orientated towards the main group bridgehead.[87] The ligands **69** adopt a typical tridentate coordination mode to Cu(I) in the CuPF$_6$ complexes [(**69**)Cu(MeCN)PF$_6$], with some degree of interaction between the PF$_6^-$ anions and the bridgehead atoms being found for increasingly Lewis acidic Sb and Bi. However, more radical structural changes were found for the CuCl complexes of the Sb and Bi ligands, Although the Sb complex is monomeric in the solid state (with tridentate coordination of **69** to Cu(I)) the Bi complex forms a dimeric

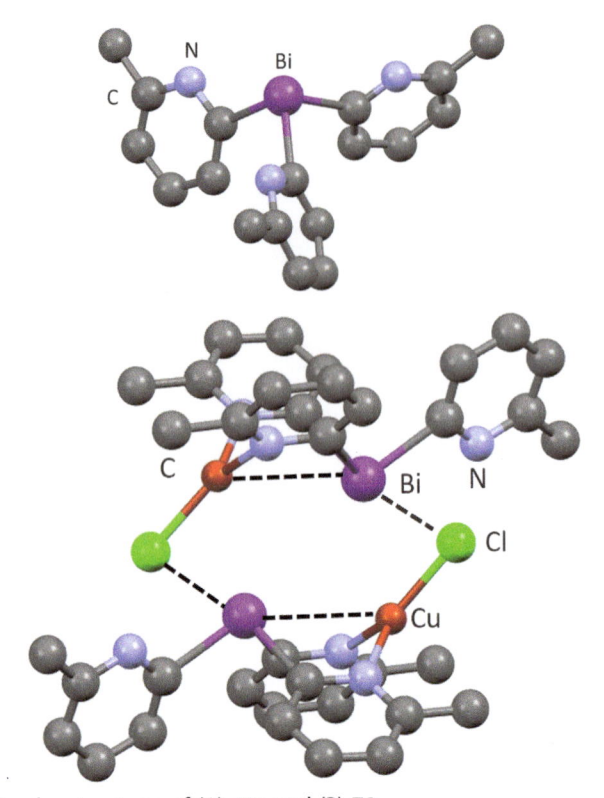

Fig. 44 Molecular structures of (A) **69c** and (B) **70₂**.

arrangement [CuCl(6-Me-2-py)₂Bi(6-Me-py)]₂ (**70₂**) in which one of the expected Cu—N bonds of each monomer unit is sacrificed in favor of an intermolecular Bi•••Cl interaction (Fig. 44B). DFT calculations indicated that this interaction involves Cl lone pair to Bi—C σ* donation. In solution, a monomer-dimer equilibrium was identified, and VT-NMR data indicate a net enthalpic gain for dimerization. Electrochemical measurements supported an increase in the donor ability of the ligands **69** as Group 15 is descended, in line with the increase in the ionic character of the bridgehead E—C bonds. However, UV-visible spectroscopy suggests that there is little change in π-acceptor character. Further indication of the electronic influence of the bridgehead atom was found by using these ligands in the Cu-mediated aziridination of aryl alkenes (Scheme 28), wherein Hammett analysis of the polar *vs* radical contribution to aziridination indicated a progressive shift towards the latter going down the group, again consistent with

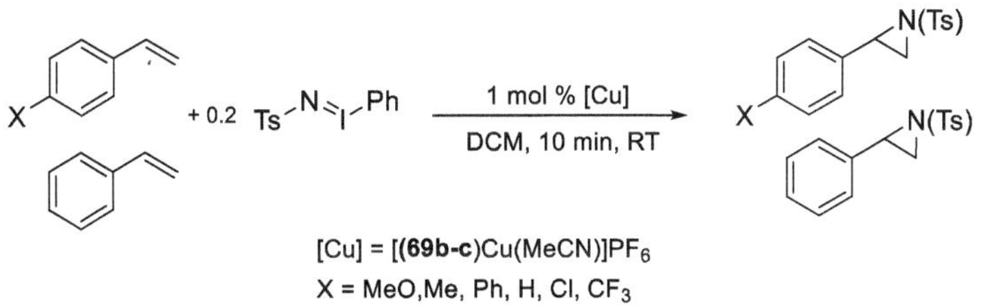

Scheme 28 Cu-mediated aziridination of aryl alkenes.

the increase in donor character of the ligands **69** (which results in more electron rich Cu(I) centres, favoring a radical pathway).

The chemistry of tris(2-pyridyl)bismuthines has recently attracted further attention and a more comprehensive study of the impact of the electronic and steric effects of pyridyl substitution has been undertaken on the series of ligands Bi(py')$_3$ (py' = substituted pyridyl, see Scheme 29) (some being isolated as their lithium halide complexes).[117] In the case of [Bi(py')$_3$LiX] (X = Cl, Br), clear evidence of anion bridging via X•••Bi interactions is seen from polymeric association in the solid-state (see, for example, [LiClBi(2-py)$_3$] (**71**) in Fig. 45). Spectroscopic measurements indicated that ligands containing 6-R-2-py groups do not coordinate LiCl, whereas ligands containing substitution at other positions (or with no substituents) do. Studies on the coordination of silver salts indicated that 6-Me- and 6-Ph-substituted Bi(2-py')$_3$ were competent ligands for Ag(I), whereas the presence of more electron withdrawing groups within the

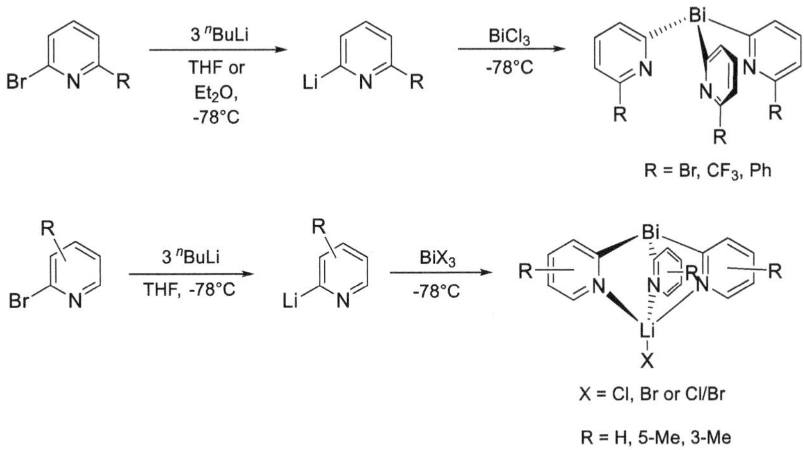

Scheme 29 Synthesis of tripodal pyridyl(bismuth) species, isolated as either free ligands or lithium halide complexes.

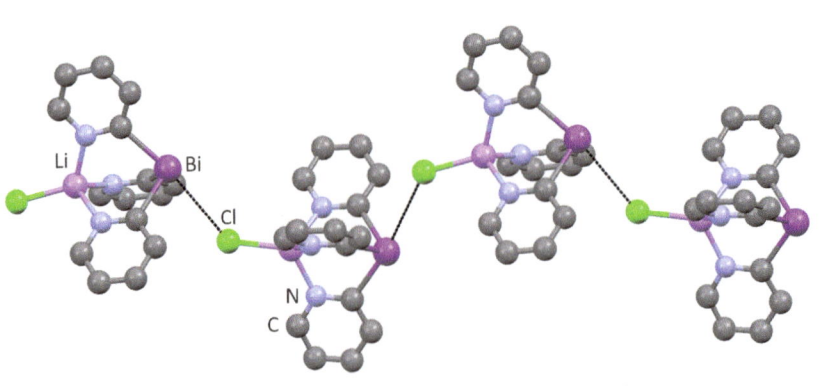

Fig. 45 Extended structure of **71**, showing intermolecular Bi•••Cl interactions.

2-pyridyl rings (Br or CF$_3$) disfavoured coordination as a result of the poor donor properties. Simple complexes containing tripodal coordination of Ag(I) are formed, rather than aggregates (as occurs with P(2-py)$_3$[93]). Some differences in stability were noted for different coinage metal complexes, with Cu(I) complexes being less thermally stable than those of Ag(I). For example, the prolonged reaction of Bi(6-Me-2-py)$_3$ with [Cu(MeCN)$_4$]BF$_4$ gave a mixture of [Cu(Me-bpy)$_2$]BF$_4$ and [{Bi(6-Me-2-py)$_3$}(MeCN)Cu(Me-bpy)](BF$_4$)$_2$ **[72]**(BF$_4$)$_2$ (Me-bpy = 6,6′-dimethyl-2,2′-bipyridine) (Fig. 46), as a result of reductive coupling of the 6-Me-2-py substituents and the formation of Bi(0) and/or Cu(0). This pyridyl coupling reaction proved highly dependent on the anion present. Using CuCl instead of CuBF$_4$ resulted in much faster coupling to the bipyridine products. This observation suggests an anion-triggered mechanism involving hypervalent Bi intermediates and is supported by the observation that reactions of CuCl$_2$ with Bi(6-R-2-py)$_3$ (R = Me, CF$_3$) in MeCN gives the Cu(I) complexes containing hypervalent [Bi(6-Me-2-py)$_2$Cl$_2$]$^-$ anions [for example, [Bi(6-R-2-py)$_2$Cl$_2$Cu(MeCN)] **(73)** (Fig. 47)], along with the expected bipyridine compounds (Scheme 30). The ability of pyridyl bismuthines to act as soft pyridyl transfer agents (as a result of their more polar bridgehead Bi—C bonds) was confirmed by the isolation of a phosphine-supported organogold complex [(6-Me-2-py)AuPPh$_3$] from the reaction of Bi(6-Me-2-py)$_3$ with ClAuPPh$_3$ (which ostensibly associates into dimers *via* aurophilic interactions).

Turning to other pyridyl isomers, a low-yielding synthesis of tris(3-pyridyl)bismuth was reported in 2012 by the reaction of an *in situ* prepared organozinc reagent with BiCl$_3$.[118] By contrast, 2- and 4-pyridyl isomers

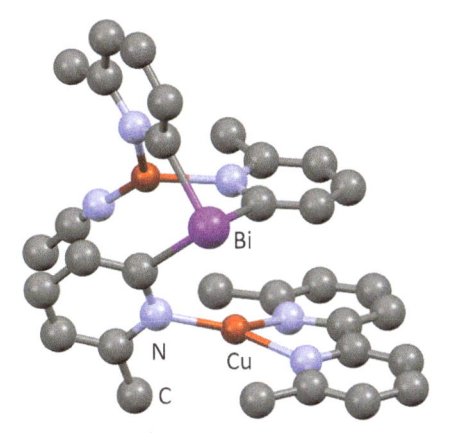

Fig. 46 Molecular structure of [**72**]$^{2+}$.

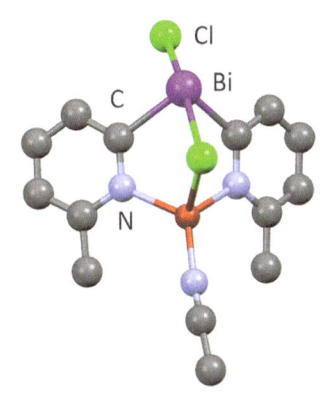

Fig. 47 Molecular structure of **73**.

Scheme 30 Formation of hypervalent bismuthines and substituted bipyridines.

R = Me X = Cl
R = CF$_3$ X = Cl
R = Me X = Br

could not be obtained effectively by this route. Bi(3-py)$_3$ was subsequently applied to Pd-catalyzed cross-coupling with a substituted pyrazine.[119] There have been no other recent structural reports on heavier tris(3-pyridyl)compounds and the same is true for the 4-py isomers, save for a report in 2006

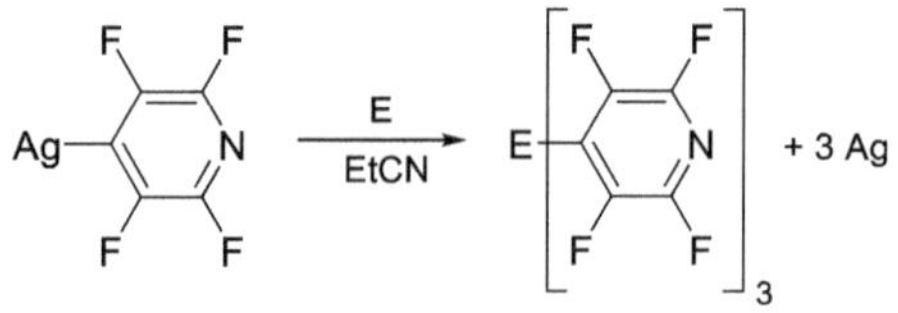

Scheme 31 Synthesis of fluorinated tris(4-pyridyl)pnictogens (As, Sb, Bi).

on the synthesis of a range of perfluorinated derivatives of the type E(C$_5$F$_4$N)$_3$ (E=As, Sb, Bi) which were prepared *via* an unusual route involving the elemental pnictogens (Scheme 31).[120] Only the arsenic derivative could be characterized in the solid state, where a π•••π stacking motif was observed between the heterocyclic rings. While Sb and Bi homologues could not be crystallized, NMR spectroscopy and mass spectrometry confirmed that the corresponding (moisture sensitive) antimony and bismuth compounds had been prepared.

5. Outlook

Great strides have been made in preparing new tripodal ligands bearing pyridyl donors over the past decade or so. Focus on the design of their steric and electronic properties by bridgehead modification and heteroaryl ring substitution of 2-pyridyl species has resulted in a foray to more remote areas of the p-block elements. Aluminates have taken centre stage in introducing a metallic bridgehead to give new analogues of ubiquitous tris(pyrazolyl)borate ligands. These have found applications in catalysis, while Al—C bond lability and solution aggregation properties have led to an unprecedented way of determining enantiomeric excess in scalemic mixtures of chiral alcohols. Meanwhile, access to gallates and indates has gone hand in hand with pioneering work on the chemistry of divalent heavy Group 14 species. Tetravalent silanes and stannanes have proved to be useful ways of introducing less investigated 3- and 4-pyridyl-based tripodal ligands and the ability of these to support both polymeric and discrete supramolecular cages has led to newfound applications. Nonetheless, the importance of systematically investigating the influence of bridgehead atom character is so far limited to 2-pyridyl species and is yet to be extended to 3- and 4-pyridyl isomers. Strong indication of the metal-bridging behavior of these isomers, in which the N-donor is more remote from the bridgehead atom, is likely to result in extended structures, the topology of which could be modulated by variations in the inter-ligand angle afforded by a change in identity of the bridgehead atom. It is therefore expected that interest in application of tripodal ligands of this type will continue to grow, especially in the arenas of supramolecular and materials chemistry.

Acknowledgments
We thank the Leverhulme Trust (grant RPG-2017-146, A. J. Peel), the Royal Commission for the Exhibition of 1851 (A. J. Plajer), Walters-Kundert Scholarship, Selwyn College, Cambridge (J. E. Waters) and the Spanish MINECO-AEI and the EU (ESF) for a Ramon y Cajal contract (R.G.-R., RYC-2015–19035).

References
1. Gispert JR. *Coordination Chemistry*. Wiley-VCH; 2008.
2. Gade LH, Hofmann P. *Molecular Catalysts: Structure and Functional Design*. Weinheim, Germany: Wiley VCH Verlag; 2014.
3. de la Lande A, Salahub D, Moliner V, Gérard H, Piquemal JP, Parisel O. *Inorg Chem*. 2009;48:7003–7005.
4. Gonzalez MA, Yim MA, Cheng S, Moyes A, Hobbs AJ, Mascharak PK. *Inorg Chem*. 2012;51:601–608.
5. Kodera M, Katayama K, Tachi Y, et al. *J Am Chem Soc*. 1999;121:11006–11007.
6. Kitajima N, Tolman WB. *Prog Inorg Chem*. 1995;43:419–531.
7. Hamacek J, Vuillamy A. *Eur J Inorg Chem*. 2018;2018:1155–1166.
8. Deshmukh MS, Yadav A, Pant R, Boomishankar R. *Inorg Chem*. 2015;54:1337–1345.
9. Reglinski J, Spicer MD. *Coord Chem Rev*. 2015;297–298:181–207.
10. Jeong SY, Lalancette RA, Lin H, et al. *Inorg Chem*. 2016;55:3605–3615.
11. Canty AJ, Minchin NJ, Healy PC, White AH. *Dalton Trans*. 1982;1795–1802.
12. Canty AJ, Minchin NJ, Engelhardt LM, Skelton BW, White AH. *Dalton Trans*. 1986;645–650.
13. Richard Keene F, Stephenson PJ, Snow MR, Tiekink ERT. *Inorg Chem*. 1988;27:2040–2045.
14. Keene FR, Tiekink ERT. *Acta Crystallogr*. 1990;C46:1562–1563.
15. Canty AJ, Minchin NJ, Skelton BW, White AH. *Aust J Chem*. 1992;45:423–427.
16. Lee FW, Chan MCW, Cheung KK, Che CM. *J Organomet Chem*. 1998;563:191–200.
17. Andersen PA, Astley T, Hitchman MA, et al. *Dalton Trans*. 2000;3505–3512.
18. Canty AJ, Skelton BW, Traill PR, White AH. *Acta Crystallogr*. 2004;C60:m305–m307.
19. Canty AJ, Chaichit N, Gatehouse BM, George EE. *Inorg Chem*. 1981;20:4293–4300.
20. Keene FR, Wilson TA, Szalda DJ. *Inorg Chem*. 1987;26:2211–2216.
21. Jonas RT, Stack TDP. *Inorg Chem*. 1998;37:6615–6629.
22. Brunner H, Maier RJ, Zabel M. *Synthesis*. 2001;2484–2494.
23. Arnold PJ, Davies SC, Dilworth JR, et al. *Dalton Trans*. 2001;567–573.
24. McWhinnie WR, Kulasingam GC, Draper JC. *J Chem Soc A*. 1966;1199–1203.
25. Kucharski ES, McWhinnie WR, White AH. *Aust J Chem*. 1978;31:53–56.
26. Anderson PA, Keene FR. *Z Krist*. 1993;206:275–278.
27. Mosny KK, de Gala SR, Crabtree RH. *Transition Met Chem*. 1995;20:595–599.
28. Yang W, Schmider H, Wu Q, Zhang YS, Wang S. *Inorg Chem*. 2000;39:2397–2404.
29. Xie Y, Liu X, Ni J, Jiang H, Liu Q. *Appl Organomet Chem*. 2003;17:800.
30. Xie YS, Liu XT, Yang JX, et al. *Collect Czech Chem Commun*. 2003;68:2139–2149.
31. Xie Y, Ni J, Jiang H, Liu Q. *J Mol Struct*. 2004;687:73–78.
32. Astley T, Headlam H, Hitchman MA, et al. *Dalton Trans*. 1995;3809–3818.
33. Ke-Wu Y, Yuan-Qi Y, Zhong-Xian H, Yun-Hua W. *Polyhedron*. 1996;15:79–81.

34. Astley T, Hitchman MA, Keene FR, Tiekink ERT. *Dalton Trans*. 1996;1845–1851.
35. Wajda-Hermanowicz K, Pruchnik F, Zuber M. *J Organomet Chem*. 1996;508:75–81.
36. Adam KR, Anderson PA, Astley T, et al. *Dalton Trans*. 1997;519–530.
37. Gregorzik R, Wirbser J, Vahrenkamp H. *Chem Ber*. 1992;125:1575–1581.
38. Xie Y, Lee C-L, Yang Y, Rettig SJ, James BR. *Can J Chem*. 1992;70:751–762.
39. Espinet P, Hernando R, Iturbe G, Villafañe F, Orpen AG, Pascual I. *Eur J Inorg Chem*. 2000;2000:1031–1038.
40. Casares JA, Espinet P, Martín-Alvarez JM, Espino G, Pérez-Manrique M, Vattier F. *Eur J Inorg Chem*. 2001;2001:289–296.
41. Boggess RK, Zatko DA. *J Coord Chem*. 1975;4:217–224.
42. Boggess RK, Hughes JW, Chew CW, Kemper JJ. *J Inorg Nucl Chem*. 1981;43:939–945.
43. Louie BM, Rettig SJ, Storr A, Trotter J. *Can J Chem*. 1985;63:2261–2272.
44. Sambade D, Parkin G. *Polyhedron*. 2017;125:219–229.
45. Tredget CS, Lawrence SC, Ward BD, Howe RG, Cowley AR, Mountford P. *Organometallics*. 2005;24:3136–3148.
46. Steiner A, Stalke D. *Chem Commun*. 1993;1702–1704.
47. Steiner A, Stalke D. *Inorg Chem*. 1995;34:4846–4853.
48. Simmonds HR, Wright DS. *Chem Commun*. 2012;48:8617–8624.
49. Cui C, Lalancette RA, Jäkle F. *Chem Commun*. 2012;48:6930–6932.
50. Shipman PO, Cui C, Lupinska P, Lalancette RA, Sheridan JB, Jäkle F. *ACS Macro Lett*. 2013;2:1056–1060.
51. Pawar GM, Lalancette RA, Bonder EM, Sheridan JB, Jäkle F. *Macromolecules*. 2015;48:6508–6515.
52. García F, Hopkins AD, Kowenicki RA, McPartlin M, Rogers MC, Wright DS. *Organometallics*. 2004;23:3884–3890.
53. Szczepura LF, Witham LM, Takeuchi KJ. *Coord Chem Rev*. 1998;174:5–32.
54. Alvarez CS, García F, Humphrey SM, et al. *Chem Commun*. **2005**;198–200.
55. García F, Hopkins AD, Kowenicki RA, et al. *Organometallics*. 2006;25:2561–2568.
56. Bullock TH, Chan WTK, Eisler DJ, Streib M, Wright DS. *Dalton Trans*. 2009;1046–1054.
57. García-Rodríguez R, Bullock TH, McPartlin M, Wright DS. *Dalton Trans*. 2014;43:14045–14053.
58. García-Rodríguez R, Wright DS. *Dalton Trans*. 2014;43:14529–14532.
59. García-Rodríguez R, Simmonds HR, Wright DS. *Organometallics*. 2014;33:7113–7117.
60. García-Rodríguez R, Wright DS. *Chem A Eur J*. 2015;21:14949–14957.
61. García-Rodríguez R, Hanf S, Bond AD, Wright DS. *Chem Commun*. 2017;53:1225–1228.
62. García-Rodríguez R, Kopf S, Wright DS. *Dalton Trans*. 2018;47:2232–2239.
63. García-Romero Á, Plajer AJ, Álvarez-Miguel L, Bond AD, Wright DS, García-Rodríguez R. *Chem Eur J*. 2018;24:17019–17026.
64. Plajer AJ, Kopf S, Colebatch AL, Bond AD, Wright DS, García Rodríguez R. *Dalton Trans*. 2019;48:5692–5697.
65. Zeckert K. *Dalton Trans*. 2012;41:14101–14106.
66. Zeckert K, Fuhrmann D. *Inorg Chem*. 2019;58:16736–16742.
67. García F, Hopkins AD, Humphrey SM, McPartlin M, Rogers MC, Wright DS. *Dalton Trans*. 2004;40:361–362.
68. Parks JE, Wagner BE, Holm RH. *J Organomet Chem*. 1973;56:53–66.
69. Plajer AJ, Colebatch AL, Enders M, et al. *Dalton Trans*. 2018;47:7036–7043.
70. Deshmukh MS, Mane VS, Kumbhar AS, Boomishankar R. *Inorg Chem*. 2017;56:13286–13292.
71. Kumar KS, Mane VS, Yadav A, Kumbhar AS, Boomishankar R. *Chem Commun*. 2019;55:13156–13159.

72. Cotton FA, Dikarev EV, Petrukhina MA, Schmitz M, Stang PJ. *Inorg Chem.* 2002;41:2903–2908.
73. Beswick MA, Belle CJ, Davies MK, et al. *Chem Commun.* 1996;2619–2620.
74. Morales D, Pérez J, Riera L, Riera V, Miguel D. *Organometallics.* 2001;20:4517–4523.
75. Beswick MA, Davies MK, Raithby PR, Steiner A, Wright DS. *Organometallics.* 1997;16:1109–1110.
76. Reichart F, Kischel M, Zeckert K. *Chem Eur J.* 2009;15:10018–10020.
77. Zeckert K, Zahn S, Kirchner B. *Chem Commun.* 2010;46:2638–2640.
78. Zeckert K. *Organometallics.* 2013;32:1387–1393.
79. Zeckert K, Griebel J, Kirmse R, Weiß M, Denecke R. *Chem Eur J.* 2013;19:7718–7722.
80. Schrader I, Zeckert K, Zahn S. *Angew Chem Int Ed.* 2014;53:13698–13700.
81. Zeckert K. *Inorganics.* 2016;4:19.
82. Yang ES, Plajer AJ, García-Romero Á, et al. *Chem Eur J.* 2019;25:14003–14009.
83. Kharat AN, Bakhoda A, Hajiashrafi T, Abbasi A. *Phosphorus Sulfur Silicon Relat Elem.* 2010;185:2341–2347.
84. Trofimov BA, Artem'Ev AV, Malysheva SF, et al. *Tetrahedron Lett.* 2012;53:2424–2427.
85. Artem'Ev AV, Gusarova NK, Shagun VA, et al. *Polyhedron.* 2015;90:1–6.
86. Gneuß T, Leitl MJ, Finger LH, Rau N, Yersin H, Sundermeyer J. *Dalton Trans.* 2015;44:8506–8520.
87. Hanf S, García-Rodríguez R, Bond AD, Hey-Hawkins E, Wright DS. *Dalton Trans.* 2016;45:276–283.
88. Artem'ev AV, Doronina EP, Rakhmanova MI, et al. *New J Chem.* 2016;40:10028–10040.
89. Dubován L, Pöllnitz A, Silvestru C. *Eur J Inorg Chem.* 2016;2016:1521–1527.
90. Solyntjes S, Neumann B, Stammler HG, Ignat'ev N, Hoge B. *Eur J Inorg Chem.* 2016;2016:3999–4010.
91. Hanf S, García-Rodríguez R, Feldmann S, Bond AD, Hey-Hawkins E, Wright DS. *Dalton Trans.* 2017;46:814–824.
92. Artem'Ev AV, Kashevskii AV, Bogomyakov AS, et al. *Dalton Trans.* 2017;46:5965–5975.
93. Artem'Ev AV, Bagryanskaya IY, Doronina EP, et al. *Dalton Trans.* 2017;46:12425–12429.
94. Artem'ev AV, Eremina JA, Lider EV, Antonova OV, Vorontsova EV, Bagryanskaya IY. *Polyhedron.* 2017;138:218–224.
95. Trofimov BA, Gusarova NK, Artem'Ev AV, et al. *Mendeleev Commun.* 2012;22:187–188.
96. Suter R, Sinclair H, Burford N, McDonald R, Ferguson MJ, Schrader E. *Dalton Trans.* 2017;46:7681–7685.
97. Wang X, Nurttila SS, Dzik WI, Becker R, Rodgers J, Reek JNH. *Chem Eur J.* 2017;23:14769–14777.
98. Hu X, Sun T, Zheng C. *Phosphorus, Sulfur Silicon Relat Elem.* 2018;193:300–305.
99. Zheng C, Hu X, Tao Q. *Mendeleev Commun.* 2018;28:208–210.
100. Artem'ev AV, Samsonenko DG. *Inorg Chem Commun.* 2018;93:47–51.
101. Artem'ev AV, Pritchina EA, Rakhmanova MI, et al. *Dalton Trans.* 2019;48:2328–2337.
102. Li CP, Ai JY, He H, Li MZ, Du M. *Cryst Growth Des.* 2019;19:2235–2244.
103. Artem'ev AV, Shafikov MZ, Schinabeck A, et al. *Inorg Chem Front.* 2019;6:3168–3176.
104. Zaffaroni R, Dzik WI, Detz RJ, van der Vlugt JI, Reek JNH. *Eur J Inorg Chem.* 2019;2019:2498–2509.
105. Baranov AY, Berezin AS, Samsonenko DG, et al. *Dalton Trans.* 2020;49:3155–3163.
106. Huber W, Linder R, Niesel J, Schatzschneider U, Spingler B, Kunz PC. *Eur J Inorg Chem.* 2012;2012:3140–3146.

107. Liu CY, Chen XR, Chen HX, et al. *J Am Chem Soc*. 2020;142:6690–6697.
108. Hanf S, Colebatch AL, Stehr P, García-Rodríguez R, Hey-Hawkins E, Wright DS. *Dalton Trans*. 2020;49:5312–5322.
109. Jongkind LJ, Reek JNH. *Chem Asian J*. 2020;15:867–875.
110. Artem'Ev AV, Gusarova NK, Malysheva SF, et al. *Mendeleev Commun*. 2012;22:294–296.
111. Kharat AN, Jahromi BT, Bakhoda A. *Transition Met Chem*. 2012;37:63–69.
112. Bocokić V, Kalkan A, Lutz M, Spek AL, Gryko DT, Reek JNH. *Nat Commun*. 2013;4:1–9.
113. Jacobs I, Van Duin ACT, Kleij AW, et al. *Cat Sci Technol*. 2013;3:1955–1963.
114. Walden AG, Miller AJM. *Chem Sci*. 2015;6:2405–2410.
115. Artem'ev AV, Gusarova NK, Malysheva SF, et al. *Mendeleev Commun*. 2015;25:196–198.
116. Plajer AJ, Colebatch AL, Rizzuto FJ, et al. *Angew Chem Int Ed*. 2018;57:6648–6652.
117. García-Romero Á, Plajer AJ, Miguel D, et al. *Inorg Chem*. 2020;59:7103–7116.
118. Urgin K, Aubé C, Pichon C, et al. *Tetrahedron Lett*. 2012;53:1894–1896.
119. Urgin K, Aubé C, Pipelier M, et al. *Eur J Org Chem*. 2013;2013:117–124.
120. Tyrra W, Aboulkacem S, Hoge B, Wiebe W, Pantenburg I. *J Fluorine Chem*. 2006;127:213–217.

CHAPTER SIX

Reactivities of N-heterocyclic carbenes at metal centers

Thomas P. Nicholls, James R. Williams, and Charlotte E. Willans*

School of Chemistry, University of Leeds, Leeds, United Kingdom
*Corresponding author: e-mail address: c.e.willans@leeds.ac.uk

Contents

1. C—H, C—C, and C—N bond activation	245
2. Reductive elimination	271
3. C—C reductive elimination	272
4. C-X reductive elimination	281
5. C—H reductive elimination	286
6. Insertion into a metal-C$_{NHC}$ bond	289
7. Hydrolytic ring opening	294
8. Ring expansion reactions	298
9. Miscellaneous ring openings	306
10. Miscellaneous NHC reactivity	313
11. Dissociation reactions	316
12. Summary	318
Acknowledgment	319
References	319

1. C—H, C—C, and C—N bond activation

C—H activation is one of the most common types of non-innocent behavior displayed by NHC ligands. C—H activation often occurs when a C—H bond of a sterically bulky N-substituent is proximal to a reactive metal center. The first example of an NHC ligand being involved in a C—H activation was reported by Hitchcock and co-workers as early as 1977.[2] In this case an imidazolin-2-ylidene NHC ligand with p-tolyl N-substituents was coordinated to a Ru metal center through both the C2 carbon and one of the *ortho*-carbon atoms of the p-tolyl N-substituent to form a five-membered metallacycle (Scheme 1A). Dimer **1** was heated to 140 °C in the presence of [RuCl$_2$(PPh$_3$)$_3$] which delivered complex **2** as the sole product of the reaction. The elevated temperature was required to produce the

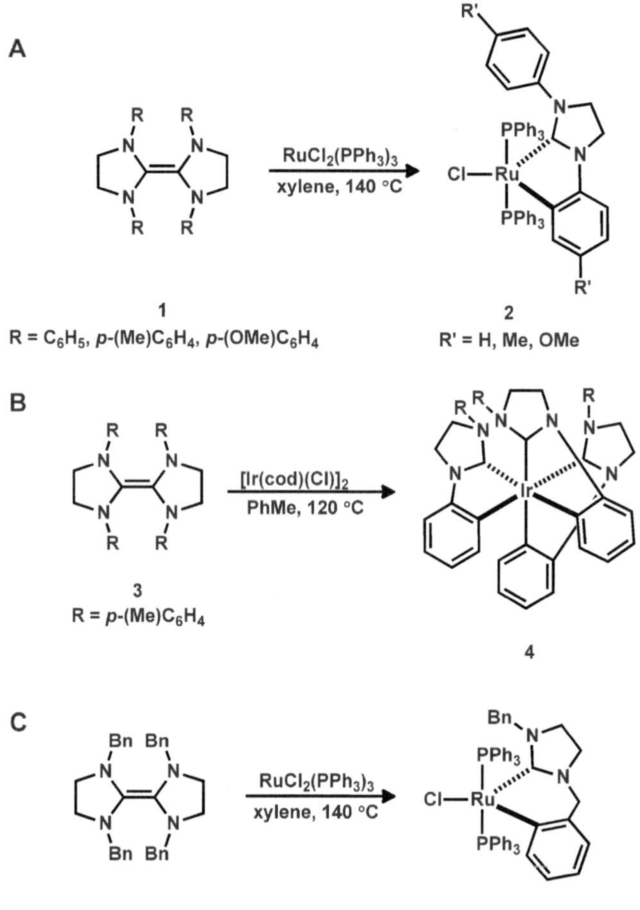

Scheme 1 Early examples of C—H activations resulting in cyclometalated Ru(II) and Ir(III) complexes.

carbene via the Wanzlick equilibrium and may not be necessary for the C—H activation step. This work was extended to include a small range of other aryl N-substituents and further continued to provide an Ir-based C—H activation product (Scheme 1B).[3] Dimer **3** was heated in the presence of [Ir(cod)Cl]$_2$ (cod = 1,5-cyclooctadiene) and octahedral complex **4** was isolated. Complex **4** features three identical cyclometalated ligands which have all undergone C—H activation and cyclometalation.[4]

A related example is represented by *ortho*-metalation of dimer **5** which incorporates benzyl N-substituents (Scheme 1C). After heating dimer **5** in the presence of [RuCl$_2$(PPh$_3$)$_3$], complex **6** was isolated which features a six-membered metallacycle rather than five-membered.[5] The reaction was performed under identical conditions to those described in Scheme 1A, however, the reaction was found to be more reliable when using [RuClH(PPh$_3$)$_3$]. It was suggested that this is likely due to the formation of H$_2$ as the by-product rather than HCl which would produce an insoluble imidazolium salt. This is an interesting example as the *N*-benzyl group has reduced steric capacity compared to *N*-aryl groups which do not have a methylene linker and are forced into closer proximity to the metal center. This suggests that factors other than sterics should also be considered to explain these *ortho*-C-H activations.

In 2000, Nolan and co-workers reported a C—H activation reaction to produce Rh complex **7**.[6] In this reaction, [Rh(coe)$_2$Cl]$_2$ (coe = cyclooctene) was treated with 1,3-bis(2,4,6-trimethylphenyl)imidazol-2-ylidine (IMes) and the cyclometalated complex **7** was formed (Scheme 2A). This represents the first example of a C(sp^3)-H activation involving an NHC ligand. Furthermore, this reaction proceeds at room temperature suggesting that the C—H activation step is more facile than originally thought and that elevated temperatures are not necessary. This is derived from the use of the carbene IMes rather than olefin dimers such that the concentration of the carbene is not limited by the Wanzlick equilibrium.

A related C—H activation was reported by Morris and co-workers which also involved the activation of the *ortho*-methyl group of IMes or SIMes (1,3-bis(2,4,6-trimethylphenyl)imidazolin-2-ylidine), however in this instance the metal used was Ru (Scheme 2B).[7] Complexes **8** and **9** were synthesized which feature C—H activation of an *ortho*-methyl *N*-substituent. The reactivity of complex **8** with excess CO was subsequently investigated. When complex **8** was exposed to CO under mild conditions (20 °C), the NHC ligand remained cyclometalated. However, using more forcing conditions (68 °C) the *ortho*-methyl group of the mesityl *N*-substituent dissociates and the NHC ligand is no longer cyclometalated indicating the reversibility of NHC-based C—H activations under certain conditions.

Another instance of C—H activation involving the *ortho*-methyl group of IMes was reported by Sola and co-workers in 2009.[8] Ir dihydride complexes **12** and **13** were treated with ethylene, propylene or diphenylacetylene and this resulted in the formation of cyclometalated complexes **16**, **17** and **18** (Scheme 2C). The hydrogenation of these olefins or internal alkynes presumably initiates a C—H activation and insertion sequence. Monohydride intermediates **14** and **15** were also isolated along with ethane, propane or diphenylethene being detected in the reaction mixture, suggesting C—H

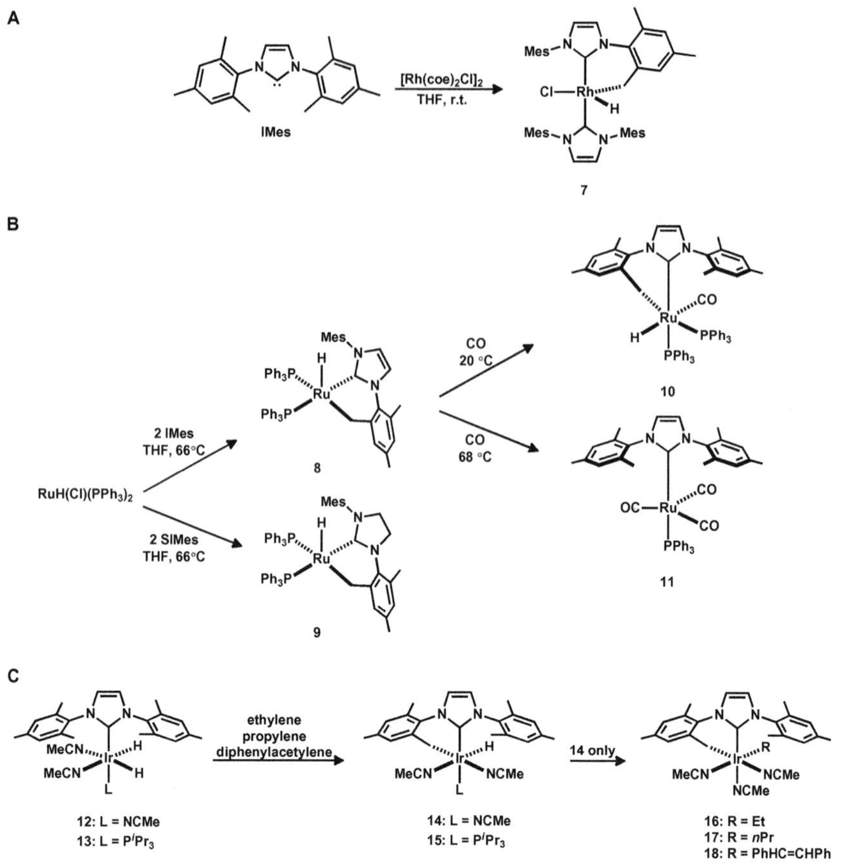

Scheme 2 C—H activations of the ortho-methyl group of the mesityl N-substituent of IMes.

activation is initiated by hydrogenation of the alkene or alkyne. All three complexes, **16**, **17** and **18**, would revert to complex **12** when treated with hydrogen gas.

While C—H activation is more common, C—C activation has also been observed within an IMes ligand.[9] When IMes was subjected to a high temperature and long reaction time in the presence of [Ru(CO)(H)$_2$(PPh$_3$)$_3$], a C—C activation product **21** was observed in which the metal center has inserted into the C—C bond of an ortho-methyl group of the mesityl N-substituent (Scheme 3). The mechanism of the formation of this product **21** was examined extensively both experimentally and by DFT.[10] Both intermediates **19$_{mono}$** and **19$_{bis}$** were isolated, and when **19$_{mono}$** was treated with trimethylsilylethene the C—H activation product **20** was produced.

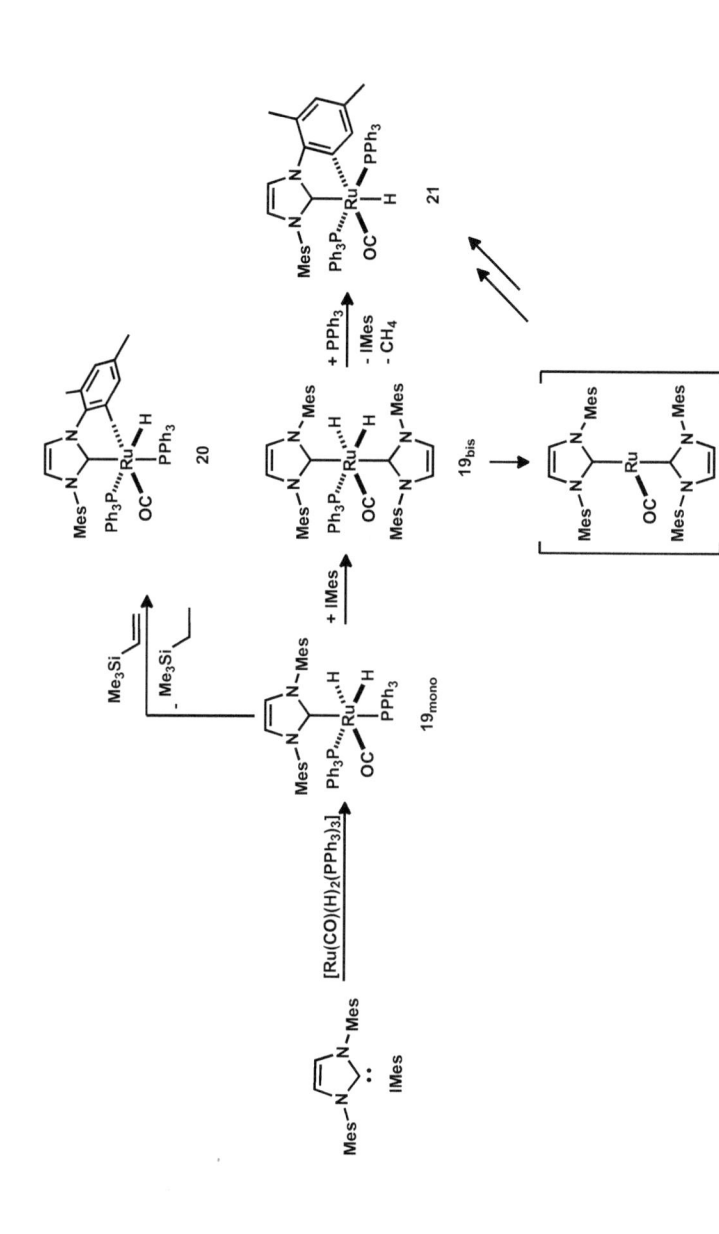

Scheme 3 C—H and C—C activation of an IMes ligand at Ru.

However, complex **20** could not be converted to complex **21** under any conditions tested suggesting that the C—H activation product **20** is not an intermediate in the formation of complex **21**. Experimental evidence and an in-depth DFT study suggested an equilibrium between **19$_{mono}$** and **19$_{bis}$**, and that the C—H activation product **20** and the C—C activation product **21** could both be formed under the reaction conditions, but that the formation of complex **20** was reversible while formation of complex **21**, which involves the loss of methane, is not. The intermediary complex **22** which is a highly reactive, unsaturated, 14-electron complex was proposed as an intermediate that leads to complex **21**.

C(sp^3)-H activation was expanded by Nolan and co-workers,[11,12] using I^{t}Bu (1,3-bis(*tert*-butyl)imidazol-2-ylidene) and [Rh(coe)$_2$Cl]$_2$. In this case the solvent played a primary role in determining the complex formed, with no C—H activation occurring in pentane, hexane delivering a single C—H activation product and benzene producing a double C—H activation congener (Scheme 4A). Perhaps most interesting is that by changing the reaction conditions slightly, complex **23** could be transformed to complex **24** and then to complex **26**, suggesting that complexes **23** and **24** are intermediates in the formation of complex **26**. However, this was not the case when reactions were performed at Ir. Ir analogues **25** and **27** were isolated, however, the Ir congener of Rh complex **23** was not isolable (Scheme 4A). The authors suggested that the formation of the Ir analogue of **23** is slower than it is for Rh and subsequent C—H activation steps are faster meaning that isolation of this complex is not possible. These crystallographically characterized intermediates provide a better understanding of these C—H activation processes. The authors were also able to abstract the chloride ligand from complexes **26** and **27** to deliver 14-electron complexes, which are proposed intermediates in many catalytic cycles (Scheme 4B). DFT studies indicate that the remarkable stability of the complexes can be attributed to enhanced π-electron donation from the NHC rings. NHC ligands are often thought of as solely σ-donors, but this study showcases a more flexible bonding mode. When these 14-electron complexes (**26** and **27**) were treated with CO in C$_6$H$_6$, coordinatively saturated complexes **28** and **29** were obtained and X-ray quality crystals were analyzed by diffraction experiments. Interestingly, both the I^{t}Bu ligands remained cyclometalated to the metals.

Finally, the 14-electron Ir complex **28** could be reacted with H$_2$ to produce dihydride complex **32** (Scheme 4C).[13] When this complex was treated with CO a double C—H activation complex **33** was isolated, showing how the ligand environment drastically affects whether these ligands undergo C—H activation and the lability of the C—H activation in this case.

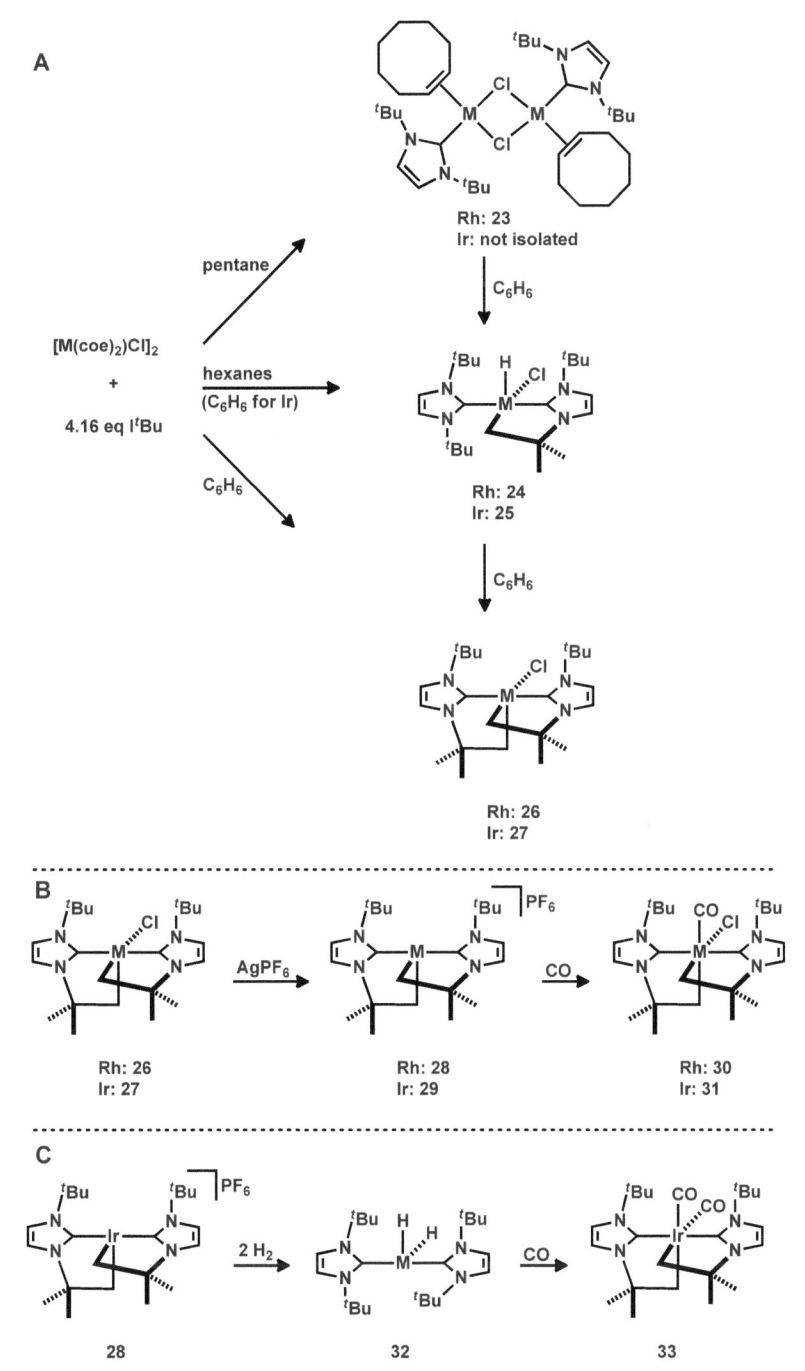

Scheme 4 (A) Solvent dependent synthesis of Rh- and Ir-NHC complexes. (B) Halide abstraction leading to 14-electron complexes. (C) Reversible C—H activation within an Ir complex.

Another C(sp^3)-H activation was reported in 2000 by Herrmann and co-workers.[14] In this example, Ir complex **34** was treated with triflic acid which led to the cyclohexyl N-substituent undergoing a C—H activation followed by a β-H migration to deliver complex **37** (Scheme 5A). The proposed mechanism involves elimination of methane upon treatment with triflic acid. This leads to a 16-electron complex **35**, which engages in C—H activation with elimination of another molecule of methane and produces complex **36**. A β-H migration delivers hydride complex **37**.

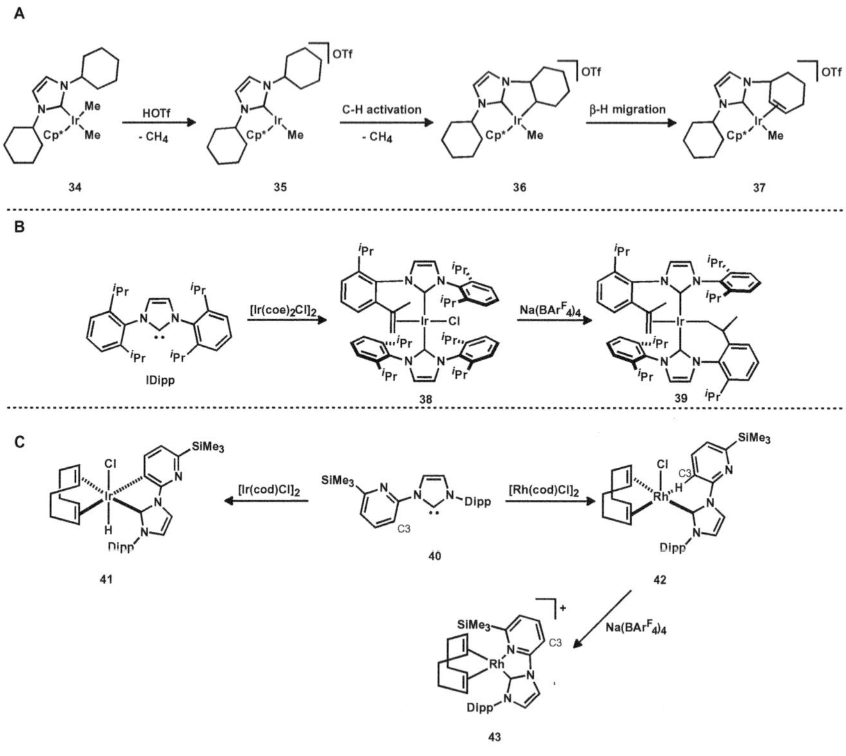

Scheme 5 (A, B) C—H activations followed by β-H eliminations. (C) C—H activation versus pyridyl coordination.

A related C—H activation followed by β-H elimination was reported by Aldridge and co-workers in 2009.[15] In this example, IDipp (1,3-bis(2,6-diisopropylphenyl)imidazol-2-ylidene, also often represented as IPr) undergoes a C—H activation, β-H elimination sequence to deliver complex **38** (Scheme 5B). Complex **38** was then reacted with Na(BAr$^F_4$) (BAr^F = 3,5-bis(trifluoromethyl)phenylborate). Abstraction of the halide initiates another C—H activation to deliver complex **39** which features an unusual seven-membered metallacycle. As such, two IDipp ligands have both undergone C—H activations which showcases the lability of these processes.

A further development of C—H activation in NHC ligands was described by Danopoulos in 2002.[16] In this example, a pyridyl-based *N*-substituent underwent C—H activation upon complexation with Ir. This is noteworthy as the C—H activation product was preferred over the pyridyl-coordinated congener (Scheme 5C). Upon exposure of NHC **40** to [Ir(cod)Cl]$_2$ in THF at −78 °C, the C—H activation product **41** was observed in which C3 is coordinated to the metal. The reaction was then performed under identical conditions but using [Rh(cod)Cl]$_2$, with the product of this reaction (**42**) being absent of any C—H activation but does display hydrogen bonding between the metal and the C3 hydrogen. This is possibly an analogue of an intermediate for the C—H activation product **41**. The authors investigated whether the steric bulk of the trimethylsilyl group excluded the possibility of the pyridyl function coordinating the metal center. However, when complex **42** was treated with Na(BAr$_4^F$), this delivered complex **43** featuring a coordinated pyridyl function.

Further investigation of C(sp^3)-H activation reactions of NHCs at Ir centers was conducted by Yamaguchi and co-workers, in which CpIr (Cp=cyclopentadienyl) analogues were synthesized and their reactivity examined.[17] The NHC ligand used throughout was ^{Me}I^{i}Pr (1,3-diisopropyl-4,5-dimethylimidazol-2-ylidene). A C—H activation of the isopropyl *N*-substituent and cyclometalation reaction was observed when complex **44** was treated with the strong base NaOMe to deliver complex **45** (Scheme 6A). Conversely, when complex **44** was treated with AgOTf to induce a halide abstraction, no C—H activation product was produced. The authors then synthesized complexes **46** and **48**. Complex **46** features a basic amine group which could conceivably bind to the metal center, and complex **48** features no similar functionality. Each of these complexes were treated with AgOTf and, in the case of complex **46**, a C—H activation of one of the methyl groups of the ^{i}Pr *N*-substituent occurred to deliver the cyclometalated product **47**. However, only halide abstraction occurred in the case of complex **48** to deliver complex **49**. The authors suggested that either an internal or external basic function is required for C—H activation to occur.

This was extended to the chemistry of IrCp*(NHC) (Cp*= pentamethylcyclopentadienyl) complexes.[18] In this case, the NHC ligands IEt (1,3-bis(ethyl)imidazol-2-ylidene), I^{i}Pr (1,3-bis(isopropyl)imidazol-2-ylidene), and I^{n}Bu (1,3-bis(butyl)imidazol-2-ylidene) were added to the

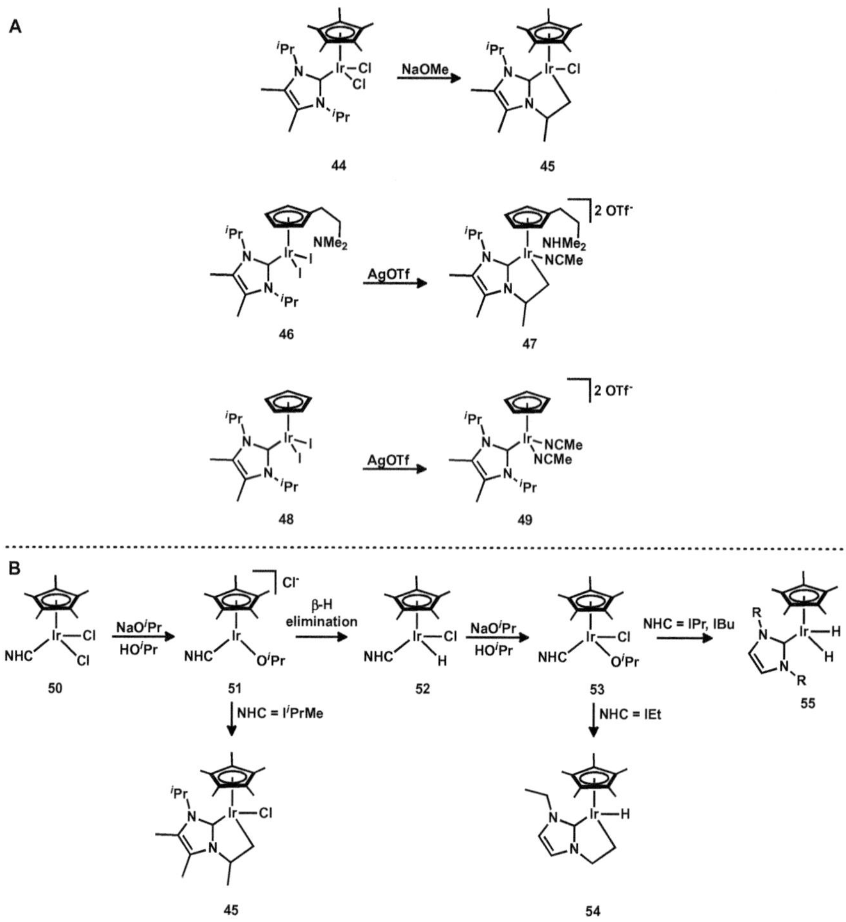

Scheme 6 C—H activations requiring either an internal or external base.

Cp*Ir framework and different products were observed. Only the IEt ligand participated in a C—H activation to deliver a cyclometalated product **54** (Scheme 6B). A mechanistic investigation revealed that two equivalents of NaO^{i}Pr were required for the cyclometalation to proceed. If only one equivalent was used a hydrido complex **52** was the sole reaction product. The different reaction products when using different NHC ligands and different amounts of base, as well as the relative rates of C—H activation compared to β-H elimination, were important factors in determining which complexes were produced.

Imidazolium salts **56** and **59** were used in competitive C—H activations in which either a C—H bond of the phenyl *N*-substituent, or the pyridyl or picolyl *N*-substituent, could be activated (Scheme 7).[19] When treated with

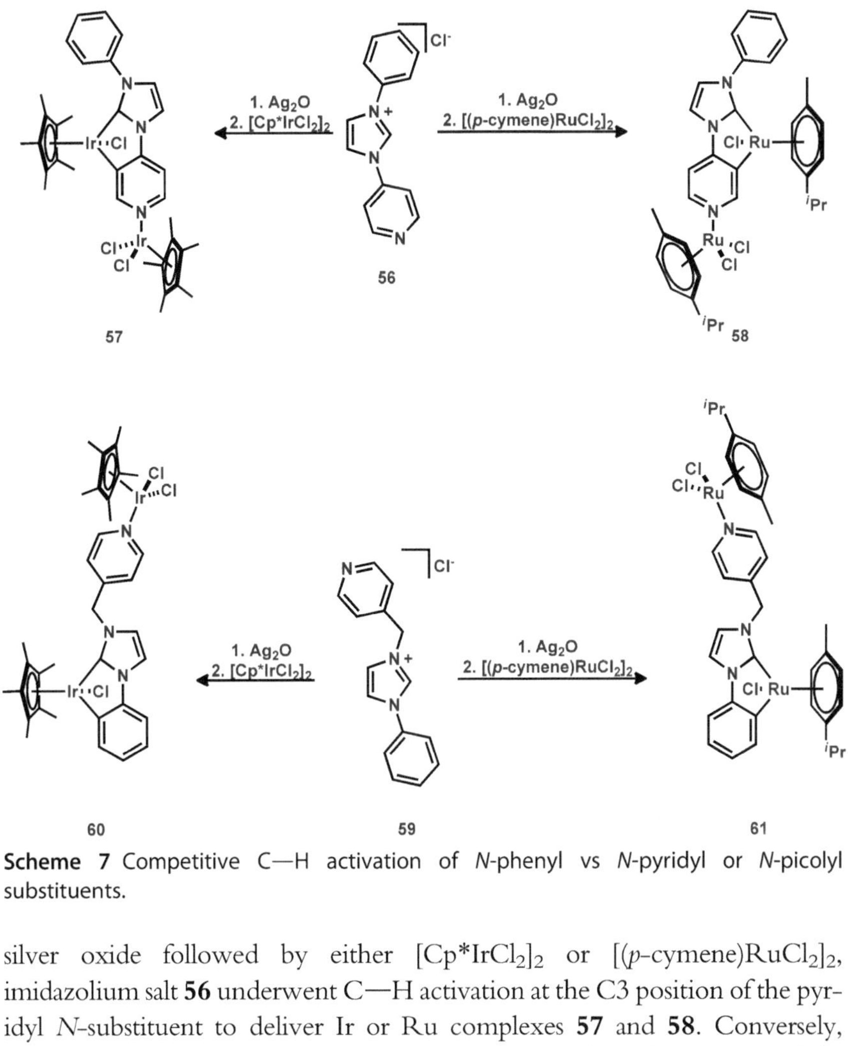

Scheme 7 Competitive C—H activation of *N*-phenyl vs *N*-pyridyl or *N*-picolyl substituents.

silver oxide followed by either [Cp*IrCl₂]₂ or [(*p*-cymene)RuCl₂]₂, imidazolium salt **56** underwent C—H activation at the C3 position of the pyridyl *N*-substituent to deliver Ir or Ru complexes **57** and **58**. Conversely, imidazolium salt **59** underwent C—H activation at the *ortho*-position of the phenyl *N*-substituent. The different reactivity can be rationalized by the ring size of the metallacycle, with the five-membered metallacycle preferred over the six-membered. A second equivalent of [Cp*IrCl₂] or [(*p*-cymene)RuCl₂] also coordinated to the free pyridyl or picolyl *N*-substituent to form bimetallic complexes.

A study investigating aliphatic and aromatic intramolecular C—H activations within NHC ligands using Cp*Ir complexes was undertaken by Peris and co-workers in 2006.[20] They initially synthesized Cp*Ir(NHC) complexes **62** and **63** which were designed such that only aromatic (**62**) or only aliphatic (**63**) C—H activations could occur (Scheme 8A). When

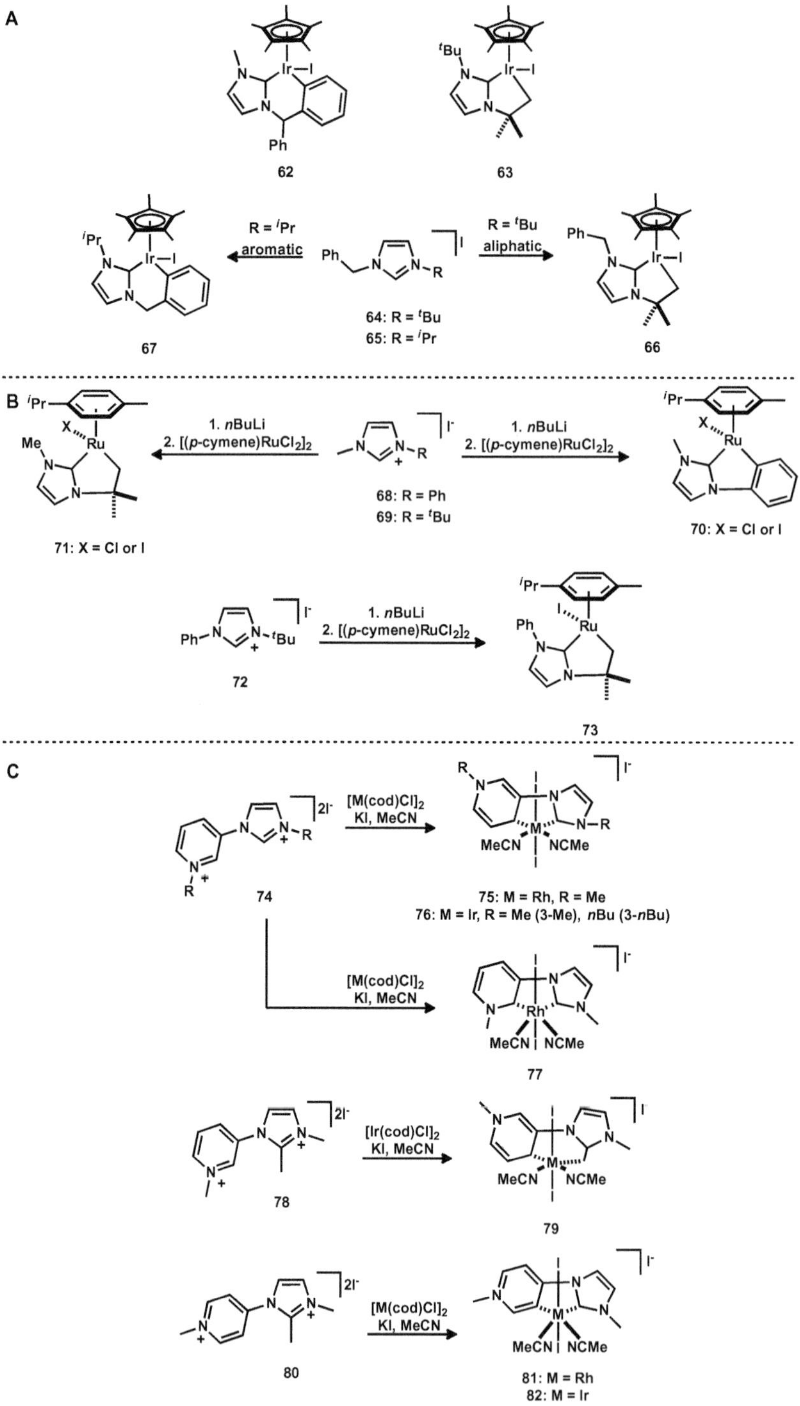

Scheme 8 Examples of competitive C(sp^3)-H and C(sp^2)-H activations.

imidazolium salt **64**, featuring a *t*Bu *N*-substituent, was used in the preparation of an Ir complex (via transmetalation of the NHC from Ag), the only product of this reaction was complex **66** in which the aliphatic C—H activation product **66** is favored. Conversely, when imidazolium **67**, featuring a less bulky *i*Pr *N*-substituent, was subjected to these reaction conditions no C—H activation product was observed. However, when strong base was added, complex **67** could be isolated which features aromatic C—H activation but no aliphatic C—H activation. The authors suggest that in the absence of substantial steric hindrance, aromatic C—H activation may be favored. Conversely, when considerable steric hindrance is present this may lead to aliphatic C—H activation as the only way to relieve congestion around the metal center.

Another example of competitive $C(sp^2)$-H versus $C(sp^3)$-H activation with [(*p*-cymene)Ru(NHC)] complexes was described by Wang and co-workers.[21] Using the imidazolium salt 1-methyl-3-phenyl imidazolium iodide (**68**), $C(sp^2)$-H activation was observed at the *ortho*-position of the phenyl *N*-substituent to deliver complex **70** which features a five-membered metallacycle (Scheme 8B). However, using 1-methyl-3-*tert*-butylimidazolium iodide (**2**), activation of the $C(sp^3)$-H bond of the *tert*-butyl substituent is observed and delivers complex **71** which also features a five-membered metallacycle. The competition experiment was performed using 1-phenyl-3-*tert*-butylimidazolium iodide (**72**), and this resulted in selective $C(sp^3)$-H activation on the *tert*-butyl substituent to deliver complex **73**. The authors suggest that steric factors are dominant in determining which group is cyclometalated as the *tert*-butyl group must have a C—H bond pointing toward and proximal to the metal center.

Aliphatic $C(sp^3)$-H activation was preferred over aromatic $C(sp^2)$-H activation in a report by Peris and co-workers.[22] Several imidazolylidene/pyridylidene Rh or Ir complexes were synthesized in which the imidazolylidene initially coordinates the metal followed by $C(sp^2)$-H activation of the pyridinium *N*-substituent, as it is proximal to the metal center (Scheme 8C). Interestingly, imidazolium/pyridinium salt **74** undergoes C(sp^2)-H activation exclusively at the *para*-position with the metal precursor [M(cod)Cl]$_2$ (M=Rh or Ir). However, with [Rh(nbd)Cl]$_2$ (nbd= norbornadiene), $C(sp^2)$-H activation occurs at the *ortho*-position to produce complex **77**. Imidazolium salt **78** was prepared with a methyl group blocking the C2-position of the imidazolium ring. When reacted with [Ir(cod)Cl]$_2$ this was expected to lead to the abnormal NHC complex metalated through C4. However, a $C(sp^3)$-H activation took place at the aliphatic methyl group blocking the C2-position rather than the aromatic C4 position to deliver complex **79**. Further, C—C bond cleavage was observed when imidazolium **80** was subjected to identical reaction conditions leading to complexes **81** and **82**.

There are several examples of C—H activation within NHC-based pincer ligands. A C—H activation occurred for an NHC pincer ligand **83** with a central pyridine moiety.[23] The reaction between NHC **83** and [Ir(cod)Cl]$_2$ produced a bimetallic complex **84** in which both the *meta*-positions of the pyridyl moiety have coordinated to Ir centers to produce a cyclometalated complex (Scheme 9A). This cyclometalation could be prevented by introducing methyl substituents at the *meta*-position of the pyridyl moiety to produce complex **85**.

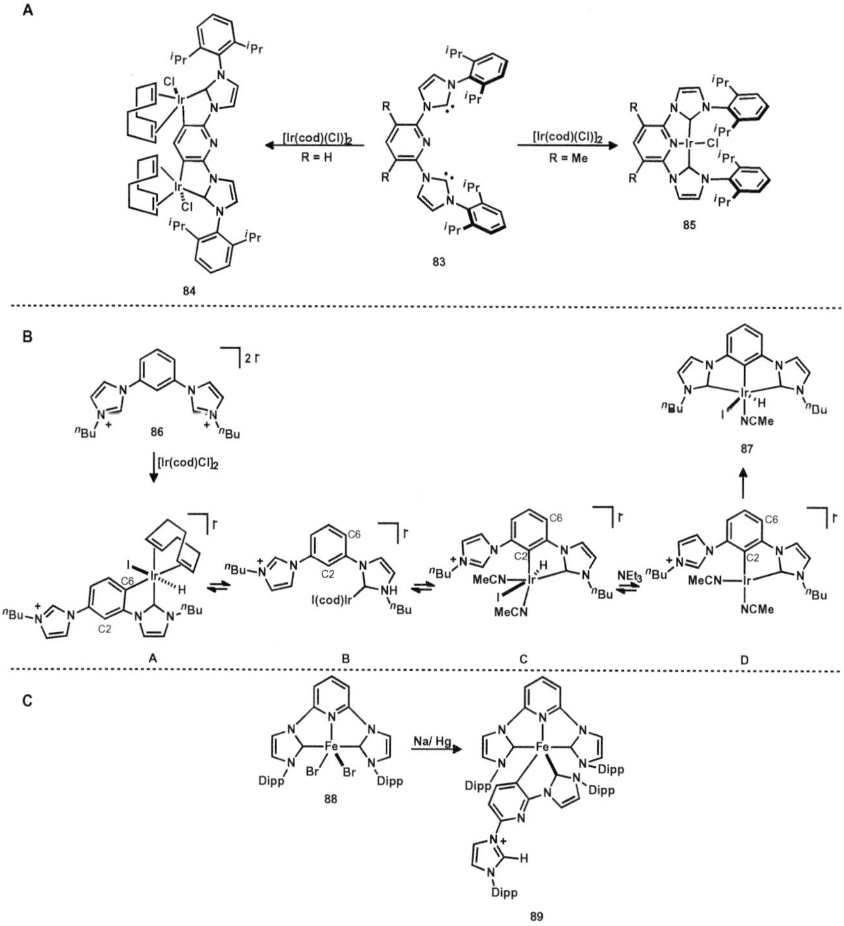

Scheme 9 C—H activations of pincer-type NHC ligands.

The metalation of pincer-type bis(imidazolium) salt **86**, which features an aryl linker, with [Ir(cod)Cl]$_2$ using a large excess of NEt$_3$ resulted in the formation of pincer complex **87**, which is the result of a C—H activation of the aryl linker (Scheme 9B).[24] The use of a large excess of base is necessary to avoid production of a mixture of complexes. The formation of these product mixtures gave some insight into the mechanism of the formation of complex **87** and some intermediates were isolated which allowed a potential mechanism to be proposed. Initial deprotonation and coordination of one of the imidazolium salts brings the aryl linker proximal to the metal center and C—H activation/cyclometalation occurs at C6 and delivers intermediate **A** which could be isolated. De-coordination of the aryl function provides intermediate **B** which could also be isolated. The cyclometalation observed in the product occurs at C2 rather than C6 so an equilibrium must exist between intermediates **A**, **B**, and **C**. Formation of intermediate **C** brings the pendant imidazolium closer to the metal and a base-assisted elimination of HI results in the Ir(I) intermediate **D**, which is set up for oxidative addition of the second imidazolium and delivers the product. This mechanism contrasts with related PCP pincer ligands, where it has been suggested that coordination of both phosphines is necessary for the C—H activation of the aryl linker to proceed.[25]

Another example of a bis(NHC) pincer ligand undergoing C—H activation to coordinate Fe in a similar manner was reported by Saßmannshausen, Danopoulos and co-workers.[26] In this example, a five-coordinate, square pyramidal Fe(II) complex **88** containing one bis(NHC) pincer ligand is treated with Na/Hg which produces a zwitterionic, five-coordinate Fe(0) complex **89**. This complex contains two bis(NHC) pincer ligands, including one which has undergone de-coordination of one of the NHCs and C—H activation at the *meta*-position of the pyridine ring to produce cyclometalated complex **89** with a pendant imidazolium arm (Scheme 9C).

Several Ru-NHC complexes are known to be efficient olefin metathesis catalysts. However, C—H activations that result in cyclometalated NHC ligands can be detrimental and result in lower activity of the catalysts. As such, understanding these decomposition processes is desirable and the development of strategies to prevent C—H (de)activation have been developed and studied both experimentally and by DFT.[27]

A double C—H insertion reaction was reported by Grubbs and co-workers and a plausible mechanism was postulated (Scheme 10A).[28] De-coordination of a phosphine ligand from complex **90**, which is the initiating step for olefin metathesis, leads to the coordinatively unsaturated complex **91**. C—H insertion of the phenyl *N*-substituent of the NHC

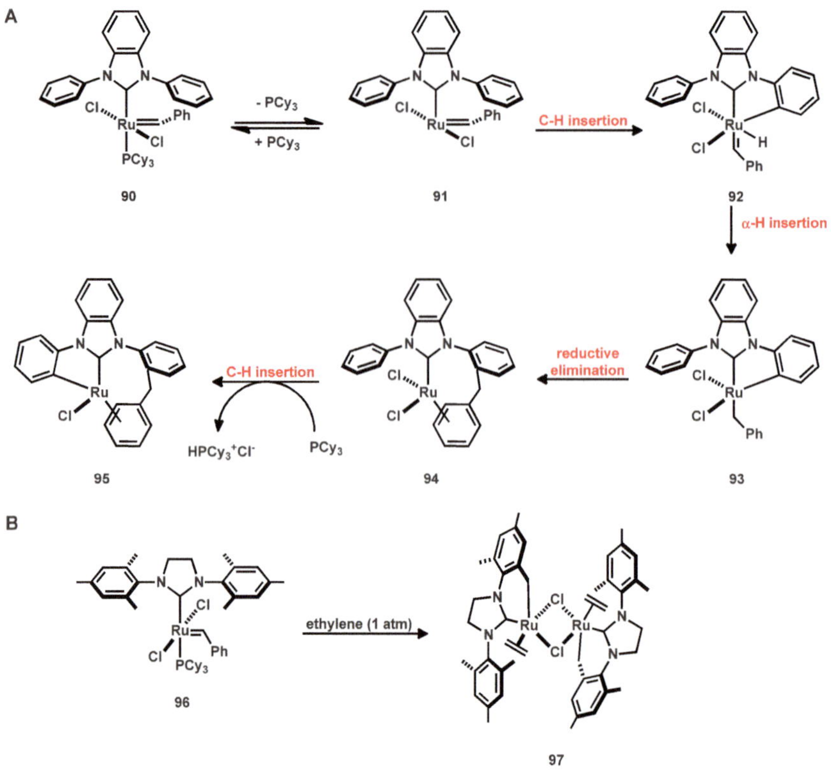

Scheme 10 C—H (de)activations of olefin metathesis catalysts **90** and **96**.

ligand produces complex **92**. The alkylidene ligand in complex **92** then undergoes α-H insertion, and reductive elimination leads to complex **94** which could be isolated. A second C—H insertion, promoted by the basic free phosphine, delivers complex **95**. This proposed mechanism was later supported by DFT calculations, with the rate limiting step being the transformation of **94** to **95**.[29]

Another decomposition product was discovered when metathesis catalyst **96** was treated with ethylene.[30] Complex **97** was isolated in 70% yield along with a quantitative amount of methyltriphenylphosphonium chloride (Scheme 10B). It is suggested that the presence of basic methylenetriphenylphosphine allows the C—H activation to proceed.

Further investigation of decomposition pathways for Ru-NHC metathesis catalysts led Piers and co-workers to examine low energy decomposition pathways of catalytic intermediates starting from the phosphonium alkylidene complexes **98** and **99** (Scheme 11).[31] When these complexes were subjected to metathesis conditions using 1,1-dichloroethene, a poor substrate for

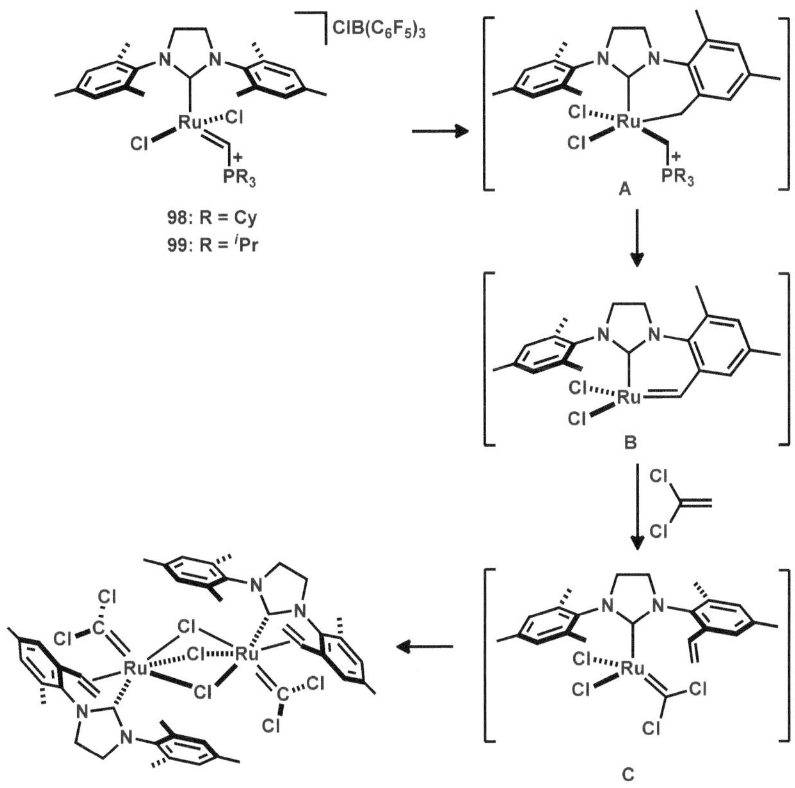

Scheme 11 C—H activation of olefin metathesis catalysts **98** and **99**.

metathesis, the final product in both cases was bimetallic complex **100**. A plausible mechanism was postulated which begins with C—H activation of one of the *ortho*-methyl groups of the mesityl N-substituent on the NHC. This results in the cyclometalated intermediate **A**, which rapidly eliminates the methylphosphonium salt via α-H elimination to give the chelate **B**. This highly reactive complex is trapped by the olefin and forms both a dichlorocarbene ligand and the styrenic vinyl group. Finally, loss of chloride leads to dimerization to deliver complex **100**.

The above examples highlight C—H (de)activations which are detrimental to catalyst performance, however, the metal-carbon bond produced by C—H activations in Ru-based Z-selective olefin metathesis catalysts improves Z-selectivity. Grubbs and co-workers reported chelated Ru catalysts which outperform their non-chelated analogues in terms of Z-selectivity.[32] Complexes **101** and **103** were treated with silver pivalate which induces C—H activation and produces complexes **102** and **104**

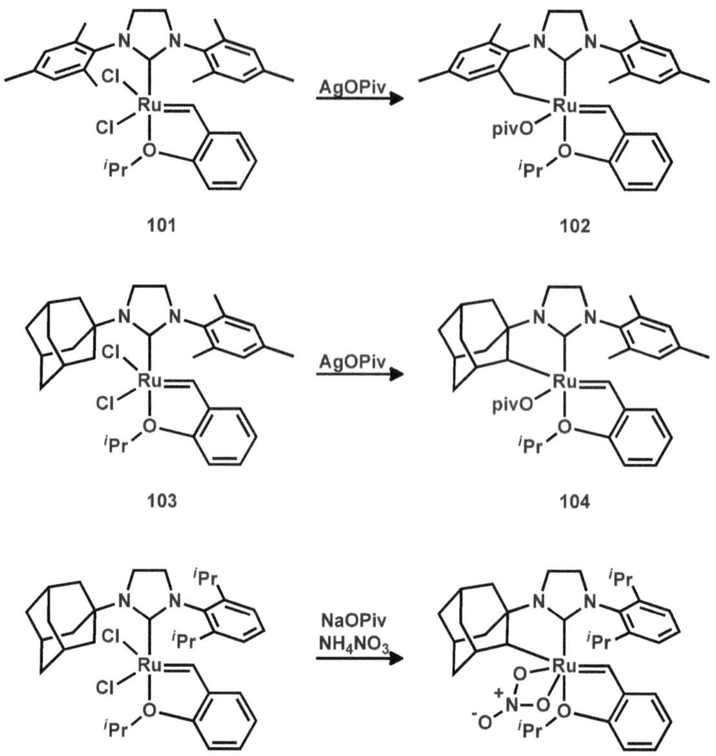

Scheme 12 C—H activation to produce Z-selective olefin metathesis catalysts **102**, **104** and **106**.

(Scheme 12). A similar complex **105** which features an unsaturated imidazol-2-ylidene NHC ligand also underwent C—H activation under similar conditions to deliver the chelated analogue **106** after treatment with ammonium nitrate.[33]

A reversible C—H activation reaction that was found to enhance the catalytic activity of a Ru-NHC complex was reported by Whittlesey, Williams and co-workers.[34] The C—H activation product **107** was synthesized by the reaction between [Ru(PPh$_3$)(CO)H$_2$] and ^{Me}I^{i}Pr, but when hydrogen was bubbled through a solution containing complex **107**, a dihydride complex **108** was formed and the Ru—C bond of the N-^{i}Pr arm had been broken (Scheme 13A). This reversible C—H activation was beneficial in tandem catalytic reactions, such as transfer hydrogenation, as complex **108** was the most active catalyst from a range of related complexes. The authors

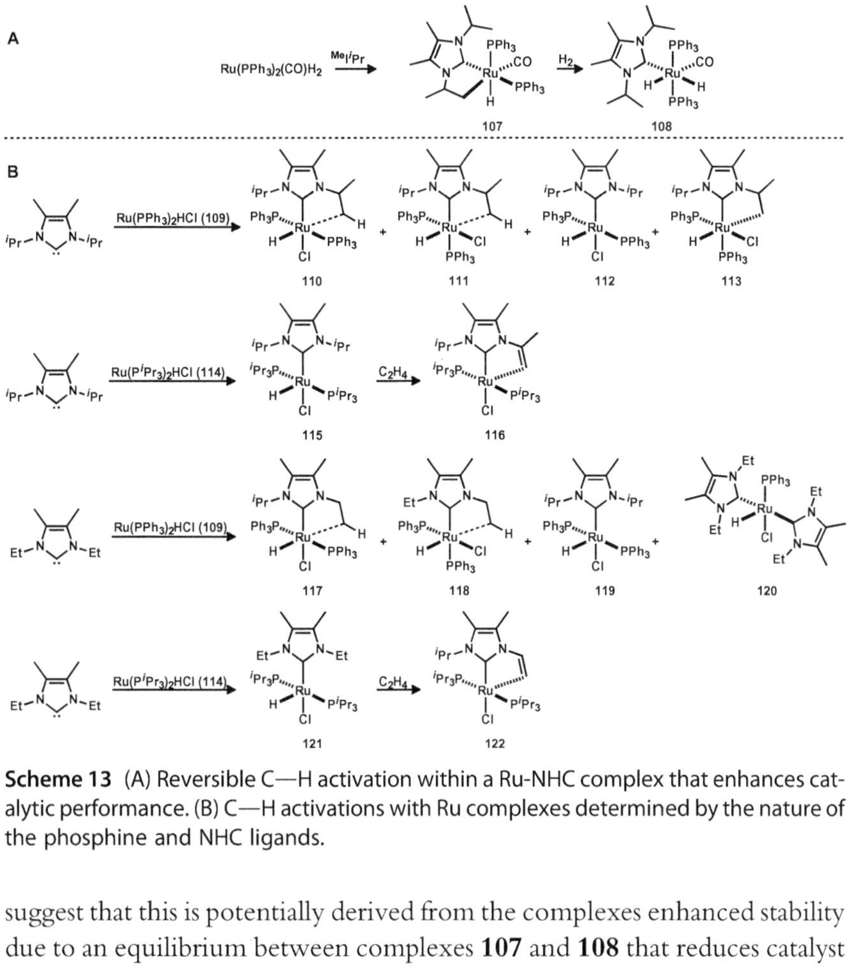

Scheme 13 (A) Reversible C—H activation within a Ru-NHC complex that enhances catalytic performance. (B) C—H activations with Ru complexes determined by the nature of the phosphine and NHC ligands.

suggest that this is potentially derived from the complexes enhanced stability due to an equilibrium between complexes **107** and **108** that reduces catalyst deactivation.

Further investigation of this C—H activation process was reported in 2009.[35] First, the reactivity of $^{Me}I^iPr$ was investigated using two different phosphine-containing Ru complexes, [Ru(PPh$_3$)$_2$HCl] **109** and [Ru(P^{i}Pr$_3$)$_2$HCl] **114** (Scheme 13B). Reaction with the PPh$_3$ analogue **109** resulted in a mixture of products, including complexes **110** and **111** which exhibit agostic interactions, as well as complex **113** in which the N-^{i}Pr arm of the NHC has undergone C—H activation and formed a new Ru—C bond. When this mixture of complexes was subjected to an atmosphere of ethene the sole product of the reaction was C—H activated complex **113**. This reactivity pattern when using the PPh$_3$ analogue **109** contrasts with that of P^{i}Pr$_3$ containing complex **114**. When $^{Me}I^iPr$ was reacted with complex **114**, under similar conditions, the only product observed was complex **115** which is the analogue of complex **112**. This complex does not exhibit any agostic interactions or undergo C—H activation. However, when treated with ethene, under more forcing conditions to those used to convert **110–112** to complex **113**, a double C—H activation delivered complex **116** as the sole product of the reaction. This divergent reactivity was explained by the increased π-basicity and steric bulk of the P^{i}Pr$_3$ ligand compared to PPh$_3$.

Similar reactions of ^{Me}IEt (1,3-diethyl-4,5-dimethylimidazol-2-ylidene) with complexes **109** and **114** were performed. Complexes **117** and **118** which feature agostic interactions were observed but no C—H activation product was detected for the PPh$_3$ analogue **109**, even when the mixture of products was treated with ethene. Conversely, when ^{Me}IEt was reacted under similar conditions using the P^{i}Pr$_3$ analogue **114**, complex **121** which is the analogue of complex **115** was observed and could be cleanly converted to C—H activation product **122** under an atmosphere of ethene. These observations highlight the increased propensity of ^{Me}I^{i}Pr to undergo C—H activations compared to ^{Me}IEt.

An Ir-NHC catalyzed dehydrogenative silylation reaction that implicates a reversible C—H activation within the complex was developed by Mashima and co-workers.[36] The silylation reaction was catalyzed by Ir-NHC complex **123** and the hemilability of the NHCs xylyl *N*-substituent was crucial for an efficient transformation (Scheme 14A). A plausible reaction mechanism was proposed in which addition of 2-phenylpyridine initiates C—H activation of the xylyl *N*-substituent. Addition of triethylsilane establishes an equilibrium between intermediates **A**, **B**, and **C**. Each of these complexes are on-cycle intermediates and the metalation/de-metalation of the xylyl *N*-substituent and the *ortho*-position of 2-phenylpyridine are both necessary. De-metalation of the xylyl *N*-substituent allows coordination of norbornene to deliver complex **D**. Reduction of norbornene and de-coordination of norbornane promotes metalation of the xylyl *N*-substituent to produce complex **E**. Reaction with another equivalent of Et$_3$SiH leads to complex **F** which undergoes ligand exchange with the substrate to complete the catalytic cycle and release the product.

An unexpected product that resulted from the reaction of IMes with LiBH$_4$ in the presence of [Ir(coe)$_2$Cl]$_2$ led to a detailed investigation of the mechanism of this C—H activation.[37] In the absence of [Ir(coe)$_2$Cl]$_2$ the reaction product is IMes•BH$_3$ (**125**) (Scheme 14B). However, with a

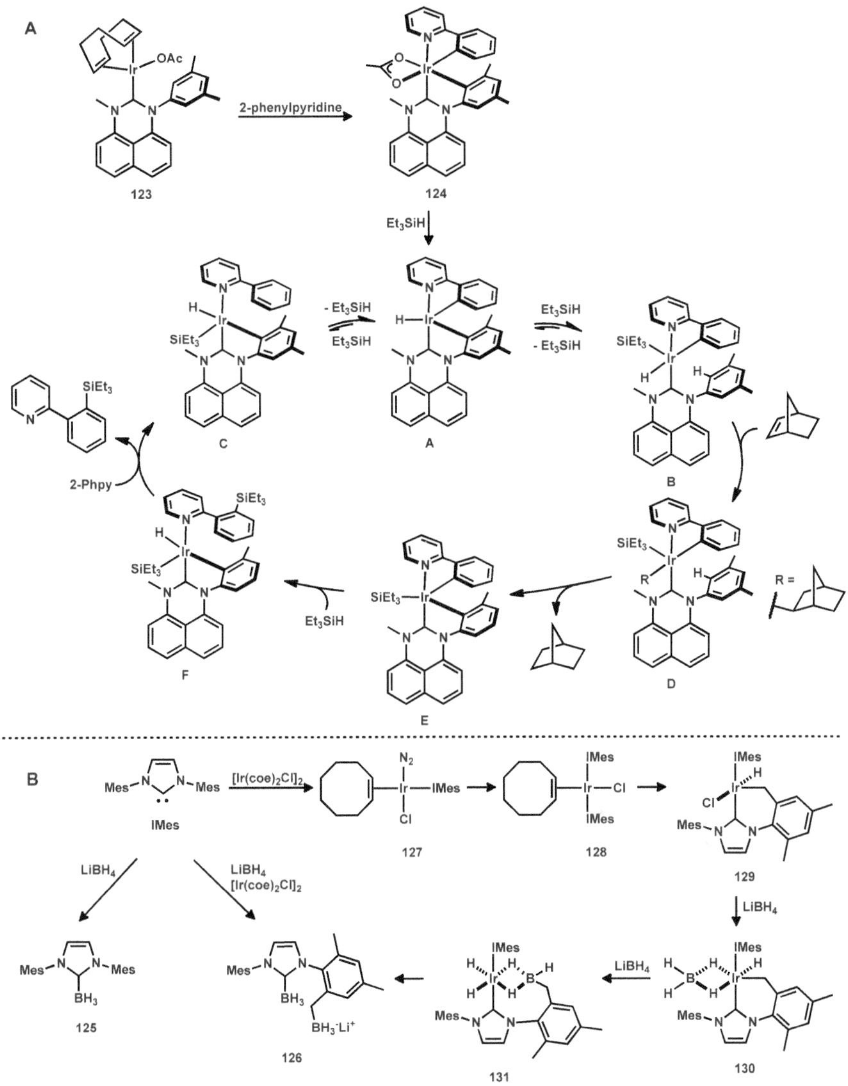

Scheme 14 Catalytic reactions with C—H activation steps implicated.

sub-stoichiometric amount of [Ir(coe)$_2$Cl]$_2$ present a diboration product **126** is observed in which the *ortho*-methyl group of one of the mesityl *N*-substituents on IMes has been furnished with a second BH$_3$ group (**126**). The authors then synthesized and isolated several complexes which are implicated in the reaction mechanism and proposed a detailed mechanism.

Initial coordination of IMes under an atmosphere of nitrogen leads to complex **127**. De-coordination of the dinitrogen ligand allows another IMes ligand to coordinate the Ir(I) center and *cis/trans*-isomerization leads to complex **128**. The C—H activation step occurs via an oxidative addition of the *ortho*-methyl group of the mesityl *N*-substituent which delivers the cyclometalated complex **129**. Salt metathesis with BH_4^- leads to intermediate **130** which reacts with another equivalent of BH_4^- to deliver complex **131**, and reductive elimination gives the diboration product **126**.

A catalytic example of C—H activation and H/D exchange was developed by Peris and co-workers.[38] In this instance, the benzyl *N*-substituent of imidazolium salt **132** underwent *ortho*-metalation to provide cyclometalated complex **133** (Scheme 15A). This was achieved using a transmetalation strategy of the NHC from Ag to Ir complex **134**, with subsequent C—H activation at room temperature to deliver complex **133**. Complex **133** could also be synthesized directly from $[Cp*IrCl_2]_2$ with the addition of NaOAc and NaI. When complex **133** was treated with CD_3OD the *ortho*-position of the benzyl *N*-substituent was quantitatively deuterated within 24 h, suggesting an equilibrium in which metalation/de-metalation of the benzyl

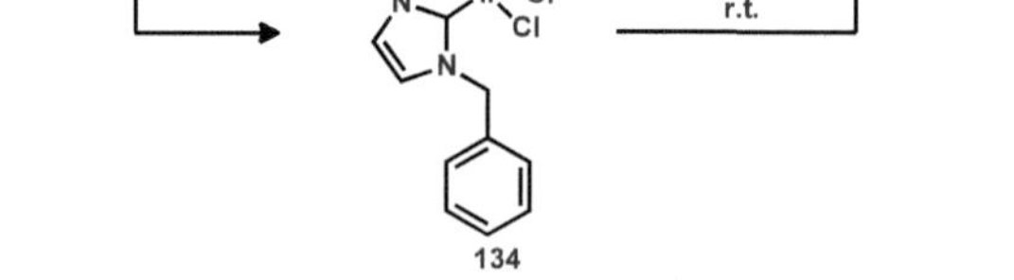

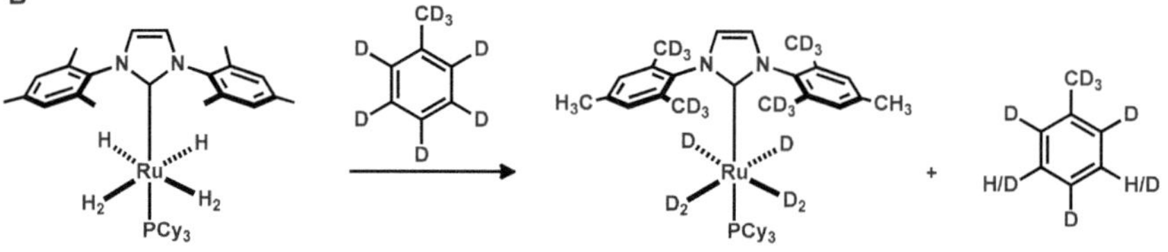

Scheme 15 Intermolecular C—H activation and H/D exchange.

N-substituent occurs. A catalytic H/D exchange reaction was developed using complex **133** and other non-cyclometalated NHC analogues for the deuteration of simple organic molecules. Each of the complexes tested were found to be capable of catalyzing these reactions.

Another example of intermolecular C—H activation is provided by Leitner and co-workers.[39] They describe the H/D exchange between Ru complex **135** and d_8-toluene (Scheme 15B). Each of the H and H_2 ligands are exchanged and, interestingly, the *ortho*-methyl protons on the mesityl N-substituents are also exchanged to deliver complex **136**. Furthermore, complex **136** could then be used to deuterate several simple aromatic molecules.

A series of cyclometalated Pd-NHC complexes was reported by Danopoulos in 2007.[40] This was the first reported example of cyclometalated Pd-NHC complexes produced by C—H activation. The nature of the products was determined by the substituents on the aryl N-substituents and the Pd source (Scheme 16). When Pd(OAc)$_2$ was used with imidazolinium salts with

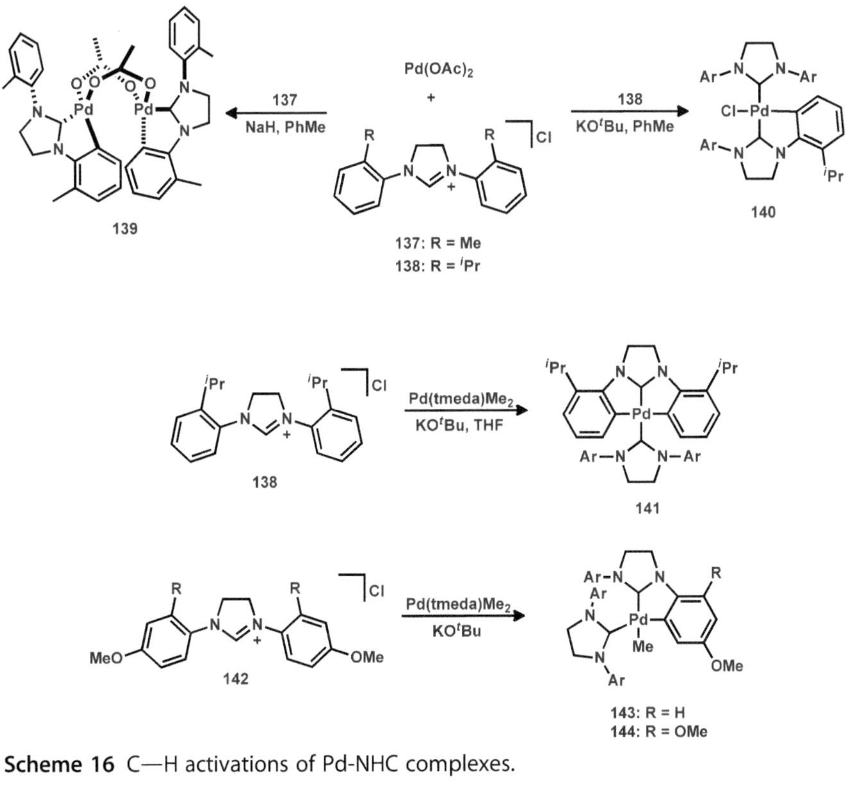

Scheme 16 C—H activations of Pd-NHC complexes.

ortho-alkyl substituents on the aryl *N*-substituents, either an acetate-bridged dimer **139**, or cyclometalated complex **140** was isolated. Conversely, when [Pd(tmeda)Me$_2$] (tmeda = *N*,*N*,*N'*,*N'*-tetramethylethylenediamine) was used a double C—H activation product **141** was formed. This contrasts with imidazolinium salts bearing *ortho*-alkoxy groups on the aryl *N*-substituents. In this case, only single C—H activation products **143**–**144** were produced. The authors suggested that the varying electronics of the ligand may result in differing mechanisms which deliver different products.

A dehydrogenative C—H activation of Ni(I) complex **145** to deliver Ni(II) complex **146** was reported by Sigman and co-workers (Scheme 17).[41] Upon exposure to O$_2$, complex **145** is oxidized to the Ni(II) dimer as a 1:1 mixture of complex **146** and the non-C-H activated complex **147**. The authors attempted similar dehydrogenative C—H activations with a range of related NHC ligands (I^{t}Bu, IMes, IAd (1,3-bis(adamantyl)imidazol-2-ylidene)) and found that this transformation was specific to IDipp, with none of the other ligands undergoing the dehydrogenation.

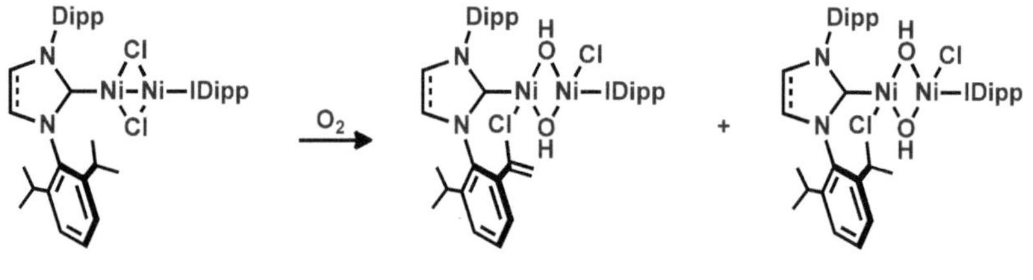

Scheme 17 Dehydrogenative C—H activation of an Ni-IDipp complex.

An interesting reaction sequence involving [Ni(cod)$_2$] and I^{t}Bu involving C—H activation followed by C—N bond cleavage was described by Caddick, Cloke and co-workers.[42] When [Ni(cod)$_2$] was treated with excess I^{t}Bu in sunlight for 5 days, complex **148** which features an I^{t}Bu ligand and two 1,3-cyclooctadiene ligands could be isolated (Scheme 18A). The isomerization of 1,5-cyclooctadiene to 1,3-cyclooctadiene was derived from irradiation by sunlight. Complex **148** undergoes a rapid C—H activation of one of the ^{t}Bu *N*-substituents to deliver complex **149**, and C—N bond cleavage and dimerization leads to complex **150** with loss of isobutene and cyclooctene.

Another C—H activation which was followed by a C—N activation was reported by Whittlesey and co-workers.[43] In this case, [Ru(PPh$_3$)$_3$(CO)HCl] **151** was heated in the presence of ^{Me}I^{i}Pr and a mixture of complexes

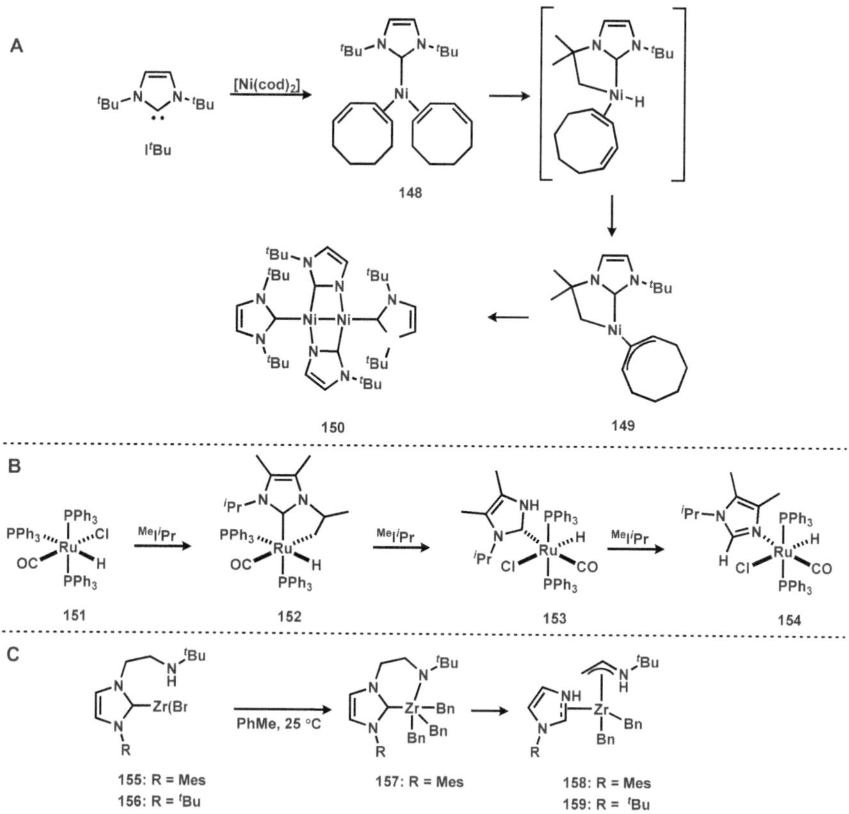

Scheme 18 (A), (B) C—H activation followed by C—N bond cleavage. (C) C—N cleavage within a Zr complex.

152, **153** and **154** were produced (Scheme 18B). An experiment in which complex **152** was treated with excess $^{Me}I^iPr$ and heated resulted in the formation of both complexes **153** and **154**. Further, when complex **153** was heated, isomerization led to complex **154**. As such, the authors suggested that, after coordination of $^{Me}I^iPr$, complex **151** undergoes a C—H activation to deliver complex **152**. This is followed by C—N bond cleavage accompanied with the loss of propane and/or propene to produce complex **153**. Finally, isomerization leads to complex **154**. It is noteworthy that both the above examples of C—N activation are preceded by a C—H activation.

A C—N activation involving a Zr metal center was reported by Ong and co-workers in 2010.[44] This was initiated by Zr complexes **155** and **156** by stirring in toluene (Scheme 18C). The NHC ligands feature a pendant amino group which coordinates the metal center (**157**) and this leads to C—N cleavage to produce an azaallyl ligand and an η^2-N,C-imidazolyl carbene moiety.

C—H activation within a Co(IV) imido complex featuring IMes was observed by heating the complex in benzene.[45] The Co(IV) complex **160** undergoes a C—H amination to produce Co(II) complex **161** (Scheme 18A). However, the Co(V) analogue **162**, which is oxidized from complex **160** with [Cp$_2$Fe](BAr$_4^F$), does not undergo C—H amination under similar conditions. The authors postulated that complex **161** is formed from complex **160** via H-atom abstraction of a benzylic C—H bond by one imido ligand, and benzyl radical rebound with the other imido ligand.

An interesting C—H activation reaction of an NHC ligand involving a Ta complex was reported by Hall, Fryzuk and co-workers.[46] Several Ta complexes were synthesized starting from a three-coordinate Li-pincer complex **163** (Scheme 19). The reaction takes place at −30 °C in THF and delivered several Ta complexes with C—H activation occurring on the ethylene linker. An in-depth DFT study of the reaction mechanism was performed. This ruled out the possibility of a concerted σ-bond metathesis pathway and confirmed α-H abstraction from an alkyl ligand, and subsequent alkylidene-mediated C—H activation.

An enantiopure IrCp*(NHC) complex (**168**) was synthesized by Peris, Fernandez and co-workers.[47] The reaction of (*S,S*)-1,3-di(methylbenzyl) imidazolium chloride **167** with [Cp*IrCl$_2$]$_2$ in the presence of NaOAc

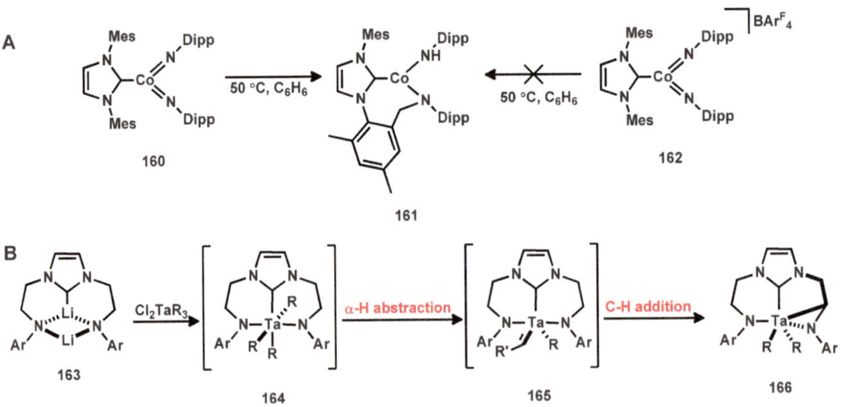

Scheme 19 (A) C—H amination of high oxidation state Co(IV) complex. (B) C—H activation within a Ta complex.

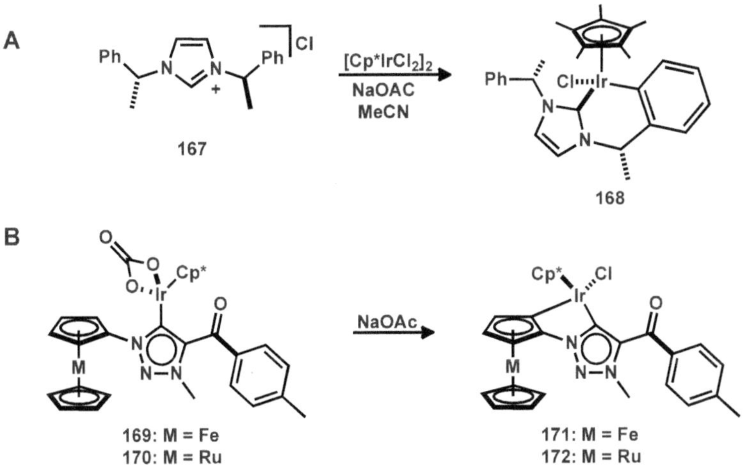

Scheme 20 (A) Synthesis of an enantiopure Cp*Ir(NHC) complex. (B) C—H activation of the Cp rings of ferrocene and ruthenocene.

delivered (S_Ir,S_C,S_C)-**168** which featured a cyclometalated benzyl ring derived from C—H activation (Scheme 20A). This complex was then used in the catalytic diboration of olefins using di(catechol)diboron and delivered modest enantioselectivity.

C—H activation of a ferrocenyl or ruthenocenyl *N*-substituent of an abnormal 1,2,3-triazolyl NHC ligand in complexes **169** and **170** occurred in the presence of NaOAc to deliver complexes **171** and **172** (Scheme 20B).[48] The enantiopure complexes are examples of bimetallic planar chiral metallocenes.

2. Reductive elimination

In the past two decades, the importance of reductive metal-NHC bond cleavage has been highlighted and its implications in NHC containing metal catalysts has been explored extensively.[49] Initially, reductive elimination was labeled to be a major pathway for catalyst deactivation and decomposition, however, later studies have also illuminated the usefulness of this type of reaction in selected catalytic cycles. The type of bond formation (C—C, C—H, C-X) has been used to categorize these reactions and here we account the emergence and propagation of the investigations focusing on, or that are inclusive of, this chemistry.

3. C—C reductive elimination

Reductive elimination of NHC ligands from metal complexes containing a range of ligated R groups (R = alkyl, aryl, alkenyl, acyl, etc.) to form new C—C bonds is the most extensively and well-studied class of reductive NHC elimination,[50,51] with initial reports focusing on NHC-alkyl coupling.[52]

The first example of C—C bond formation through reductive elimination of an NHC came about in 1998, when Cavell and co-workers were investigating Pd-NHC complexes of the type [Pd(CH$_3$)(NHC)(chelate)].[53] One such complex featuring a cod chelating ligand (**173**) was found to undergo rapid reductive elimination to form 1,2,3-trimethylimidazolium tetrafluoroborate (**174**), Pd(0) nanoparticles and uncoordinated cod (Scheme 21). The decomposition was found to occur even at low temperatures (−20 °C), with the formation of the imidazolium product detected through use of NMR spectroscopy and mass spectrometry. The same decomposition pathway was observed for other complexes under investigation by Cavell featuring different ligand environments around the Pd center.[54] The authors noted that the reductive elimination decomposition pathway was more facile with cationic complexes than for analogous neutral complexes.[55]

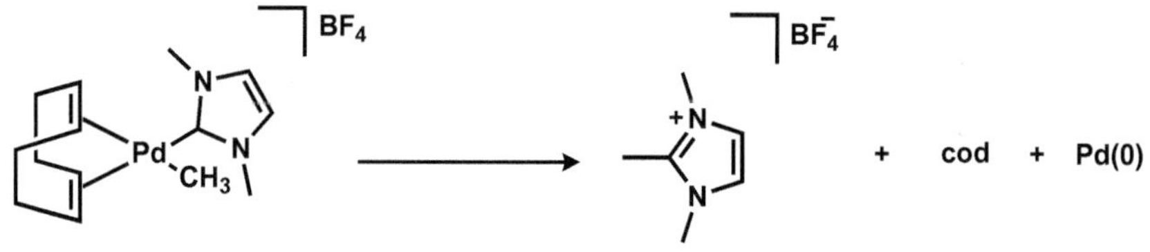

Scheme 21 C—C bond formation through reductive elimination of an NHC and methyl group from a Pd(II) complex.

Initial combined kinetic and DFT studies sought to probe the mechanism for the formation of C2-alkylated imidazolium salts from Pd(II)-NHC complexes.[56] These investigations showed that the mechanism is one of concerted reductive elimination as it is the most energetically favorable pathway, and the bond formation is not a migratory process.

The elimination of an imidazolium salt from Ni complexes was also observed during the synthesis of [Ni(CH$_3$)I(NHC)$_2$], where formation of the desired Ni complex was followed swiftly by rapid decomposition and release of C2-methylated imidazolium salt.[53] Even when the synthesis and workup was conducted at low temperatures (−50 °C), the major product was the methylated imidazolium salt. The first instance of C2-aryl imidazolium salt formation was also observed during this investigation following halide abstraction from [Pd(aryl)I(NHC)$_2$], forming the cationic Pd complex. It was again noted that the cationic Pd species decomposition to form the arylated imidazolium salt was much more rapid than for the neutral Pd complex. Later it was shown that Ni-catalyzed olefin dimerization could be performed with similar catalysts in an ionic liquid of imidazolium salt. The inevitable reductive elimination of C2-alkyl imidazolium **176** was followed swiftly by oxidative addition of a solvent imidazolium salt to reform an active catalyst, **177** (Scheme 22).[57]

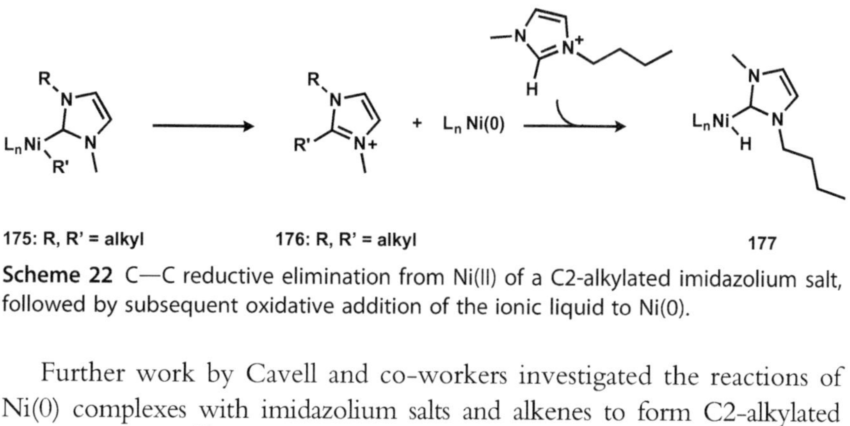

175: R, R' = alkyl   176: R, R' = alkyl   177

Scheme 22 C—C reductive elimination from Ni(II) of a C2-alkylated imidazolium salt, followed by subsequent oxidative addition of the ionic liquid to Ni(0).

Further work by Cavell and co-workers investigated the reactions of Ni(0) complexes with imidazolium salts and alkenes to form C2-alkylated imidazolium salts.[58] The reactions of the imidazolium salts with [Ni(PPh$_3$)$_n$] was proposed to follow a reaction mechanism comprising of oxidative addition of the imidazolium on to the Ni(0) center, forming the corresponding Ni(II)-NHC hydride species. Alkene coordination followed by insertion into the ensuing Ni—H bond results in the formation of a Ni-alkyl intermediate, which then undergoes reductive elimination of the C2-alkylated imidazolium salt, and reformation of the starting complex (Fig. 1). Later studies explored this mechanism in more detail, eluding that multiple possible rate determining steps exist, each of which can become dominant depending on the ligands used.[59,60]

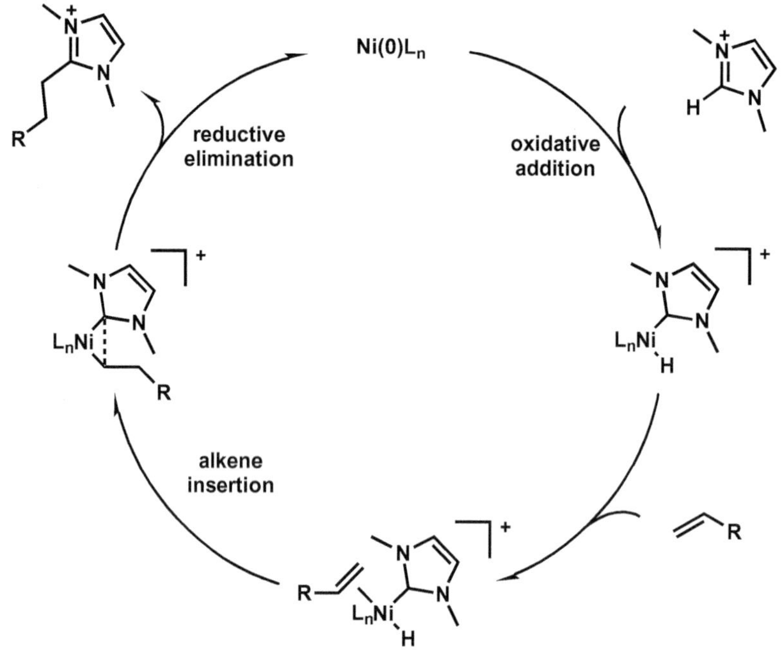

Fig. 1 Proposed mechanism of the Ni(0)-catalyzed formation of C2-alkylated imidazolium salts, involving the formation of an intermediate Ni(II)-NHC.

As an extension to this work, Normand, Cavell and co-workers showed that Ni complexes featuring NHCs with allyl *N*-substituents can undergo intramolecular annulation to form five- or six-membered rings (e.g. **179**) upon reductive elimination (Scheme 23).[61] The mechanism for these ring formations proceeds analogously to the intermolecular coupling described above (Fig. 1).[58,59]

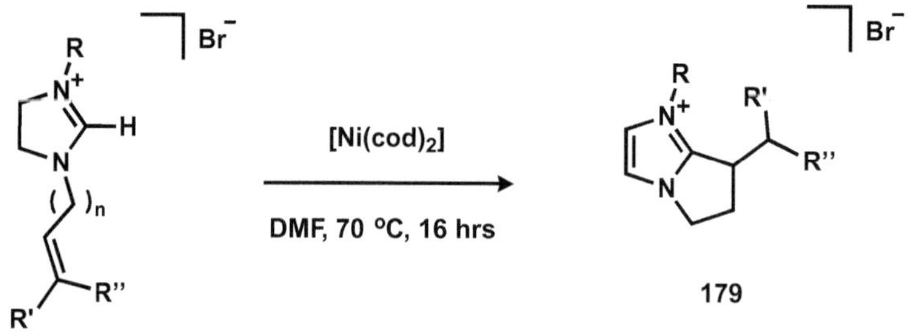

Scheme 23 Intramolecular annulation of an imidazolium salt proposed to proceed through the formation of a Ni(II)-NHC complex.

Following on from their previous observations, Cavell and co-workers investigated the possible implications of the reductive elimination of an NHC with an acyl group in catalytic carbonylation processes.[62] [Pd(CH$_3$)Cl(NHC)]$_2$ type complexes were treated with AgBF$_4$ in the presence of CO. This led to rapid decomposition of the complex **180** at $-20\ °C$ with the C2-acylated imidazolium salt **181** being the major organic component following the decomposition (Scheme 24). Small amounts of the methylated imidazolium salt **182** were also observed. Other such eliminations have been shown with different ligand environments.[63]

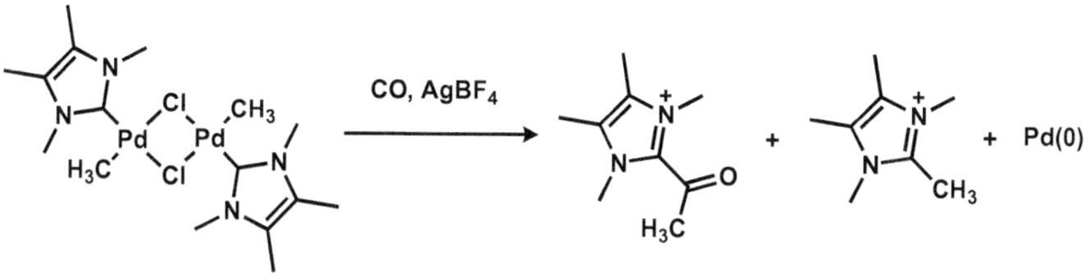

Scheme 24 Reductive elimination from Pd(II) of an NHC with an acyl group, with possible implications during catalytic carbonylation processes.

Reductive elimination of a C2-methylated imidazolium salt was observed as the decomposition product of a Pd complex bearing an NHC with coordinating 2-picolyl N-substituents.[64] Much harsher conditions were required for this reductive elimination to occur than for previous decompositions (100 °C over 12 h). The rationale for the complex to tolerate higher temperatures is that for the reductive elimination to occur, the methyl substituent must occupy a position *cis* to the NHC prior to elimination, thus dissociation of a picolyl ligand needs to take place so isomerization can ensue, and reductive elimination can occur. To further investigate this, a Pd(II)-NHC complex with diisopropyl-2-ethylamino N-substituents was studied. The weaker donor ability of the N-substituents was expected to enable reductive elimination to occur under relatively milder conditions by allowing the isomerization process to occur more freely. This was indeed the case and the methylated imidazolium product formed much more readily. A later example of a Pd complex with a chelating NHC ligand (N-Ar-SO$_3^-$ side group) was also found to undergo reductive elimination of an alkyl imidazolium salt.[65] Again, the reactivity was heavily dependent on the dissociation and displacement of the ancillary Ar-SO$_3^-$ group.

Fig. 2 Reductive elimination is affected by the bite angle of spectator ligands and the angle of the NHC relative to the plane.

Many factors seemingly affect the stability of complexes featuring chelating NHCs. In a further study it was noted that, for chelating NHC ligands twisted out of the coordination plane, there is a reduced ability of the rotationally constrained rings to interact with the methyl group and undergo reductive elimination.[66] A later DFT study on alkyl imidazolium reductive elimination focused on the angle of the NHC relative to the plane of the complex and bite angle of the spectator ligand.[67] Small ligand bite angles together with minimalization of the planarity with respect to the NHC ligand showed the highest resistance to decomposition (Fig. 2), which correlates strongly with Cavell's experimental findings.

Following on from this, further DFT studies found that the electron-donating capacity of the NHC is an important stabilizing factor for complexes.[68] Complexes ligated with NHCs featuring electron withdrawing N-substituents were found to be much more prone to reductive elimination than those featuring electron donating N-substituents. Other factors such as the steric environment of the coordinating NHC ligand have been highlighted, where results show that very bulky N-substituents can destabilize the complex.[69] The oxidation state of the metal within the complex plays an important role, with reductive elimination from complexes featuring high oxidation state metals being more facile than for analogous lower oxidation state metal complexes.[55,69,70]

Upon studying the oxidative addition of aryl chlorides with Pd(0) bis-NHC complexes to form amination pre-catalysts, successful oxidative addition to form a four-coordinate Pd(II) pre-catalyst was demonstrated.[71,72] However, the attempted oxidative addition of 4-chlorotoluene with Pd(0) bis-NHC **183**, featuring Dipp N-substituents, resulted in the exclusive formation of a C2-aryl imidazolium salt **184** (Scheme 25). These results parallel the observations by Cavell[53] and likely proceed analogously through initial oxidative addition of the aryl chloride, followed by rapid reductive elimination to afford the C2-aryl imidazolium salt.

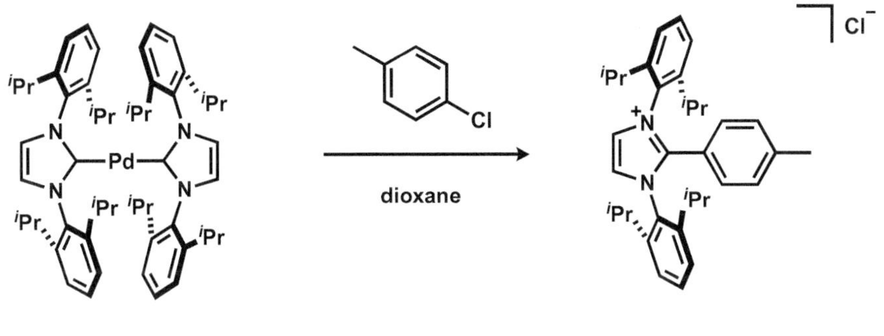

183 → **184**

Scheme 25 Attempted oxidative addition of 4-chlorotoluene to a Pd(II) bis-NHC, resulting in formation of a C2-arylated imidazolium salt.

Albéniz and co-workers illustrated that this type of reductive elimination is not exclusive to heterocyclic NHCs but can also take place when acyclic monoaminocarbene ligands are employed (Scheme 26).[73] It was illustrated that, similar to NHC ligands, the reductive elimination of iminium salts (for example, **186** from **185**) is present and a plausible route to catalyst decomposition.

(Pf = C$_6$F$_5$)

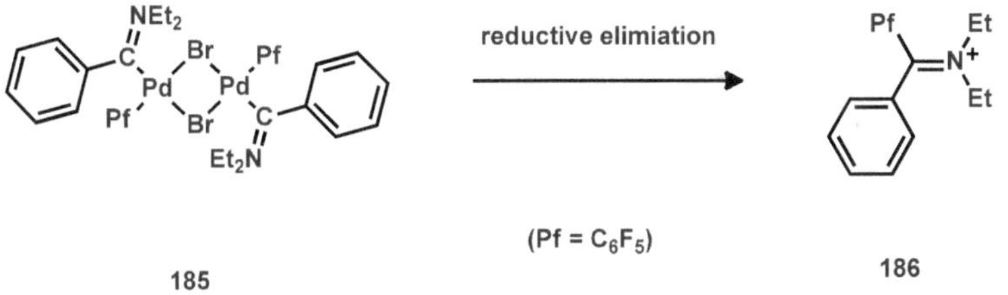

185 → **186**

Scheme 26 Reductive elimination of an iminium salt from a Pd(II)-monoaminocarbene complex.

During an investigation into direct arylation reactions with aryl chlorides using Pd-NHC catalysts, Fagnou and co-workers observed that reductive elimination of a C2-arylated imidazolium salt was the main pathway for decomposition.[74] In an attempt to limit catalyst decomposition, they used the imidazolium salts [IMes]HCl and [IPr]HCl as reaction additives and saw a marked improvement in the TONs for the target arylation. This was thought to act similarly to the systems reported by Cavell,[57] where reductive elimination of an aryl imidazolium would be followed by catalyst reformation by oxidative addition of the imidazolium additive. Interestingly, when testing this hypothesis, palladium black was treated with

[IPr]HCl which was found to catalyze the arylation reaction. This suggested that, even after palladium black formation in a reaction, catalyst recovery is somewhat possible.[49,74]

Numerous examples of the reductive elimination of NHC ligands with aryl groups from Pd-NHC complexes now exist,[49,52,53] with other metal centers such as Rh and Cu also displaying this reductive elimination process.[75–78] As with reductive elimination from metal-alkyl complexes, the reductive elimination of 2-arylimidazolium salts from aryl metal complexes has also been extensively studied computationally.[79,80] Comparative studies on reductive C—C bond formation from NHC complexes illuminate trends across the different types of coupling reactions. For example, in 2018 Gordeev and co-workers investigated R-X oxidative addition and the factors affecting subsequent reductive elimination of R-NHC coupling products.[81] The effect of the R group was investigated and the predicted reactivity in the R-NHC coupling process was found to decrease in the following order: R = vinyl > ethynyl > Ph > Me. Gordeev later noted that the reaction solvent has a larger role to play in the coupling product formation and stability than the chemical nature of the organic substituent R.[82]

The solvent dependency for reductive elimination of aryl imidazolium salts was first noted by Grushin in 2003 while studying synthetic pathways to Pd(II)-aryl complexes stabilized by NHCs.[83] Successful synthesis of the target complex [Pd(IPr)(PPh$_3$)(Ph)Cl] was achieved, however, while being stable in benzene, the complex would undergo rapid C—C reductive elimination in dichloromethane affording the corresponding 2-arylimidazolium salt. Addition of one equivalent of PPh$_3$ to the complex in dichloromethane almost completely inhibited the formation of the imidazolium elimination product, suggesting that phosphine dissociation occurs prior to reductive elimination. The increased ionizing power of the reaction medium with dichloromethane lowers the activation barrier to the polar transition state following phosphine dissociation, thus allowing reductive elimination to be more facile for these phosphine containing complexes in more polar solvents.

Many initial investigations into reductive M-NHC bond cleavage and C—C bond formation considered the implications and circumvention of this process for catalyst decomposition and complex instability.[84] This process, however, was soon exploited as a valuable tool in novel synthesis pathways and catalytic applications.[49] Ellman and co-workers described the intramolecular Rh-catalyzed annulation reactions using a variety of alkene appended benzimidazolium salts, generating new five- and six-membered rings.[85] This was followed by a series of investigations attempting to elucidate the mechanism for the C—C bond formation process.[86] Though the initial mechanism was thought to proceed via hydride migration and reductive elimination of C—H, a comprehensive study by Yates suggests a different route involving C—C reductive elimination,[87] similar to the Ni-catalyzed annulation demonstrated by Normand.[61] This chemistry has since found great use in the synthesis of many functionalized heterocycles.[51]

During the synthesis of novel Ti(III)- and V(III)-NHC complexes with pendant fluorene and indene N-substituents (**188**), Danopoulos stated that the NHC ligand would likely be acting as a reducing agent affording the target M(III) complexes (Scheme 27).[88] This was supported by the formation of a spiroimidazolium salt (**189**) isolated from the reaction solution. The authors noted that it is unlikely that **189** was formed in situ following the synthesis of the target complex **188**, therefore the NHC is likely acting as both a reagent for the reduction of the metal and a ligand following this reduction.

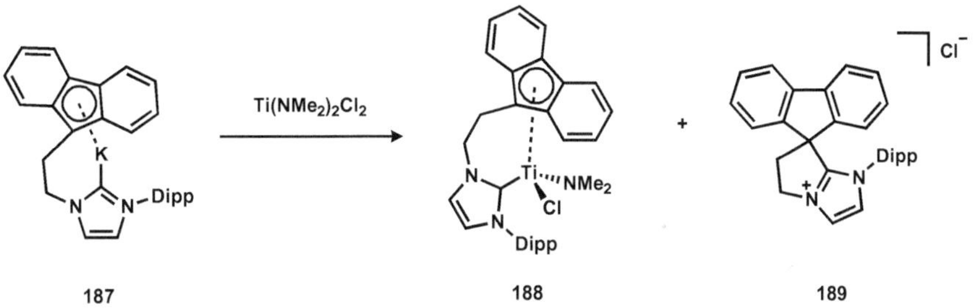

Scheme 27 Reduction of Ti(IV) by an NHC bearing fluorene N-substituents.

Astakhov described a very interesting and novel mode of operation of catalytic Pd-NHC systems, whereby NHC-aryl reductive elimination from the parent Pd(II) complex would form Pd(0) deposits of nanoparticles, however these nanoparticles would be stabilized by the C—C coupled product imidazolium salt.[89] The nanoparticles formed from this process were found to be catalytically active and were utilized in the Mizoroki-Heck reaction. The mechanism to the formation of the nanoparticles is thought to proceed through both C—C and C—H coupling to form the stabilizing salts (Scheme 28). A later computational study confirmed that this mechanism is indeed plausible but noted that this mode of action only exists for a few examples, and it was stated that this process may strongly depend on the nature of the catalytic reagents.[90]

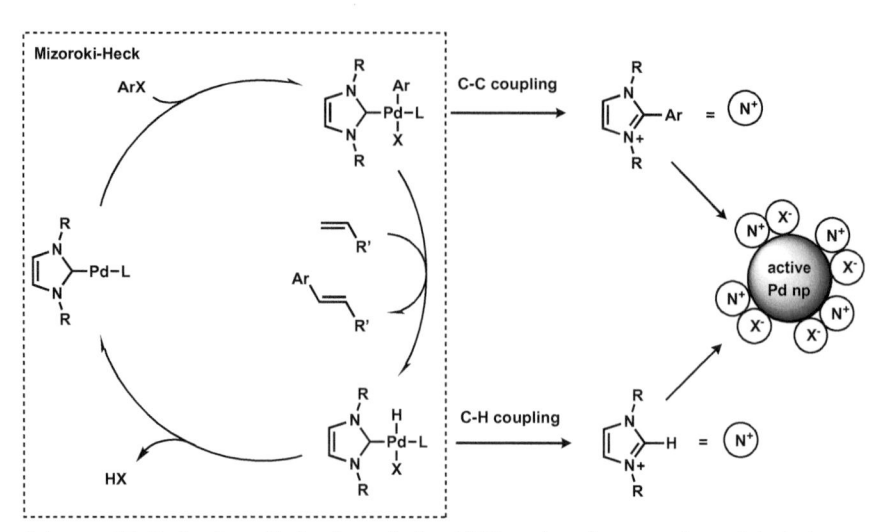

Scheme 28 Reductive elimination of the NHC and aryl group from Pd during a Mizoroki-Heck reaction, with subsequent stabilization of catalytically active Pd nanoparticles by the imidazolium salts.

Though the majority of investigations into NHC reductive elimination focused on C2—C coupling, with some examples of C2-C2' coupling of bis-NHC ligands,[70,77,91] Albrecht and co-workers described C4-C4' bond formation while investigating the electronic impact of abnormally bound NHC ligands in Pd complexes.[92] Exposure of a Pd complex **190**, featuring an abnormally bound bidentate bis-NHC, to Cl_2 led to C—C reductive elimination, generating a tricyclic bis-imidazolium dichloro salt **191** (Scheme 29A). Oxidative addition of Cl_2 to the Pd center forms an unstable Pd(IV) intermediate, where reductive elimination of the tricyclic bis-imidazolium **191** can occur. This contrasts with the analogous C2-C2' bound NHC complex which does not undergo reductive elimination, only reversion to the parent complex with halide switching. These observations agree with other findings indicating C4 reductive elimination from a metal complex is more facile than C2 elimination. Similar abnormal coupling reactions have been reported, such as the reductive elimination of a tricyclic bis-imidazolium **194**. Reductive elimination from a Pd(II) intermediate **193**, which features a bis-NHC coordinated to the Pd through the C2 position on one ring and the C4' position on the other, was speculated as the reaction pathway (Scheme 29B).[93]

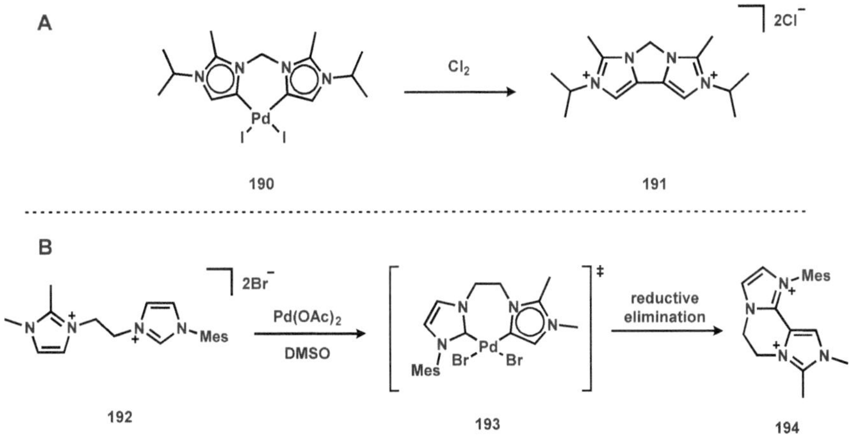

Scheme 29 C4-C4′ (A) and C2-C4′ (B) reductive elimination of bis-NHC ligands from Pd(II).

4. C-X reductive elimination

In sharp contrast to the reductive elimination of alkyl-, acyl- and aryl-azolium salts from transition metal-NHC complexes, the reductive elimination of 2-haloazolium salts is rarely covered in the literature, with examples for only a few transition metals.[49] This is despite the ubiquity of transition metal-NHC catalysts that feature X ligands (X=halide or pseudohalide), and the prevalence of investigations into reductive C-X bond formation.[94,95] This could be due, in part, to the challenges associated with gaining mechanistic insight for these reactions. 2-Haloazoliums, for example, can undergo facile oxidative addition to a transition metal complex, resulting in the rapid reformation of the metal-NHC complex following the initial reductive elimination of the halogenated azolium ion.

Initial investigations into the formation of 2-haloazolium salts from transition metal-NHC complexes were reported through reaction with molecular halogens (Cl$_2$, Br$_2$ or I$_2$),[96–99] though the mechanisms for formation were rarely probed. Albrecht and co-workers reported the formation of a 4-chloroazolium salt **196** from [PdI$_2$(bis-NHC)] (**195**) under exposure to Cl$_2$ (Scheme 30).[92] The bis-NHC in **195** is bound "abnormally" through the C4 position. The exact mechanism for this transformation was not probed, however, it was speculated that successive redox reactions with excess Cl_2 would ultimately produce the oxidative addition intermediate, [PdCl$_4$(bis-NHC)], which could then undergo reductive elimination of the C4, C4' chlorinated bis-azolium product **196**. The authors noted that this reductive elimination should be facile in order to compete with the reductive elimination of Cl_2 from the Pd(IV) intermediate. This is in accord with a recent theoretical study suggesting that Pd(IV) complexes would be very susceptible to reductive NHC-halogen coupling.[69]

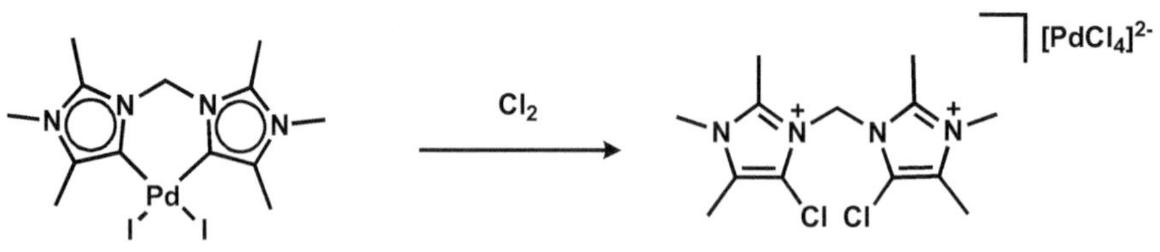

Scheme 30 Reductive elimination of C4, C4' chlorinated bis-imidazolium salt following exposure of an abnormally bound bis-NHC Pd(II) complex to Cl_2.

Stack and co-workers sought to better understand the mechanism of reductive elimination of 2-haloimidazoliums from Cu(I)-NHC complexes.[100] Oxidation of Cu(I)-NHC complexes using Selectfluor or Cu(II) salts lead to quantitative conversion to the corresponding 2-haloimidazolium salt at room temperature. DFT calculations were used to assess the mechanism of action, utilizing Selectfluor as the oxidizing agent (Fig. 3). Initially, fluoride transfer to and oxidation of the Cu(I) center of **I** would occur, yielding a 3-coordinate Cu(III) complex **II**, to which solvent MeCN could coordinate to give the 4-coordinate complex **III**. From this, a facile reductive elimination reaction would occur via **IV**, forming the 2-haloimidazolium salt **V**. The authors noted that other pathways were found but were implausible due to the scale of the associated activation barriers.

Hong and co-workers reported the quantitative formation of a 2-chloroimidazolium salt when attempting the synthesis of a Cu(II)-NHC complex via reaction of a Ag-NHC with $CuCl_2$.[101] This was again observed in a later report by Nechaev and co-workers in a diverse investigation on Cu-NHC complexes.[102] Using a similar synthetic approach to Hong, utilizing the Ag-NHC complex **197** this time with multiple equivalents Cu(II)-halides, Nechaev too observed the quantitative formation of the

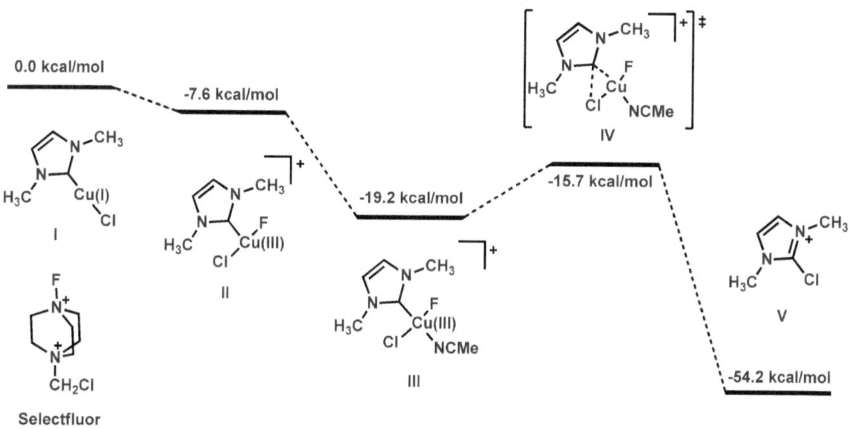

Fig. 3 Relative free energies in the reductive elimination of 2-haloimidazolium salt via a proposed Cu(III) intermediate.

C2-halogenated imidazolium salt **198** (Scheme 31). They report similar findings to Stack and co-workers, where oxidation of a Cu(I)-NHC complex with Br$_2$ leads to the formation of a C—Br coupled imidazolium salt, which were also observed using six- and seven-membered rings. A mechanism was not suggested here but one can assume a similar pathway of oxidative addition and reductive elimination.[100] In addition, the synthesis of Cu^{II}-NHC complexes was attempted via reaction of a free NHC (again with five-, six- and seven-membered rings) with CuX$_2$ (X = Cl, Br), which again leads to the formation of the NHC-halogen coupled products. A mechanism was proposed for this reaction where initial formation of a Cu^{II} bridged dimer, [(NHC)CuX$_2$]$_2$, would be followed by reductive elimination of the 2-haloamidinium salt with a CuX$_2^-$ anion. However, further mechanistic investigations were not undertaken.

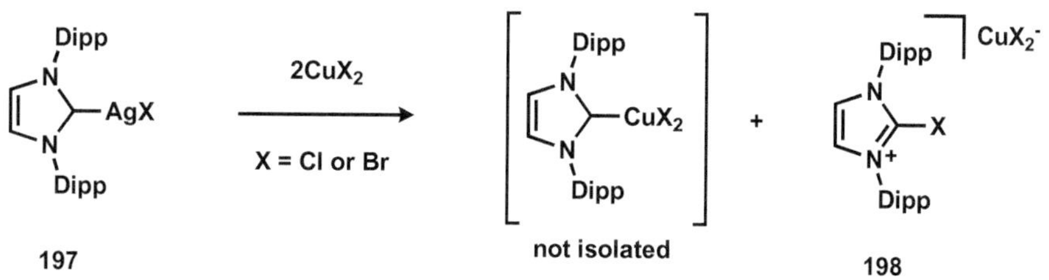

Scheme 31 Formation of a C2-halogenated imidazolium salt through reaction of a Ag-NHC with CuX$_2$.

Successful synthesis of Cu(II)-NHC complexes bearing halide ligands was demonstrated by Willans and co-workers.[77] Key to the stability of these complexes was the coordinating pyridyl *N*-substituent on the NHC, which had previously been demonstrated by Willans in an earlier study.[103] The stable Cu(II)-NHC complex **200**, of the type [Cu(II)(NHC)Br$_2$], could be isolated through the reaction of a Ag-NHC complex **199** with one equivalent of CuBr$_2$. Interestingly, however, reaction of the same Ag-NHC complex **199** with two equivalents of CuBr$_2$ would result in the formation of 2-bromoimidazolium **202** with a Cu containing anion (Scheme 32). DFT calculations indicate that, in the presence of excess CuBr$_2$, reductive elimination of the 2-bromoimidazolium from a bimetallic Cu(III)-Cu(I) intermediate **201** (where the NHC is coordinated to the Cu(III) center) would be a plausible and facile process.

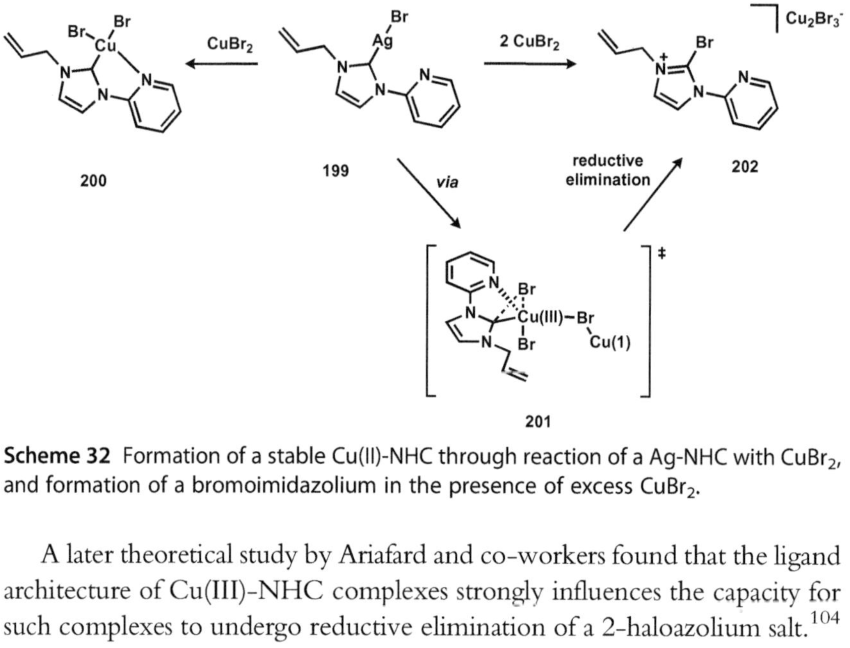

Scheme 32 Formation of a stable Cu(II)-NHC through reaction of a Ag-NHC with CuBr$_2$, and formation of a bromoimidazolium in the presence of excess CuBr$_2$.

A later theoretical study by Ariafard and co-workers found that the ligand architecture of Cu(III)-NHC complexes strongly influences the capacity for such complexes to undergo reductive elimination of a 2-haloazolium salt.[104] Inclusion of strong σ-donating anionic ligands on Cu (Ph, CH$_3$, vinyl, etc.) rendered reductive elimination of haloazolium salts energetically unfeasible, with the reverse being true for weakly σ-donating ligands (Br, Cl, CN etc.). This is in agreement with the experimental findings of Fairlamb and co-workers, where Cu(III)-NHC complexes with a Ph ligand did not undergo reductive elimination of the corresponding C-X (X=Cl, Br) coupled product.[78]

A rare example for the reductive elimination of C—N bonds from transition metal–NHC complexes is described by Gautier and co-workers, where various Ag-NHC complexes with pendant azidoaryl (-Ar-N$_3$) N-substituents reacted with CuCl to yield fused nitrogen heterocycles.[105] A mechanism for this reaction was proposed and verified by DFT calculations (Scheme 33). First, the Ag-NHC complex **202** reacts with CuCl to give AgCl and the corresponding Cu-NHC complex **203**, where the azidoaryl N-substituent on the NHC interacts with the Cu center through the N closest to the aromatic ring. Nitrogen evolution, thought to be the driving force of the reaction, results in the formation of the Cu-NHC complex **204**, from which reductive elimination can occur to give the cyclized product **205**.

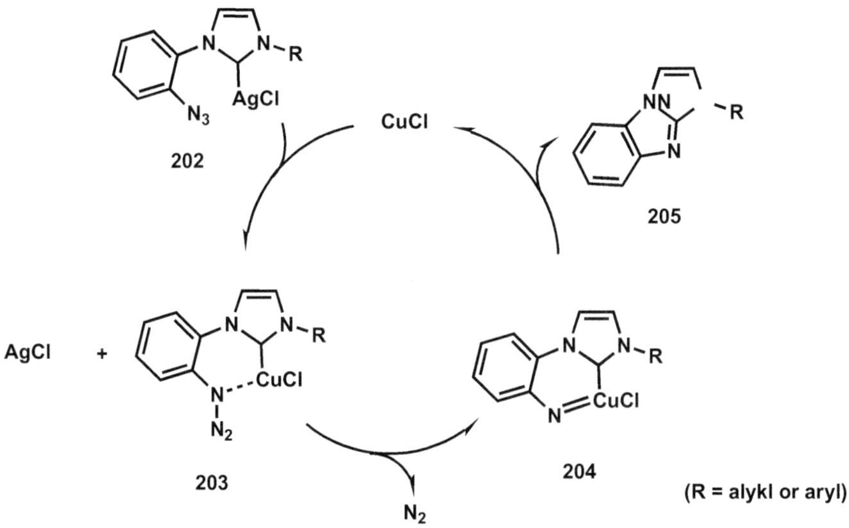

Scheme 33 Reductive elimination of C—N from a Cu-NHC resulting in fused nitrogen heterocycles.

During an investigation into the deactivation of metal–NHC catalytic systems, Ananikov and co-workers observed the formation of imidazolones directly after the addition of KOH to a solution of a Pd–NHC catalyst.[106] The authors stated that a reductive elimination pathway would be the most plausible route after performing various isotope labelling experiments. They postulated that ligand exchange on the metal with R'O⁻, to give the corresponding hydroxy Pd-NHC complex, would be followed by reductive elimination of an NHC-O coupled imidazolium salt (Scheme 34).

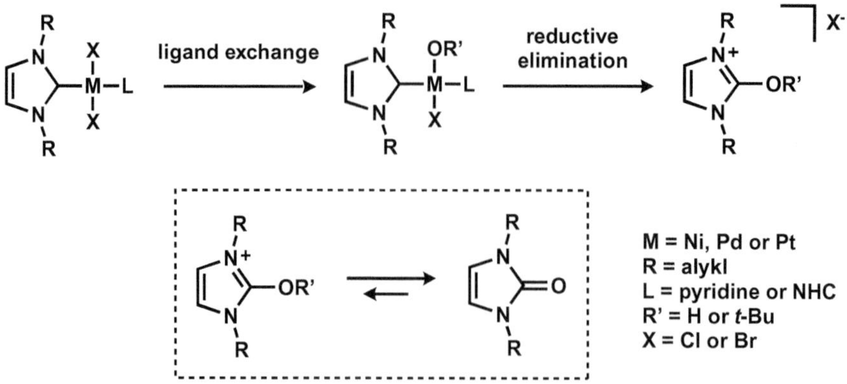

Scheme 34 Reductive elimination from metal centers to form imidazolones.

Elimination of the R' group from the imidazolium salt would then result in the formation of the product imidazolone. Other examples of reductive NHC-O coupling from transition metal complexes exist that may proceed by a similar mechanism.[107,108]

5. C—H reductive elimination

As with C-X reductive elimination, there are relatively few investigations focusing on the formation of 2-H azolium salts via reductive elimination from transition metal-NHC complexes.[49] This may be in part due to the difficulty in differentiating between true reductive elimination of an azolium salt and protonolysis of the metal-NHC bond by protic impurities in reaction mixtures.[99,109] In an example from Bullock and co-workers, imidazolium was formed during ketone hydrogenation catalysis using a W-NHC complex featuring a hydride ligand.[110] Importantly the authors made it clear that the experiments conducted could not distinguish between a reductive elimination process and other possible formation mechanisms.

While attempting to identify intermediates for a proposed olefin insertion mechanism using a halide abstracted cationic Pd-NHC complex (**206**, Scheme 35), Cavell and co-workers identified the formation of imidazolium ions.[53] The imidazolium was proposed to have formed via coordination, and migratory insertion of an olefin into the Pd—Ar bond of complex **206**, followed by β-hydride elimination to afford the unstable cationic Pd—H complex **207**. This could then undergo reductive elimination of an imidazolium salt, **208**. Ananikov and co-workers later reported similar findings while working on a multimodal Heck catalytic system, where a

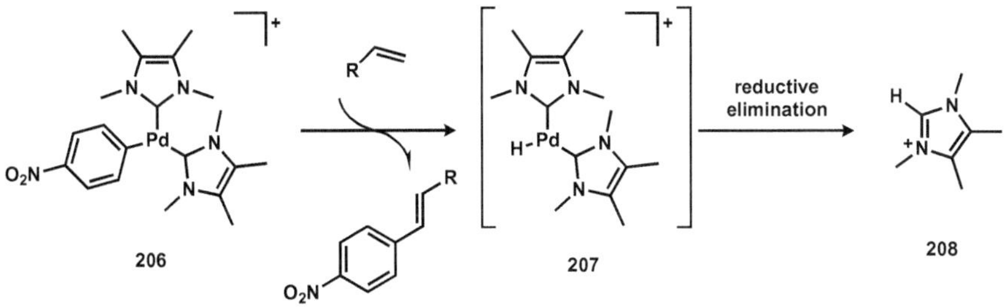

Scheme 35 Reductive elimination of an imidazolium via a proposed Pd-hydride intermediate.

Pd-NHC hydride intermediate would undergo reductive elimination of an imidazolium salt, forming catalytically active Pd(0) metal deposits.[89,111]

Cavell and co-workers later stated that, in addition to reductive elimination of 2-alkylated imidazolium salts from a Ni-NHC ethyl complex, the reductive elimination of a 2-H imidazolium salt could also occur.[57] β-hydride elimination of the coordinated ethyl ligand from complex **175** would yield a Ni—H, from which the reductive elimination of the imidazolium could take place (Scheme 21).

Reductive elimination of an imidazolium from a transition metal complex under a hydrogen atmosphere was also demonstrated for cationic Pd-NHC complexes.[112,113] Oxidative addition of H_2 onto the Pd center was thought to precede the irreversible reductive elimination of the imidazolium. Van Rosenburg observed a similar catalyst degradation pathway when exploring the use of a Co-NHC dimer, $[Co_2(CO)_6(IMes)_2]$, for use in hydroformylation.[114,115] The catalyst was exposed to an atmosphere of CO and H_2 (1:1 and 1:2) at elevated temperatures to evaluate its viability for hydroformylation, however this lead to catalyst decomposition. Reductive elimination of IMes-H from the Co dimer upon exposure to H_2 was suggested as the decomposition pathway, due to the formation of $[IMesH]^+[Co(CO)_4]^-$ and the monitored collapse of CO signals by IR spectroscopy.

Coordination of an NHC N-substituent does not always offer enhanced stability toward transition metal-NHC catalyst decomposition, as demonstrated by Albrecht and co-workers when investigating Ru(II)-NHC complexes as hydrogenation catalysts.[116] It was observed that Ru(II) complexes furnished with NHCs with pendant allyl N-substituents, e.g. **209**, underwent catalyst decomposition, affording free imidazolium salts (Scheme 36). In situ high-pressure NMR spectroscopy was used to help identify a plausible mechanism for this decomposition. It was suggested that intramolecular alkene hydrogenation of the NHC *N*-substituent under the H$_2$ atmosphere would initially occur, rendering the NHC as a monodentate ligand (**210, 211**). This would be followed by reductive elimination of the NHC and a hydride ligand from Ru complex **211** to yield an imidazolium cation **212**.

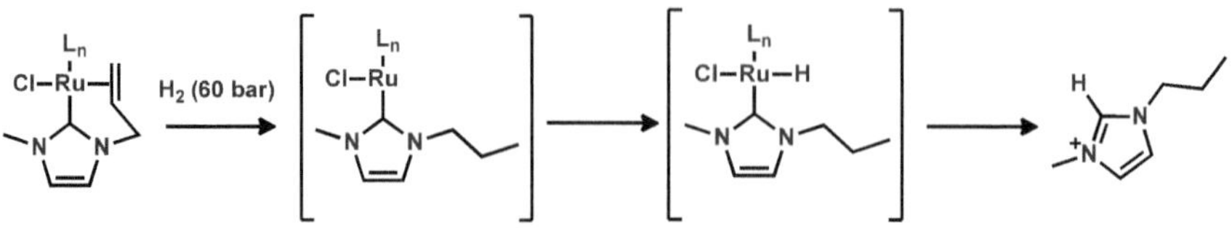

Scheme 36 Reductive elimination of an imidazolium from Ru with a proposed hydride intermediate.

In a similar fashion, Veige and co-workers investigated metal-NHC complexes as hydrogenation catalysts, using Ir and Rh complexes featuring a bidentate bis-NHC ligand, predicting that the bis-NHC would be more resistant to reductive elimination than a monodentate analogue.[117] Under catalytic hydrogenation conditions the Rh-NHC catalyst decomposed rapidly to form black deposits of Rh^0, and analysis by ^{1}H NMR spectroscopy identified the imidazolium, which was suggested to be formed by reductive elimination. Interestingly, though full catalyst decomposition had occurred, the conversion to the target hydrogenated product was >98%. The analogous Ir-NHC complex was tested under that same conditions and, along with no black deposits being observed, no conversion to the product was seen. As Ir forms stronger M-NHC bonds that Rh, it was suggested that the Ir-NHC complex was less susceptible to reductive elimination of the bis-imidazolium thus not producing the active M^0 catalytic species.

Cavell and co-workers highlighted an interesting example where the reversible reductive elimination of a single imidazolium salt from [PtBrH(NHC)$_2$] (**213**) could be achieved through addition of an alkene (styrene or dimethyl fumarate) (Scheme 37).[118] The starting complex **213** features two different NHC ligands with one bound normally through the C2 position and the second bound abnormally through the C4 or C5 positions on the heterocycle, as the C2 position is blocked. Interestingly, the reductive elimination following alkene addition was only found to occur for the

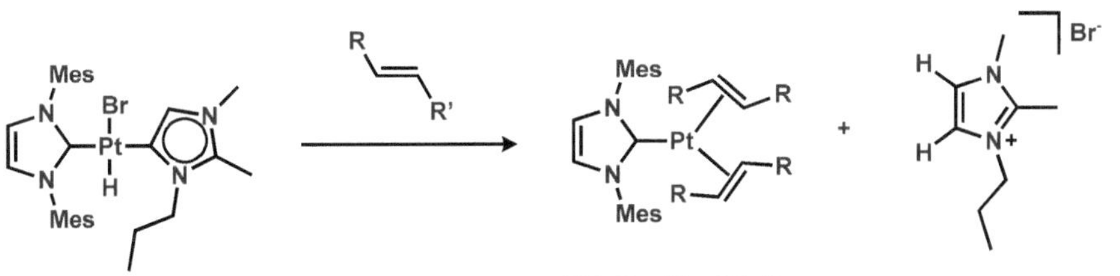

Scheme 37 Reductive elimination of an abnormally bound NHC from Pt to form an imidazolium.

abnormally bound NHC, with no detectable traces of the normal bound reductive elimination product. Cavell noted that alkylation of ionic liquids at the C2 position may not be enough to prevent involvement in reactions for which they are used as solvents, as reactions may still take place at the C4 or C5 positions.

6. Insertion into a metal-C$_{NHC}$ bond

The use of metal–NHCs in catalysis has shown that these complexes can possess remarkable activity. It used to be thought that the NHC ligands act merely as spectator ligands and remain firmly bound to the metal center during a catalytic cycle. However, as described previously, both experimental and theoretical studies have shown that reductive elimination of the NHC from a metal center can occur, with migratory insertion being a potential first step in some of these reactions. In 2002, the insertion of an alkene into a Rh-NHC bond was proposed as a key step in the catalytic intramolecular coupling of an alkene with benzimidazole.[86] In the following year, Danopoulos and co-workers reported the first experimental evidence of methyl migration from a Pd(II) center to the carbene carbon of an NHC, with the heterocyclic carbon remaining coordinated to the metal through an organo-Pd bond.[119] The reaction of a pyridine-anchored pincer bis(NHC) (**217**) with [Pd(tmeda)Me$_2$] resulted in the isolation of the major product **218** in which one of the methyl groups has migrated from Pd to the NHC carbon (Scheme 38). Computational studies indicated that the reaction proceeds via a 5-coordinate intermediate in which all three functionalities of the pincer ligand are coordinated to Pd(II), in addition to the two methyl ligands. This is in contrast with the concerted pathway proposed by Cavell for alkyl-carbene reductive elimination,[56] with the ridged chelating ligand thought to prevent reductive elimination from the metal center in this

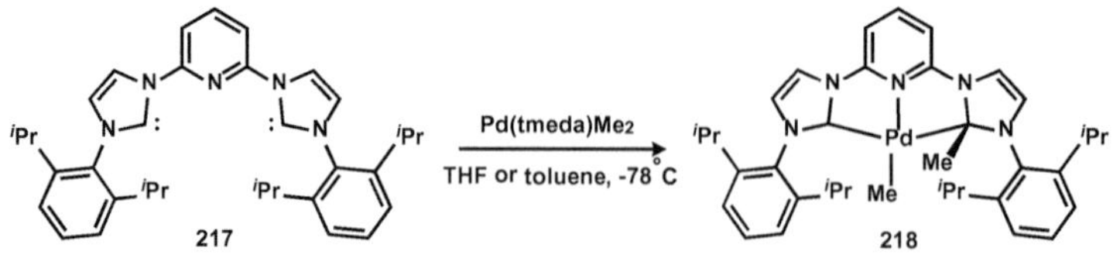

Scheme 38 Reaction of a pincer bis(NHC) with Pd(tmeda)Me₂ resulting in methyl migration.

case. In the 5-coordinate intermediate, one of the methyl groups is in the plane of the ligand, with the second being axial, which allows interaction with the empty p-orbital of the NHC carbene carbon.

A few years later, a related reaction was reported at Ni which was thought to proceed via a similar methyl migration mechanism. Reaction of the same pincer ligand with [Ni(DME)Br₂] (DME = dimethoxyethane) resulted in the expected cationic square planar Ni(II)-bis(NHC) complex.[120] However, the same reaction of the NHC with [Ni(tmeda)Me₂] resulted in ring-opening of one of the heterocycles, with the proposed mechanism for this transformation being based on a migratory insertion.

Studies on the reactions of Ru(II)-NHC complexes with alkynes have shown migration of the NHC from Ru on to a coordinated alkene. In 2006, it was reported that the reaction of [Ru(Cp)(IPr)(CH₃CN)₂]⁺ (219) with terminal alkynes led to migration of the NHC to an electrophilic alkylidene (221), formed through a metallacyclopentatriene intermediate (220) which was characterized using a combination of 1D and 2D NMR spectroscopy (Scheme 39A).[121] The C—C coupling of the terminal alkynes was highly regioselective, with the substituents being exclusively in the 1 and 3 positions, and DFT calculations supported the proposed mechanism with relatively low activation barriers. In the following year the same group reported reactions of [Ru(Cp)(IPr)(CH₃CN)₂]⁺ and [Ru(Cp)(SIPr)(CH₃CN)₂]⁺ with the parent acetylene which resulted in the formation of a (ruthenocenylmethyl)imidazolium salt 225 (Scheme 39B).[122] DFT calculations were conducted to understand the mechanism of this reaction, with the first step being, again, the formation of the metallacyclopentatriene intermediate 222. In this case, a third acetylene coordinates to the metal center (223) and is thought to insert into the Ru-C$_{NHC}$ bond to form a metallacyclopentadiene 1-metallacyclopropene intermediate (224). A series of vinyl insertion steps leads to the [2+2+1] cyclotrimerization product 225, which converts to the Ru sandwich complex 226 at −40 °C, bearing a methylimidazolium-substituted cyclopentadienyl ligand.

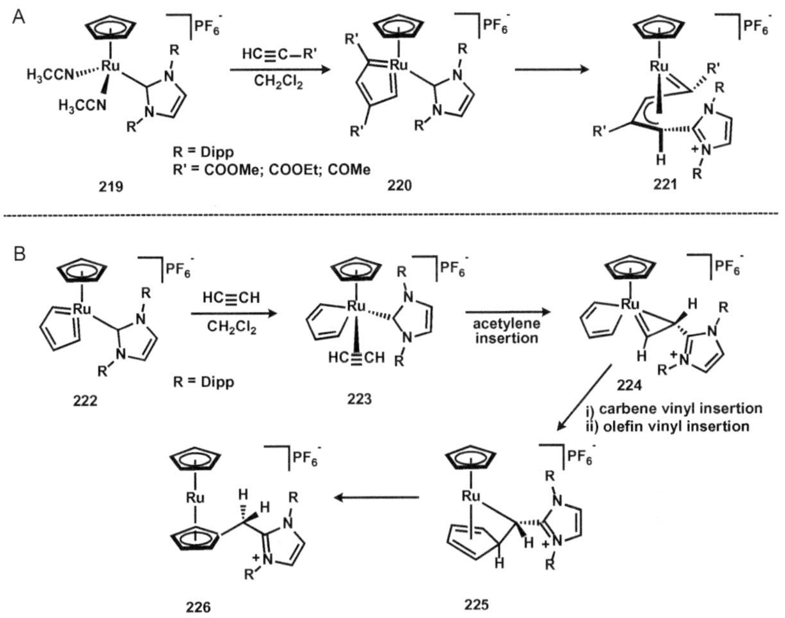

Scheme 39 Reaction of Ru(II)-NHC complexes with alkynes.

A study into the activation of internal alkynes by Ru–NHC complexes was reported in 2013, in which migratory insertion of an NHC bearing a picolyl N-substituent was found to occur (Scheme 40).[123] Following Ru–C$_{NHC}$ bond cleavage in **227** the NHC inserts into a Ru-coordinated alkyne, leading to C$_{picolyl}$–N bond cleavage and tautomerization of the heterocycle with coordination to the metal through the nitrogen (**228**).

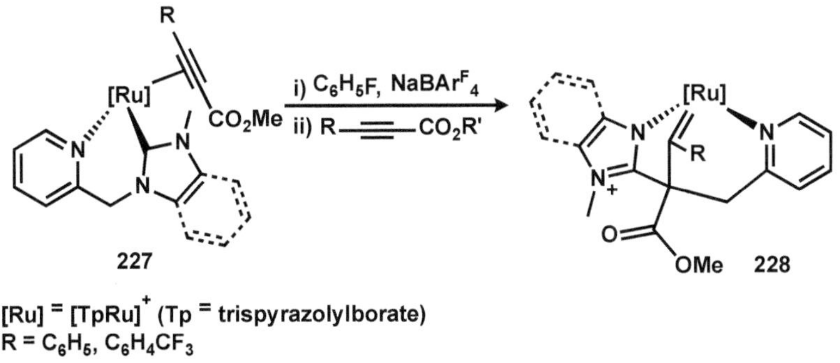

Scheme 40 Migratory insertion of an NHC into an internal alkyne resulting in C$_{picolyl}$-N bond cleavage and tautomerization.

Akin to the studies at Ru by Becker and co-workers discussed in the previous paragraph, a metallacyclopropene intermediate is proposed as an intermediate in the process, though no experimental or theoretical evidence is provided in this case.

Nolan and co-workers provided an example of the insertion of an NHC into a Pt-coordinated olefin through reaction of Pt(1,5-hexadiene)Cl$_2$ **229** with IPr.[124] A reaction intermediate (**230**), involving displacement of one of the olefins and coordination of an NHC to the Pt, was isolated and characterized (Scheme 41). A second product was observed in very low yield, which could be increased through the addition of excess IPr. This was found to be the zwitterionic imidazolium species **231**, in which an NHC had inserted into one of the olefins. DFT calculations considered both intramolecular and intermolecular pathways, with the activation energy for an intermolecular process being significantly lower than that for an intramolecular, thus external attack by an NHC at the olefin was proposed.

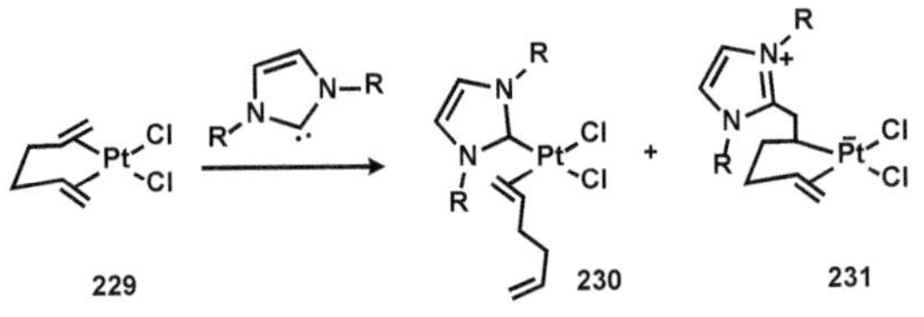

Scheme 41 Insertion of an NHC into a Pt-olefin bond.

Following previous studies in which NHCs were found to be excellent ancillary ligands to support 3- and 4-coordinate Fe(II)-alkyl, -aryl and -amido complexes,[125,126] Deng and co-workers examined the reactivity of Fe(II)-alkynyl complexes with NHC ligation.[127] The reaction of Fe(IPr$_2$Me$_2$)$_2$(C≡C^{t}Bu)$_2$ **232** with CN^{t}Bu in benzene resulted in formation of diamagnetic Fe(0) complex **233** bearing three CN^{t}Bu ligands and a zwitterionic vinyliminacyl ligand (Scheme 42). An imidazolium is bonded to the central carbon atom of the coordinated allylic group. A pathway for this process is proposed which involves the initial formation of Fe(C≡C^{t}Bu)$_2$(CN^{t}Bu)$_4$ upon reaction with CN^{t}Bu, and subsequent loss of the two NHC ligands. A series of migratory insertion reactions can lead to the formation of Fe vinylidene species, with external attack by one of the released NHC ligands producing the isolated product.

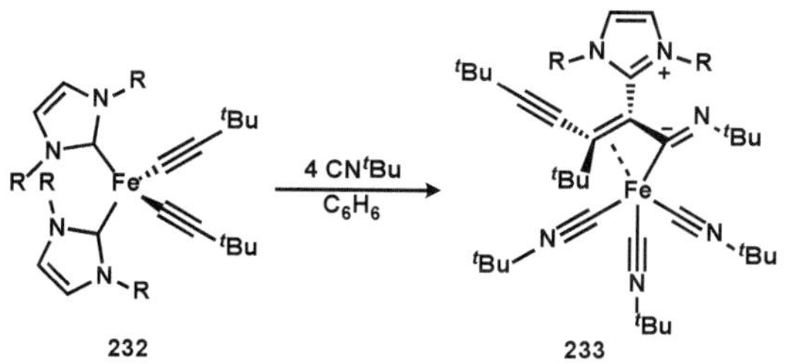

Scheme 42 Reaction of an Fe(II)-alkynyl NHC complex with CN^{t}Bu.

Ammonia-borane has attracted attention as a hydrogen storage material, with its dehydrogenation rate being one of the biggest challenges in the field. Ni-NHC complexes have been used as catalysts for this reaction,[128] with Hall and co-workers using computational methods to explore the mechanism of the Ni(NHC)$_2$-catalyzed process.[129] DFT calculations support an initial proton transfer step from the nitrogen atom of ammonia-borane to the NHC carbene carbon atom. The carbon atom remains bound to the metal center through an organo-Ni bond, with the new C—H bond being activated by the metal to transfer the H to Ni and regenerate the NHC. A second H is transferred to the metal from the boron atom and subsequently forms H$_2$. Release of a second equivalent of H$_2$ is also thought to be assisted by the NHC, through activation of a 3-center 2-electron Ni-H-B bridging bond.[130] This study demonstrates the importance of NHC ligand participation in some catalytic processes, which helps lower the barrier to activation.

The soft nature of NHC ligands and oxophilic nature of earlier transition metals renders early metal-NHC complexes less stable than late transition metals. However, ligands have been developed that contain anionic N-substituents, with the chelating ligands being beneficial to the stability of group 4 and 5 metal-NHCs. Reaction of an imidazolium salt bearing two phenol N-substituents with BuLi followed by ZrCl$_4$(THF)$_2$ results in a 6-coordinate Zr(IV) complex, in which the ligand binds tridentate, and two chlorides and a THF molecule remain coordinated to the metal center.[131] The use of a basic Zr precursor, Zr(CH$_2$Ph)$_4$, in toluene produces a 5-coordinate complex (**235**), through the tridentate ligand, a benzyl and a chloride. However, adding this complex to THF, or conducting the original reaction in THF, results in the migration of the benzyl group to the NHC

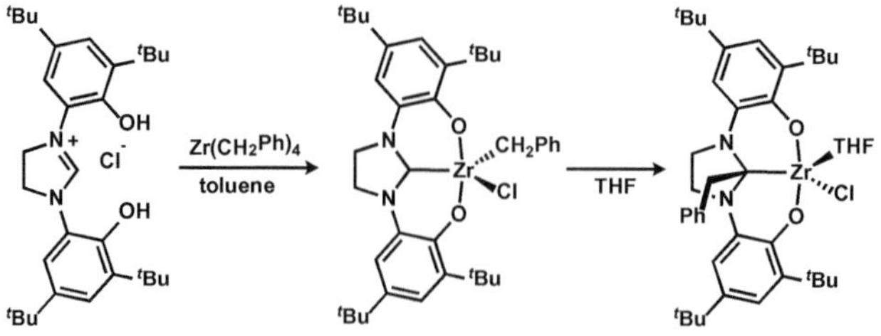

Scheme 43 Reactions of an imidazolium salt bearing two phenol N-substituents with a basic Zr precursor.

carbon to produce complex **236** (Scheme 43). Through characterization using X-ray crystallography, the structure is described as a heptacoordinate Zr(IV) complex, bearing a η^5-O,N,C,N,O-pentadentate trianionic ligand, a chloride and a THF molecule. DFT calculations support the fact that the energy of the migratory insertion step is lowered by the coordination of THF to the Zr center.

7. Hydrolytic ring opening

There are several examples of ring openings and expansions of NHCs that occur in the presence of water. Often only trace amounts are necessary for this to proceed and different mechanisms are prevalent depending on the concentration of water in the reaction mixture.[132] A ring expansion was reported by Weaver and co-workers in which an imidazolinium salt **237** was treated with KO^{t}amyl and Cu(OTf)$_2$ in THF (Scheme 44A).[133] This delivered piperazinone **238** as well as imidazolinone **239** and none of the desired Cu complex was formed. Adventitious water was suggested to initiate the ring expansion and either oxygen or water to produce the imidazolinone.

A ring opening of a pendant NHC within an Ir complex was observed by Braunstein and co-workers.[134] Complex **240** was treated with strong base and complex **241** which contains the pendant NHC was observed by NMR spectroscopy (Scheme 44B). The ring opened product **242** was also observed in the reaction mixture and complex **241** slowly decayed to complex **242** over time. The rate of this decay was accelerated if wet solvents were used in place of anhydrous solvents.

Another example involving an Ir complex was provided by Peris and co-workers.[135] In a reaction between imidazolium salt **243** and

[IrCp*Cl$_2$]$_2$ in the presence of Cs$_2$CO$_3$ and KI a mixture of complexes **245** and **246** was observed (Scheme 44C). The authors suggested that the ring opened intermediate **244** was involved in the mechanism and performed several experiments to show this. First, the hydrolysis of **243** to form **244** under basic conditions was shown to be feasible suggesting that ring opening occurs prior to complexation. When this reaction mixture was then treated with [IrCp*Cl$_2$]$_2$, an increased yield of the ring opened product **246** was obtained.

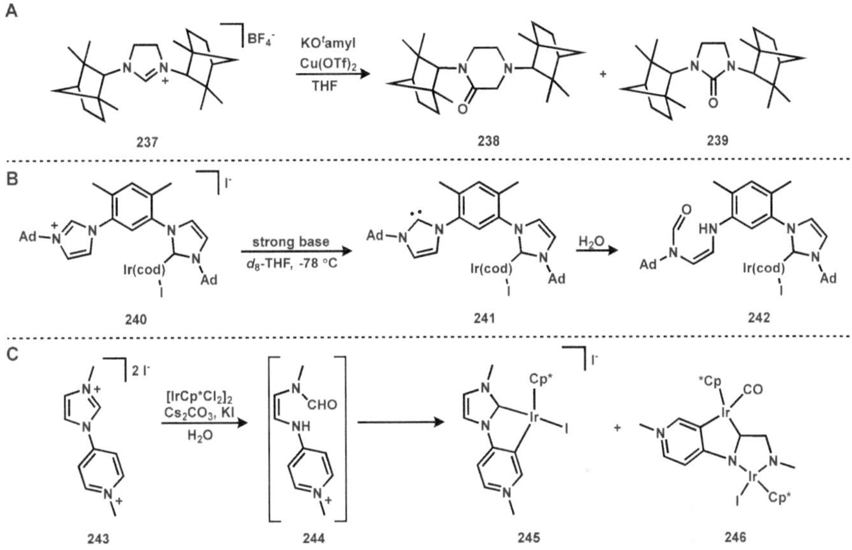

Scheme 44 (A) Hydrolysis of an imidazolinium salt promotes ring expansion. (B, C) Ring opening of a pendant NHC within Ir complexes.

The hydrolysis of a Ru sandwich complex **247** was reported by Ganter and co-workers.[136] This work provided mechanistic insights through DFT. Deprotonation of a benzimidazolium cationic ligand on Ru leads to the free carbene intermediate **A** which interacts with a molecule of water (Scheme 45A). H-Bonding is observed between the carbene carbon and a hydrogen atom on water as well as two H-bonding interactions between H atoms of the Cp* ligand and the O atom of water. Intermediate **B** is produced by O-insertion into the H bonding interaction, and C—N cleavage delivers the product **248**. If MeOH is used instead of water, an analogue of intermediate **B** could be isolated in which the OH group is replaced by OMe which further validates this mechanism.

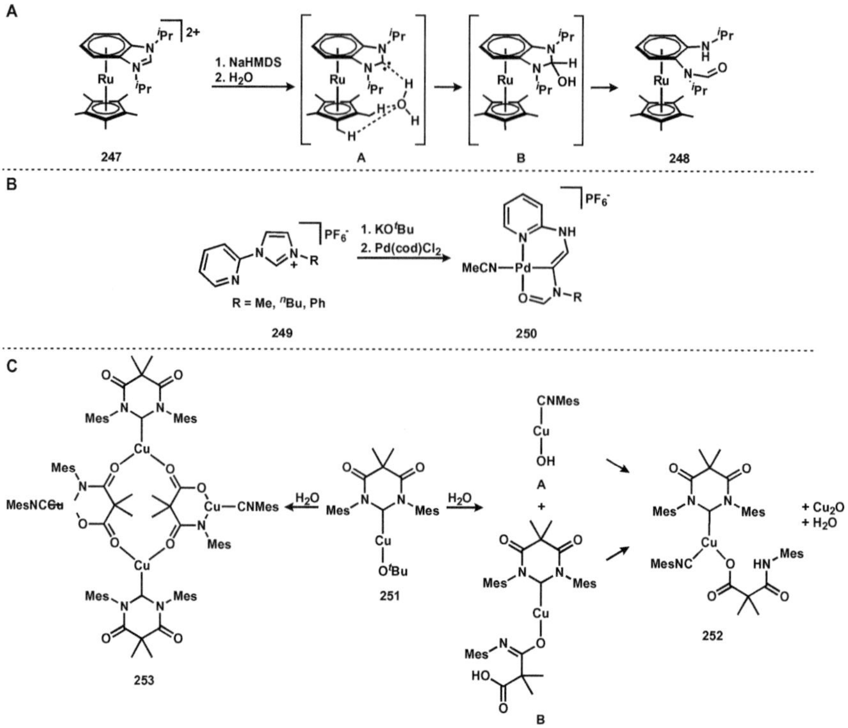

Scheme 45 (A) Hydrolysis of an NHC ligand on a Ru sandwich complex. (B) Ring opening of an NHC to produce an NCO pincer ligand. (C) Hydrolytic ring opening within Cu complexes.

A hydrolytic ring opening of imidazol-2-ylidenes to produce Pd complexes (**250**) which feature N, C, O pincer ligands was reported by Choudhury and co-workers (Scheme 45B).[137] The imidazolium salts **249** were treated with KO^{t}Bu to produce the NHCs before [Pd(cod)Cl$_2$] was added. It was suggested that the ring opening takes place before addition of Pd and is promoted by adventitious water.

A hydrolytic ring opening of an unusual NHC ligand within Cu complex **251** was observed by Whittlesey and co-workers (Scheme 45C).[138] Upon addition of water, complex **251** rapidly transformed to the ring opened product **252**. It was postulated that intermediates **A** and **B** are involved, and that the combination of these species delivers complex **252** as well as H$_2$O and Cu$_2$O which precipitates from the reaction mixture. If complex **251** was left for several weeks, another ring expanded product **253** was obtained. It is possible that complex **252** is an intermediate in the formation of **253**.

A Ag$_2$O-induced hydrolytic ring opening to produce Pd complexes was developed by Dai and co-workers.[139] The ring opening was controlled by the counter ion of the imidazolium salt **254** (Scheme 46A). Halide counter ions delivered the ring opened complex **256**, whereas PF_6^-, BF_4^-, NO_3^-, and OTf^- counter ions delivered complexes **255** which are absent of any ring opening. Ag$_2$O was found to be necessary for the ring opening to occur and it was suggested that it acts as a reservoir of hydroxide ion which is generated by the hydrolysis of Ag$_2$O. Also, AgBF$_4$ promoted a ring closing reaction to form imidazolium salt **254** from the corresponding ring opened formamide while AgI was not capable of mediating this reaction, perhaps due to insolubility. This explains the divergent reactivity provided by different counter anions. While the Ag$_2$O-induced hydrolytic ring opening is likely to occur with all counter anions, the ring closing reaction only takes place with AgBF$_4$ or other soluble Ag salts.

The hydrolytic ring opening of NHCs was exploited by Zane, Feroci and co-workers to synthesize the natural product hymeniacidin.[140] An electrochemical methodology was developed to produce the NHC from

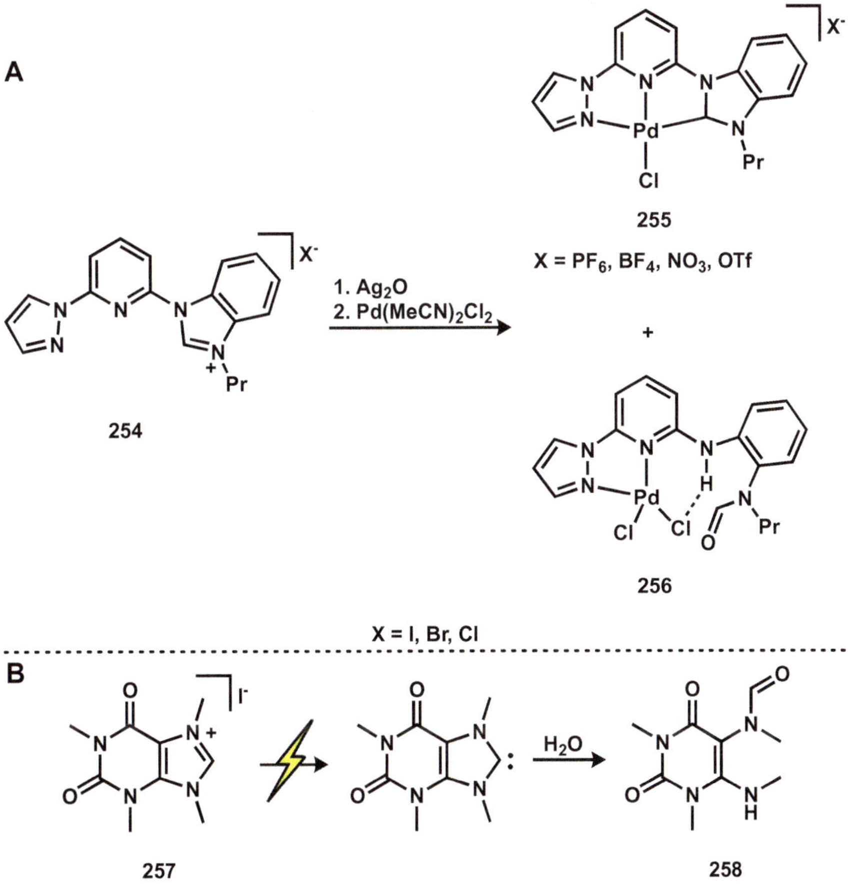

Scheme 46 (A) Ag$_2$O-induced hydrolytic ring opening to produce Pd complexes. (B) Synthesis of the natural product Hymeniacidin by hydrolytic ring opening of an NHC.

the corresponding xanthinium salt **257** which is derived from caffeine, and hydrolytic ring opening delivered hymeniacidin **258** (Scheme 46B). Hydrolysis of the xanthinium salt **257** was ruled out as it is stable in water for up to a week.

8. Ring expansion reactions

Over the last decade there has been a significant amount of research into the ring expansion of NHCs which involves C—N insertion of a transition metal or other element to produce six-membered heterocycles. The majority of these transformations have been promoted by either Be, B or Si.[141,142] However, the first report of this type of reaction was provided by Grubbs and co-workers in 2006.[143] An imidazolinium salt featuring a chelating phenoxide moiety (**259**) was subjected to KHMDS (potassium hexamethyldisilylamide) followed by [NiClPh(PPh$_3$)$_2$] (Scheme 47). The product of this reaction was a six-membered metallacycle **260** in which the phenyl group on the Ni starting material has migrated to the former carbene carbon and Ni has inserted into the C—N bond of the NHC. A related ring expansion was observed with the chloroform protected NHC **261**. When subjected to ^{n}BuLi followed by ClPPh$_2$ the NHC ring

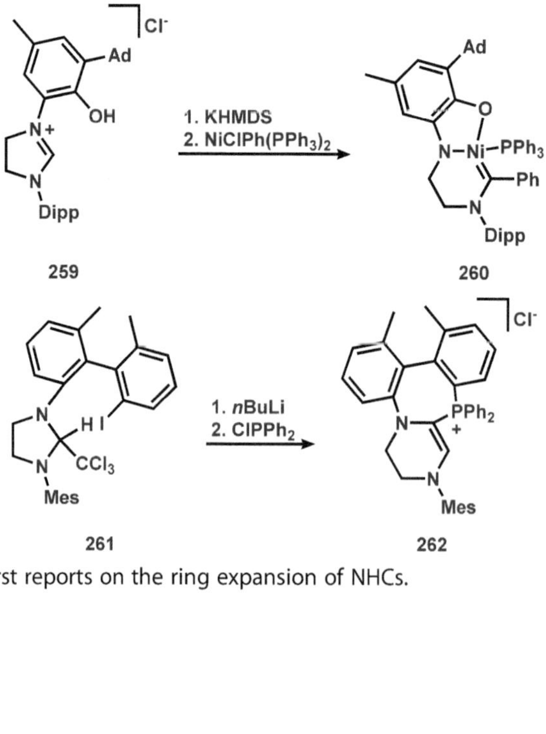

Scheme 47 First reports on the ring expansion of NHCs.

was expanded by a single carbon atom with an appended phosphine, presumably derived from the CCl$_3$ protecting group.

The first report of ring opening with Be was by Hill and co-workers in 2012, in which a BeH$_2$ unit inserted into a C—N bond of IDipp.[144] A reaction between [(IDipp)BeMe$_2$] **263** and excess PhSiH$_3$ initially led to formation of the dimer [(IDipp)BeHMe]$_2$ **264** (Scheme 48A). This poorly soluble complex reacted with another equivalent of PhSiH$_3$ when heated to 80 °C to deliver the ring expanded product **265**. Evidence of the Be-Me/Si—H metathesis products was obtained by NMR scale experiments and the authors suggested a BeH$_2$ mediated C—N insertion which leads to the ring expanded product **265**.

A related example was reported shortly afterward by Rivard and co-workers in which BH$_2$ complex **266** also underwent a C—N bond activation/ring expansion to deliver complex **267** (Scheme 48B).[145] Heating complex **266** to 100 °C in toluene delivered the product **267**. The isotopologue D$_2$-**266** was subjected to identical reaction conditions and this delivered complex D$_2$-**267** with deuteration only at the carbene carbon suggesting that the BD$_2$ moiety inserts into the C—N bond of the IDipp ligand to deliver the ring expanded product, similar to the BeH$_2$ report by Hill and co-workers.

Another example featuring NHC insertion into a Si—H bond which delivers ring expanded diazasilinanes was reported by Radius and co-workers in the same year.[146] In this instance, NHCs I^{t}Bu, I^{i}Pr and IPr were treated with either PhSiH$_3$ or Ph$_2$SiH$_2$ and heated to 100 °C for 3 days which delivered ring expanded products **268** and **269** (Scheme 48C). The NHCs I^{i}Pr, IPr and IMe were treated with Ph$_3$SiH under more forcing conditions (140 °C) and this delivered complexes **270**. A mechanistic investigation involved treatment of I^{i}Pr with Ph$_2$SiD$_2$ and the product D$_2$-**269** was formed in which deuterium was incorporated only at the carbene position. The NHC SIDipp could be converted to complex **271** and heating this complex delivered complex **272**, suggesting the intermediacy of complex **271** in the formation of **272**.

Following these three experimental reports a series of theoretical studies were performed, the first by Dutton, Wilson and co-workers.[147] They examined the ring expansion of IMe with several silanes but focused on the use of Ph$_2$SiH$_2$ (Scheme 49A). The combination of IMe and Ph$_2$SiH$_2$ leads to an adduct **A** in which the NHC coordinates the silane. Si—H activation and hydrogen atom migration to the carbene carbon

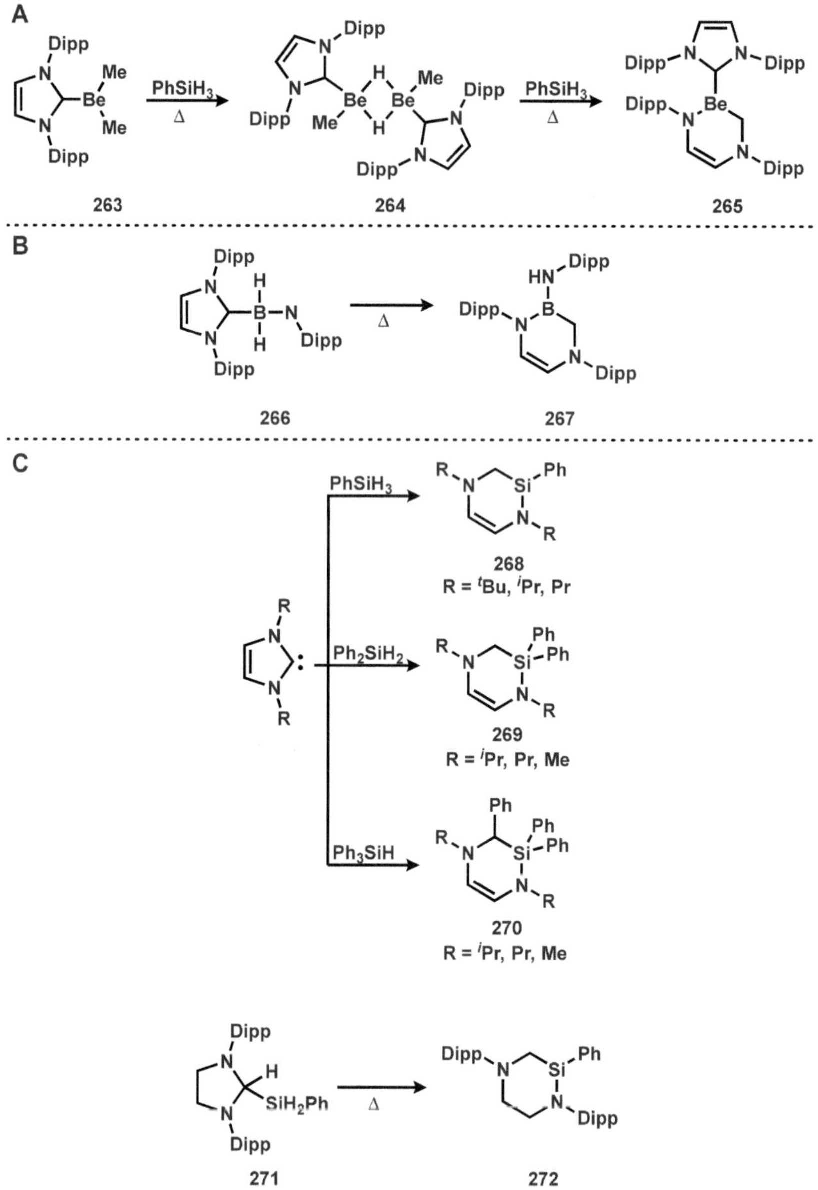

Scheme 48 Early reports of ring expansion of NHCs featuring Be, B and Si.

delivers intermediate **B**. This is the rate limiting step of the transformation. Insertion of Si into the C—N bond of the NHC leads to intermediate **C** and transfer of a second hydrogen atom to the carbene carbon produces the product **273**. This work was expanded to include theoretical investigations of both the B and Be versions of the ring expansion of NHCs.[148] All three transformations were found to proceed through similar mechanisms with each containing the fundamental steps, (i) E-H activation and hydrogen atom transfer to the carbene carbon, (ii) insertion of either Be, B or Si into the C—N bond of the NHC, and (iii) transfer of a second hydrogen atom to the former carbene carbon. The transformation featuring B required the presence of the NHDipp substituent on the starting material (theoretically modeled as NHMe) as the lone pair on the N atom stabilizes several intermediates (Scheme 49B). As such, IMe•BH$_3$ is not a viable substrate for this transformation because of the absence of the stabilizing effect provided by the N atom. The exact speciation of the starting complex for the Be analogue of the transformation is ambiguous, however, it was proposed to proceed through a (IMe)$_2$BeH$_2$ complex (**276**) and that this complex which features coordination of two NHC ligands was the most likely starting point from which the initial hydrogen atom transfer would take place (Scheme 49C).

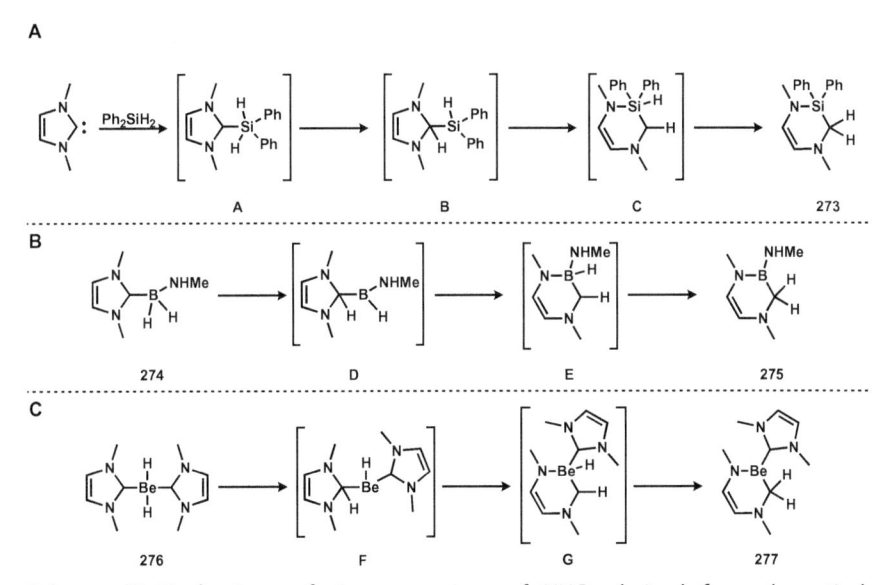

Scheme 49 Mechanisms of ring expansions of NHCs derived from theoretical investigations.

Another theoretical study concluded that cyclic alkylaminocarbenes (CAACs) would not participate in the B version of the ring expansion reaction as the barrier for C—N activation is too large. However the initial hydrogen atom transfer is feasible and the reaction would likely terminate at this point.[149] A theoretical investigation by Su examined the reactivity of group 14 elements (C—Pb) in the ring expansion reaction and concluded that the reactivity toward this reaction decreased when going down the group.[150]

Dutton, Wilson and co-workers also examined the reactivity of a range of NHC ligands in the Si and B versions of the ring expansion of NHCs. Five NHCs were examined including IMe, SIMe, ^{Me}IMe, ^{Cl}IMe and 1,3-dimethylbenzimidazol-2-ylidene. Of these five, the ring expanded product of the saturated NHC (SIMe) was found to be the most thermodynamically stable and presented the lowest barriers in the transition states. This was partially rationalized by the absence of aromatic character in the NHC so it does not need to overcome any barrier associated with the loss of aromatic character in the product.[151]

An experimental investigation of the Be version of the ring expansion of NHCs was conducted by Hill and co-workers which corroborated many aspects of the mechanism proposed by Dutton, Wilson and co-workers.[152] The requirement for two equivalents of NHC was confirmed and a metathesis product, PhSiMeH$_2$, was also observed which suggests that BeH$_2$ complex **276** is a likely starting point for the ring expansion reaction.

The requirement for an NHDipp substituent in the ring expansion of NHCs with BH$_2$ species led Inoue and co-workers to investigate iminoborane dihydride species in this transformation.[153] Iminoboranes **278** were treated with various NHCs on an NMR scale such that as the temperature was slowly increased, the ring expansion reaction could be monitored (Scheme 50A). The NHCs screened in this study were ^{Me}IMe, IMes, SIMes and IDipp. Only two combinations of iminoborane and NHC failed to deliver the ring expanded products **279**. These were iminoborane **278a** combined with ^{Me}IMe, and iminoborane **278b** combined with IDipp. This led the authors to suggest that, while sufficient steric bulk is required to produce the BH$_2$ species, too much inhibits the reaction.

A ring expansion reaction between NHCs with phosphine N-substituents and 9-borabicyclo[3.3.1]nonane (9-BBN) was reported by Stephan and co-workers.[154] The NHCs **280** were treated with 9-BBN at room temperature and boracycles **281** were produced (Scheme 50B). These frustrated Lewis pairs (FLPs) were efficient catalysts in the reduction of CO$_2$. The mechanism of the formation of boracycles **281** proceeds through a similar pathway to that described above, however, instead of a second hydrogen atom transfer, an electrophilic carbon inserts into a B—C bond to produce a new C—C bond as the final step in the mechanism.

A double ring expansion reaction was discovered during development of a Zn-catalyzed methodology for the borylation of tertiary alkyl halides.[155] In this instance, the combination of (IMes)ZnCl$_2$ (**282**), B$_2$pin$_2$ (pin=1,1,2,2-tetramethylethylene glycolate) and KO^{t}Bu lead to the formation of the double ring expansion product **283** (Scheme 50C). While the authors did not speculate on a mechanism, complex **283** includes

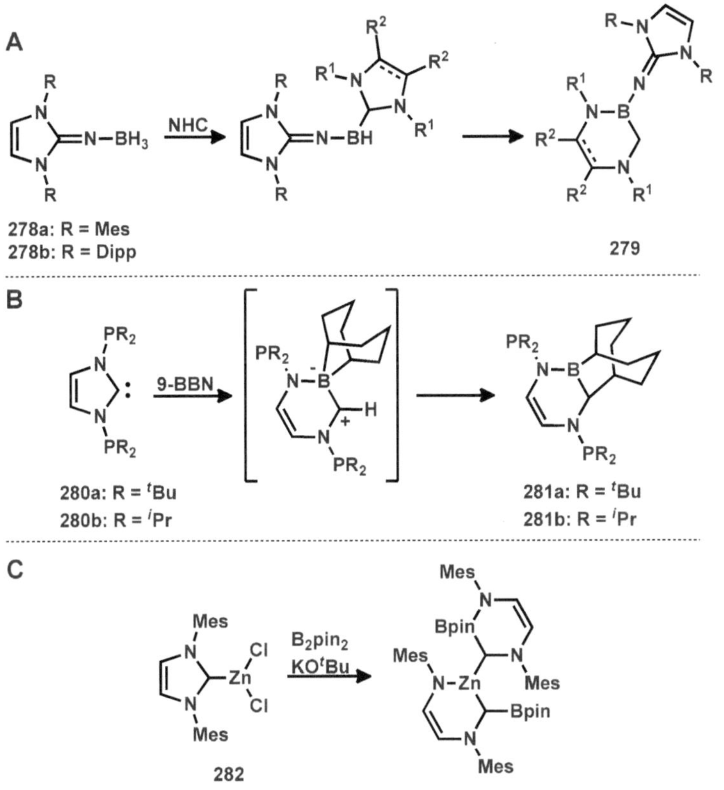

278a: R = Mes
278b: R = Dipp

279

280a: R = ^{t}Bu
280b: R = ^{i}Pr

281a: R = ^{t}Bu
281b: R = ^{i}Pr

282

283

Scheme 50 (A) Ring expansion of NHCs with iminoborates. (B) Ring expansion of NHCs containing phosphorous *N*-substituents to produce FLPs. (C) Double ring expansion featuring Zn and Bpin.

familiar features such as insertion of Zn into the C—N bond of the NHC ligand and Bpin migration to the carbene carbon instead of H atom migration. A second ring expansion of another NHC is also observed in which a second Bpin moiety inserts into the C—N bond of the NHC.

Ring expansion of NHCs was also observed when B_2cat_2 (cat = 1,2-dihydroxybenzene) complexes **284**, which feature NHCs bound to the B centers, were heated in THF and this delivered ring expanded complexes **285** (Scheme 51A).[156] Both NHC ligands were necessary in the starting complexes **284** as no reaction occurred if only one NHC ligand was present. The ring expanded products **285** feature both an endocyclic and an exocyclic Bcat unit. A similar reaction using $B_2(neop)_2$ (neop = neopentyl glycolato) and two equivalents of IPr delivered ring expanded product **286**, however this reaction proceeded at room temperature suggesting heightened reactivity of $B_2(neop)_2$ over B_2cat_2. The ring expanded product only features one endocyclic B(neop) unit and another of the neop moieties has also undergone ring expansion.

A related example features the saturated NHC SIDipp and HBcat.[157] Combining equimolar amounts of these substrates results in an interesting product **287** in which one Bcat moiety inserts into the C—N bond of the NHC and two hydrogen atoms migrate to the former carbene carbon, as in previous examples (Scheme 51B). The Bcat moiety also ring opens and one of the oxygen atoms coordinates another Bcat moiety which is further stabilized by another SIDipp ligand.

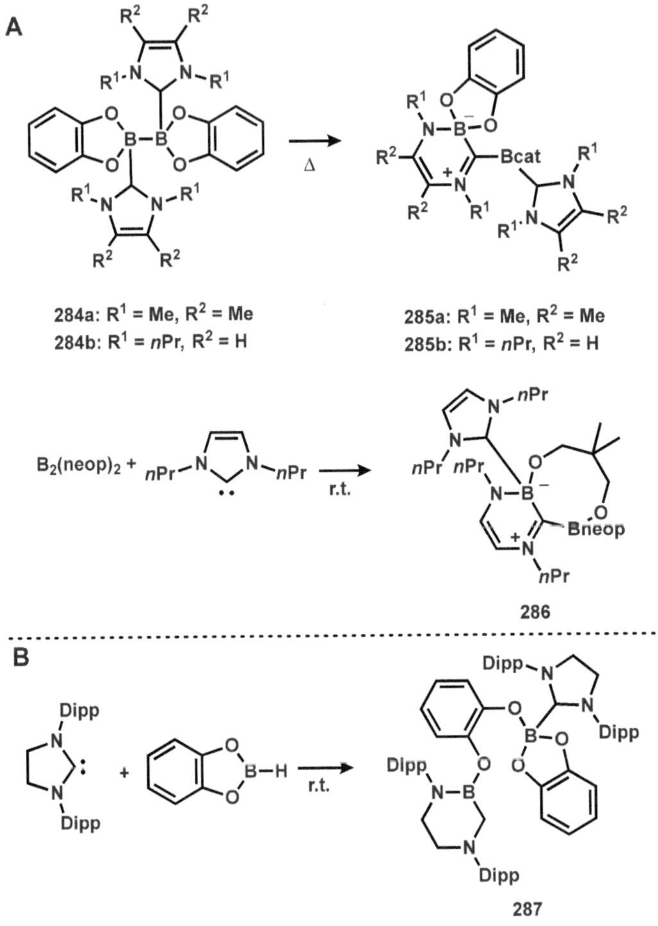

Scheme 51 Ring expansion reaction of NHCs with B compounds.

The ring expansion reaction was extended to an Al analogue by Radius and co-workers in 2017.[158] When Al complex **288** was treated with I^{i}Pr, the ring expanded product **289** was observed (Scheme 52A). The mechanism of its formation was suggested to be similar to the above examples.

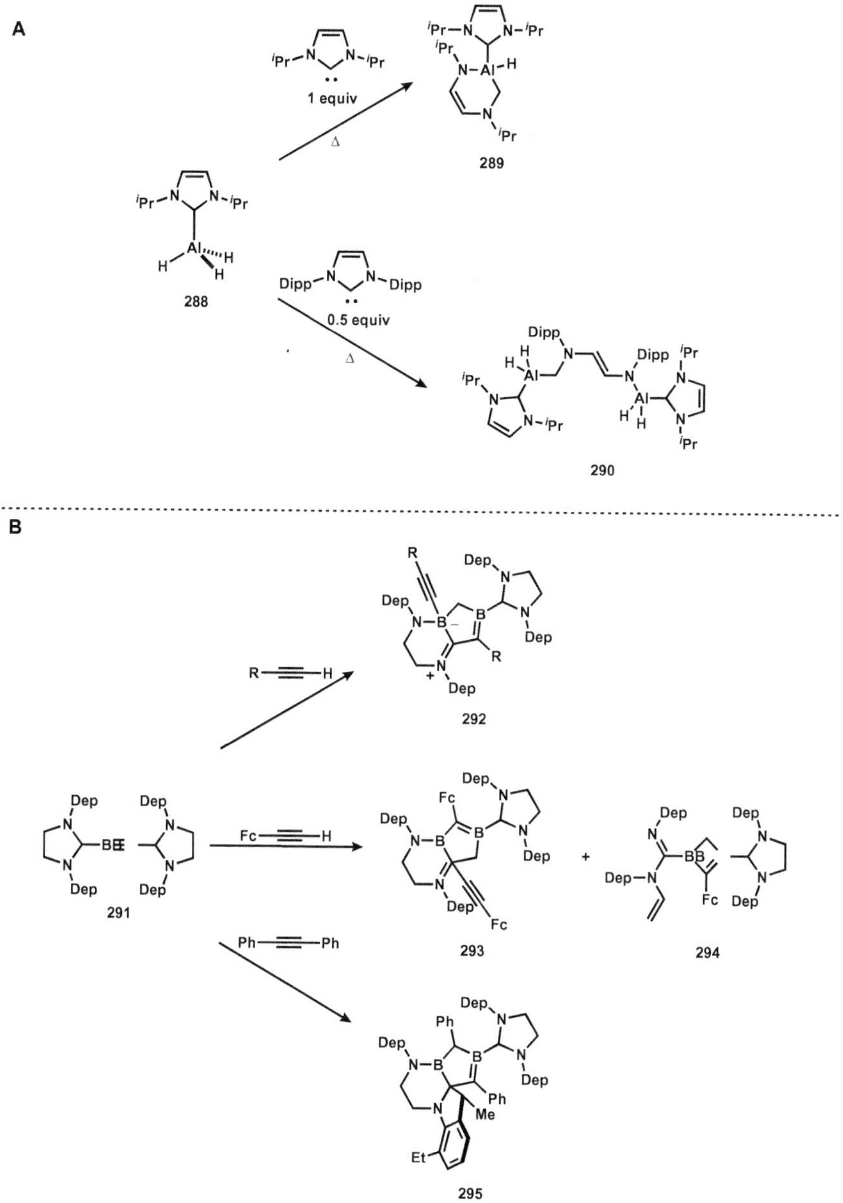

Scheme 52 (A) Ring expansion and ring opening within Al complexes. (B) Ring expansion/opening induced by a diboryne complex and alkynes.

Experimental evidence for initial adduct formation between complex **288** and the free NHC to form a bis-NHC adduct was reported, and subsequent migration of two hydrogen atoms to the former carbene carbon and C—N insertion of the Al moiety was postulated. When the bulkier NHC IDipp was used in place of I^{i}Pr, a ring opened product **290** was observed but the ring expanded product was not. Again, two hydrogen atoms have migrated to the former carbene carbon and a new Al—C and Al—N bond have been formed. The ring opening only occurs for IDipp with no opening of the I^{i}Pr ring observed.

Ring expansions/openings of NHCs in diboryne complexes induced by alkynes were reported by Braunschweig and co-workers in 2019.[159] Diboryne complex **291** was treated with a range of different alkynes and different reactivity was observed (Scheme 52B). Terminal alkynes delivered ring expanded complex **292** in which the ring expansion of the NHC has occurred as well as insertion of the alkyne into the B≡B triple bond. Other constitutional isomers were also observed such as complex **293** when a ferrocenyl-based alkyne is used. In this instance, a ring opening product **294** was also observed. When diphenylacetylene is used the ring expanded product **295** was observed but a C—H activation of one of the 2,6-diethylphenyl (Dep) substituents also occurs.

9. Miscellaneous ring openings

Ring opening reactions were observed when the sterically demanding ligand 6-Dipp (**296**) was complexed with Ir or Au (Scheme 53A).[160] When 2 equivalents of 6-Dipp were treated with [Ir(coe)$_2$Cl]$_2$, the ring opened complex **298** was formed in which the C—N bond of the heterocycle has been cleaved, resulting in a chelating alkene/amidine ligand. However, if four equivalents of 6-Dipp were used this resulted in complex **299**, in which the coe ligand has been replaced by another ring opened ligand but is only complexed to Ir through the alkene, leaving the formamidine moiety pendant. If the preformed allylformamidene **297** was treated with [Ir(coe)$_2$Cl]$_2$, complex **298** was observed suggesting that ring opening occurs prior to complexation. A related example using [(6-Dipp)AuCl] led to complex **300** (Scheme 53B). This complex was formed when using either starting materials **296** or **297**. The authors suggested that ring opening of **296** occurs to give **297**. This ligand is bound to the Au through the alkene moiety and then a ring closure occurs via attack of the imine function on the alkene. Comparisons between abnormal NHCs and complex **300** could be drawn, however, X-ray diffraction data are consistent with a secondary alkyl ligand.

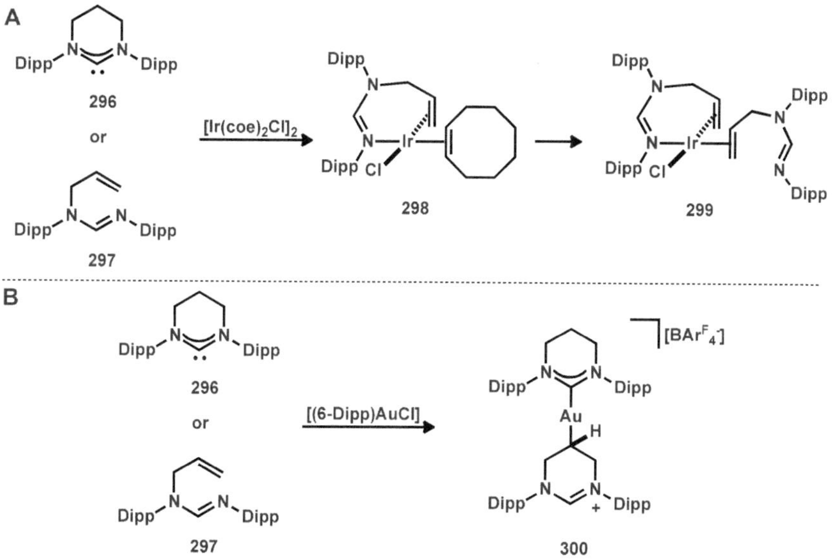

Scheme 53 Ring openings of the sterically demanding 6-Dipp ligand.

An elegant study of ring openings of SIMes with s-block elements was reported by Hevia and co-workers.[161] When SIMes was treated with MCH$_2$SiMe$_3$ (M=Li or K) along with other donor ligands such as PMDETA (pentamethyldiethylenetriamine) or THF, a ring opening of the NHC takes place and in each case (**301** and **302**) the ligand is rearranged to the same non-symmetrical amide ligand (Scheme 54A). Complex **301** contains a K metal center and the ring opened ligand is coordinated through an amido nitrogen and is further stabilized by π-interactions with the indolyl C=C bonds as well as the *ipso*-carbon of the mesityl ring. The tridentate PMDETA ligand completes the coordination sphere. Complex **302** contains a Li metal center but the ring opened ligand is only coordinated through the amido moiety. Another SIMes and a THF ligand complete the coordination sphere. A mechanism was proposed for the formation of complex **302**, and it follows that a similar mechanism was operative for the formation of complex **301** (Scheme 54B). Initial coordination of SIMes with LiCH$_2$SiMe$_3$ produces dimer **A**, followed by cyclometalation of one of the methyl groups of the mesityl *N*-substituent to deliver **B**. Subsequent coupling of the carbene carbon to the benzylic CH$_2$ group of the mesityl substituent and C—N bond cleavage of the heterocycle leads to the ring opened intermediate **C**. A 1,2-hydrogen atom shift rearomatizes the indole moiety to provide **D**, and coordination of donor

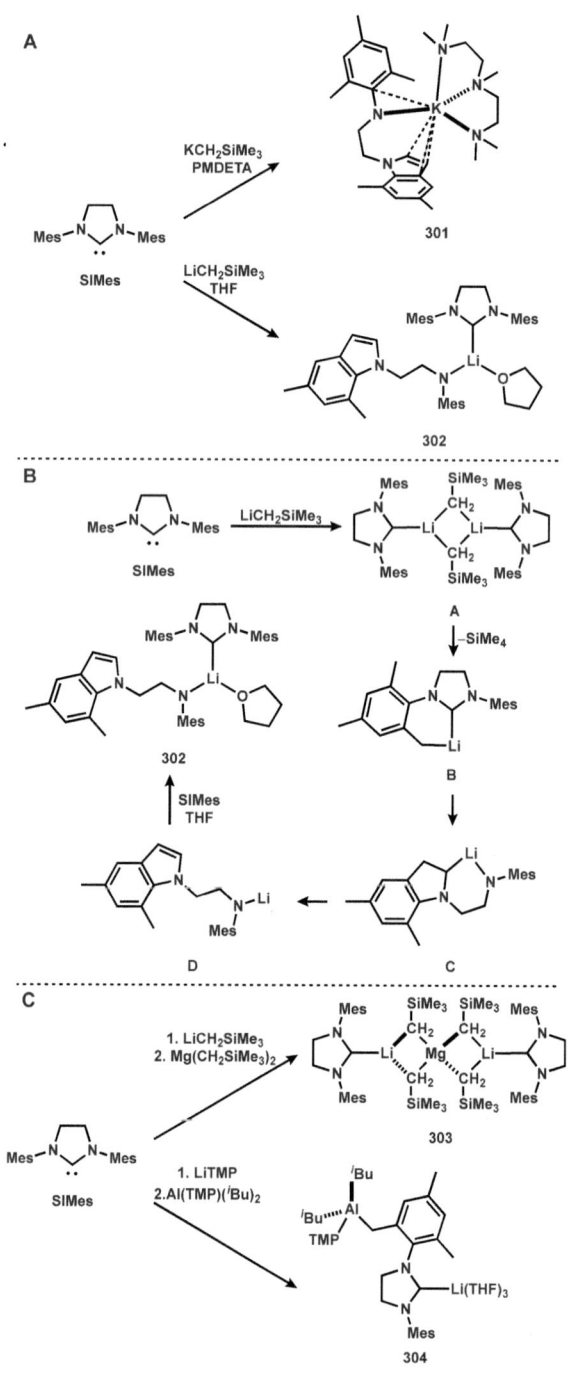

Scheme 54 (A) Ring opening reactions with s-block elements. (B) Proposed mechanism. (C) Isolated analogues of intermediates.

ligands SIMes and THF complete the sequence. While none of these intermediates could be isolated or even observed when the reaction was repeated at lower temperature, some analogues could be trapped by the addition of either Mg or Al species (Scheme 54C). Complex **303** is an analogue of intermediate **A**. The addition of Mg(CH$_2$SiMe$_3$)$_2$ allows the formation of less polar Mg—C bonds compared to Li—C bonds which renders the complex less reactive. The reaction of SIMes with LiTMP and Al(TMP)(^{i}Bu)$_2$ provides complex **304** which is an analogue of intermediate **B**, which is trapped as the bimetallic complex with Al at the benzylic position that was cyclometalated in intermediate **B**. The higher carbophilicity of the Al(TMP)(^{i}Bu)$_2$ fragment compared to Li allows transmetalation to occur and for this intermediate to be isolated.

An interesting report by Slaughter and co-workers describes a ring opening and complete removal of the carbene carbon from the heterocyclic ring of an imidazol-2-ylidene ligand.[162] When imino-NHC **305** was treated with tetrabenzylhafnium, the unexpected complex **306** was isolated (Scheme 55). Through the course of the reaction three benzyl groups have migrated to the carbene carbon and it has been completely excised from the heterocyclic ring, resulting in a tridentate eneamino-amido ligand and a tribenzylmethyl ligand. Interestingly, when imino-NHC **305** was treated with HfCl$_4$, the expected (imino-NHC)HfCl$_4$ complex was formed and no ring opening was observed.

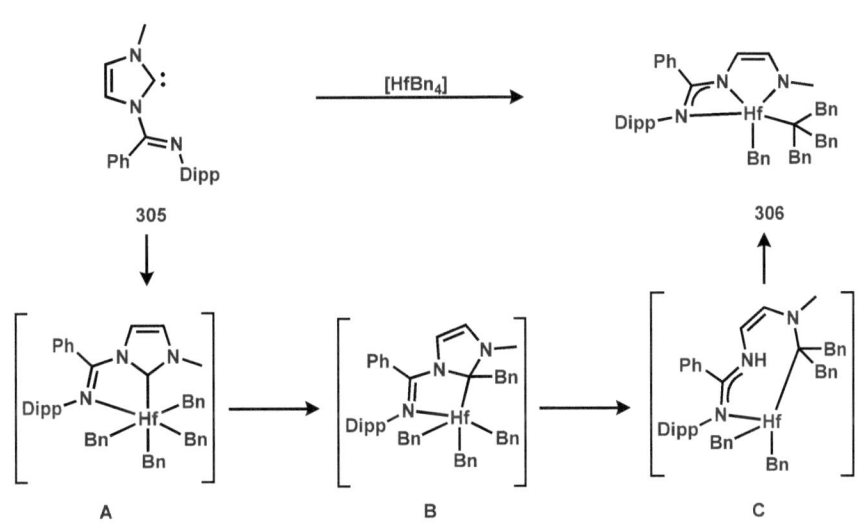

Scheme 55 Ring opening and removal of the carbene carbon from an NHC.

An interesting C—N cleavage and ring opening of a cyclometalated imidazol-2-ylidine to form a new cyclopentadienyl ligand was reported by Tatsumi, Ohki and co-workers.[163] When Fe complex **307** was treated with either phenylacetylene or diphenylacetylene, new sandwich complexes **308** were produced (Scheme 56A). Complexes **308** feature new cyclopentadienyl ligands and the cyclometalated NHC ligand has undergone ring opening to produce a diimine moiety that is pendant on the new Cp ring. C—N cleavage results in the loss of an isopropyl N-substituent as propylene which was observed by NMR spectroscopy in the reaction mixture. If the reaction between complex **307** and phenylacetylene was performed at room temperature a putative intermediate **309** could be isolated, in which loss of the isopropyl N-substituent has already taken place but ring opening of the NHC has not.

When treated with phenylacetylene, Co-NHC complexes **310** undergo migratory insertion and, depending on the N-substituents, ring expansion to deliver either complex **311** or **312** (Scheme 56B).[164] After coordination of

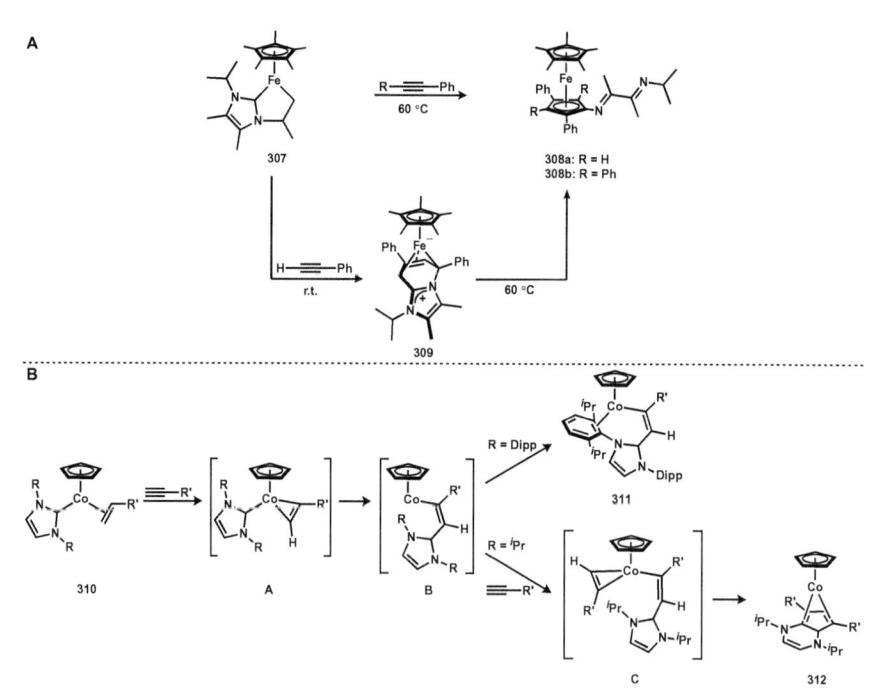

Scheme 56 (A) C—N cleavage and ring opening to produce a cyclopentadienyl ligand. (B) Migratory insertion and ring expansion within a Co complex.

the alkyne (**A**), migratory insertion occurs to deliver intermediate **B**. With Dipp *N*-substituents, this intermediate is stabilized by η^2-coordination of the aryl ring and produces complex **311**. Conversely, with isopropyl *N*-substituents, this coordination mode is not available, and the complex coordinates a second phenylacetylene (**C**) and ring expansion occurs with one carbon atom of phenylacetylene included in the heterocyclic ring. A 1,3-cyclopentadienyl ligand is produced and binds to Co in an η^4 fashion to deliver complex **312**.

An unexpected ring opening was reported by Bercaw and co-workers when attempting to synthesize early transition metal-NHC complexes.[165] Imidazolium salt **313** was expected to form an N, C, O pincer complex upon treatment with [TiCl$_2$(NMe$_2$)$_2$]. However, ring opening of the heterocycle occurred to deliver complex **315** (Scheme 57A). It is possible that this reaction may be initiated by migration of an amido group from the Ti center to the carbene carbon or attack of free amine. Complex **315** slowly rearranges to the zwitterion **316** to deliver a rare example of an anionic Ti complex. A similar rearrangement of the free imidazolium salt **313** was also observed in solution which delivers benzimidazolium salt **314**. This could be treated with [TiCl$_2$(NMe$_2$)$_2$] for a higher yielding route to complex **316**.

A ligand rearrangement derived from a ring opening/ring closure mechanism was reported by Chen and co-workers.[166] In this instance, bis-imidazolium salt **317** was reacted with either Ni(OAc)$_2$•4H$_2$O or Pd(OAc)$_2$ and the products of each respective reaction were complexes **318** which do not display any non-innocent reactivity of the NHC ligand (Scheme 57B). Conversely, if additional external bases were included (NaOAc, NEt$_3$, Na$_2$CO$_3$, Na$_2$HPO$_4$, K$_3$PO$_4$, NaOH) as well as the metal acetates, then the ligand rearranged to produce a saturated imidazolin-2-ylidene ring and a new amide ligand. The requirement for trace amounts of H$_2$O was ruled out by extensively drying all reagents and solvents. It was suggested that the transformation proceeds by deprotonation of complexes **318** and then a concerted C—N cleavage, C—N coupling step delivers the rearranged complexes **320**.

A ring opening of a bis-NHC pincer ligand with a central pyridine moiety was reported by Danopolous and co-workers in 2008.[120] When bis-NHC **321** was treated with [NiMe$_2$(tmeda)], the ring opened complex **322** was observed as the sole product of the reaction (Scheme 58A). It was suggested that the ring opening is initiated by migration of one of the methyl groups on Ni to the NHC carbon.

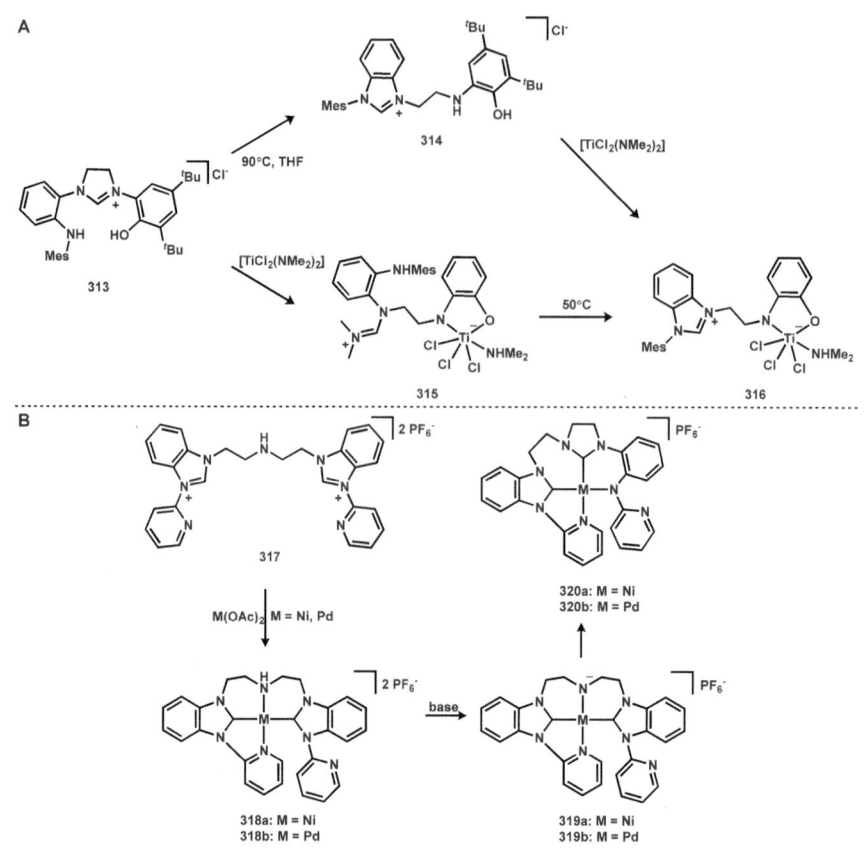

Scheme 57 (A) Rearrangement and ring opening of an NHC ligand within a Ti complex. (B) Ring opening/ring closure within a tetradentate NHC ligand on Ni and Pd.

A ring opening of SIDipp was reported by Bertrand and co-workers in 2010.[167] When SIDipp was treated with HBpin the dimer **323** was formed (Scheme 58B). There was no further work to elucidate the mechanism of this reaction, however, the authors commented that SIDipp behaved very differently from CAACs which were used extensively in this study with no evidence of ring opening.

An unusual example of a ring opening of an NHC ligand was described by Tilley and co-workers.[168] A cyclometalated imidazol-2-ylidene with 2,6-dimethylphenyl (xylyl) N-substituents which was coordinated to a Ru center (**324**) underwent ring opening upon treatment with LiCH$_2$SiMe$_3$ (Scheme 58C). The mechanism of this ring opening is unclear; however, C—N bond cleavage must occur, and it is possible that Li inserts into the C—N bond.

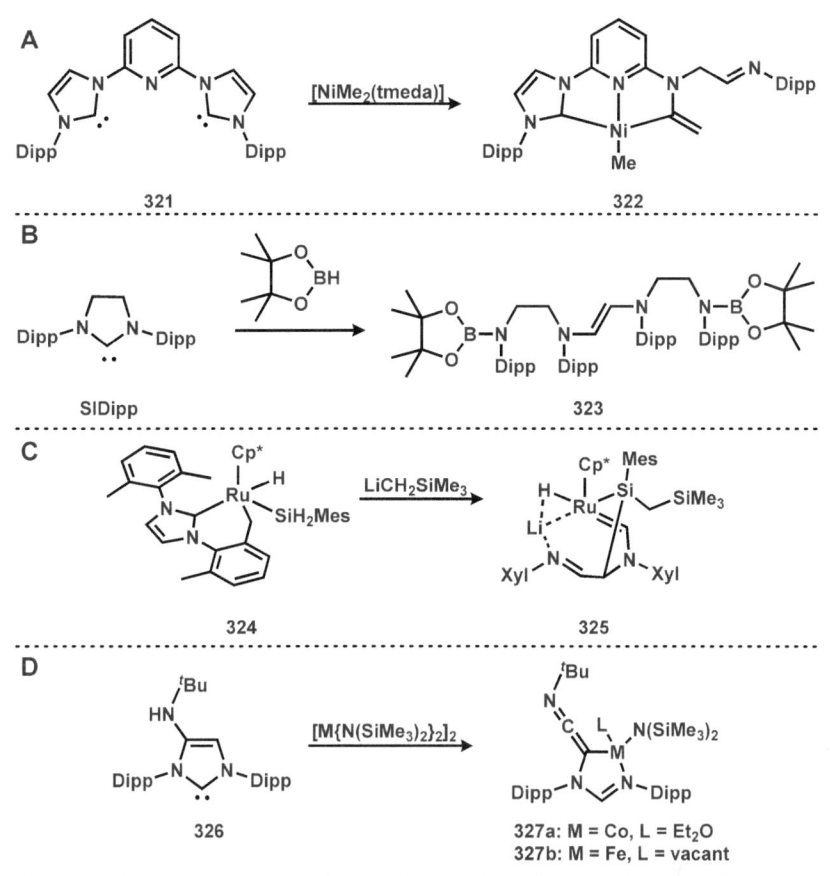

Scheme 58 (A) Ring opening of a bis-NHC pincer ligand induced by methyl migration. (B) Ring opening of SIDipp mediated by HBpin. (C) Ring opening of a cyclometalated NHC ligand. (D) Ring opening of a functionalized IDipp NHC ligand.

A ring opening of a 4-*tert*-butylamino functionalized IDipp ligand (**326**) was described by Danopoulos, Braunstein and co-workers.[169] In this instance, [M{N(SiMe$_3$)$_2$}$_2$]$_2$ (M = Co or Fe) inserted into the heterocyclic C—N bond to form a five-membered metallacycle with one of the backbone carbon atoms transferred to an exocyclic position to provide complexes **327** (Scheme 58D).

10. Miscellaneous NHC reactivity

An alkylidene insertion into the mesityl *N*-substituent of SIMes within Ru olefin metathesis catalysts was observed by Keister, Diver and co-workers.[170] Complexes **328** were treated with CO and unexpected products **329** featuring ring expansion of the mesityl *N*-substituent were produced (Scheme 59A). X-ray diffraction analysis identified the new rings as cycloheptatrienes due to the alternating long/short bond lengths. The reaction could also be promoted by isocyanides and it was suggested the strong π-acidity of these ligands may weaken π-backbonding with the alkylidene ligand making it more electrophilic and thus de-coordinate from the Ru center. After de-coordination, cyclopropanation of the closest "double bond" of the mesityl *N*-substituent occurs and rearrangement delivers the products **329**. The mechanism of this transformation was confirmed by DFT and it was found that many π-acidic ligands are able to promote this reaction.[171] This work was extended to include a broad range of related olefin metathesis catalysts and it was found that these insertion reactions were more likely to proceed when promoted by ligands with higher π-acidity. If the alkylidene has an appended heteroatom which donates electron density, the transformation becomes less likely and several other factors were also explored.[172]

A rearrangement of an NHC bearing an anionic aryloxide *N*-substituent **331** was reported by Zhang and co-workers.[173] When imidazolium salt **330** was treated with *ⁿ*BuLi at −78 °C, the expected deprotonation of the phenol moiety and the C2 position of the imidazolium ring was

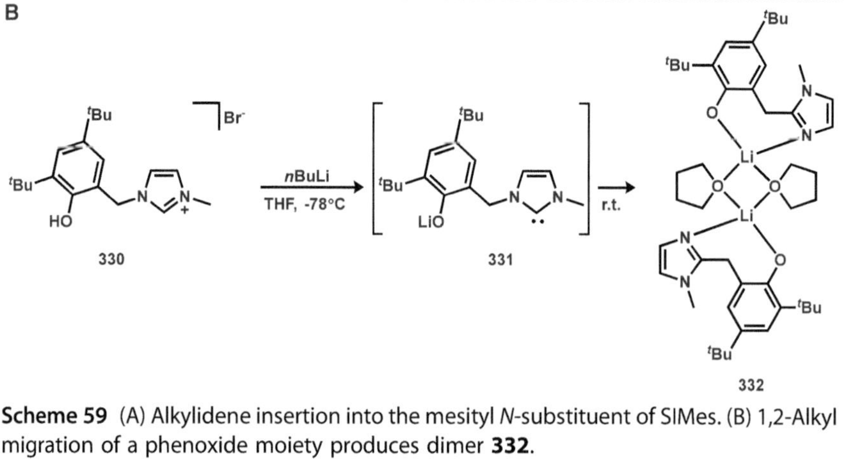

Scheme 59 (A) Alkylidene insertion into the mesityl *N*-substituent of SIMes. (B) 1,2-Alkyl migration of a phenoxide moiety produces dimer **332**.

observed. However, upon warming to room temperature a 1,2-alkyl shift had occurred to produce Li complex **332** (Scheme 59B). The migration of the aryloxide moiety to the carbene position leaves an N atom free to coordinate Li and a bidentate ligand is produced. Coordinated THF molecules complete the coordination sphere of the dimer **332**.

There have been several reports of protic NHCs which contain a simple proton as at least one of the *N*-substituents. Grotjahn and co-workers reported several Ir complexes bearing protic NHC ligands which feature ambidentate reactivity.[174] Cyclometallation of complex **333** could be achieved by heating to provide complex **334a** which features a protic NHC (Scheme 60A). The proton *N*-substituent could be removed using a combination of NaOMe and NaH to provide an imidazol-2-yl ligand present in complex **335a**. Otherwise, complex **334a** could be treated with just NaOMe and the protic NHC is retained while a chloride ligand is substituted for a hydride ligand in complex **334b**. The addition of excess NaOMe resulted in deprotonation of the protic NHC to deliver imidazol-2-yl complex **335b**. The formation of both complexes **334** and **335** showcases the ambidentate nature of this ligand class.

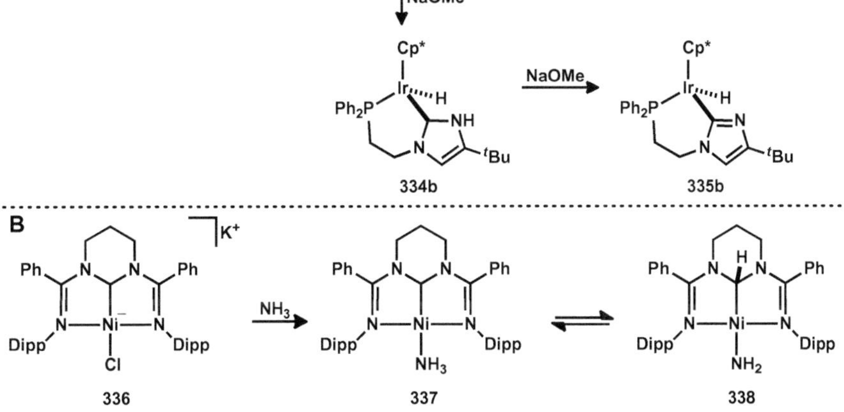

Scheme 60 (A) Ambidentate reactivity of a protic NHC. (B) Ammonia activation via ammine/amido equilibrium.

A Ni NCN-pincer complex developed by Roesler and co-workers was found to activate ammonia through a non-innocent NHC.[175] The NCN pincer ligand features a central six-membered heterocyclic ring with amidine N-substituents (**336**) (Scheme 60B). When treated with ammonia two new complexes were formed (**337** and **338**) which were in equilibrium. In the ammonia complex **337** the NHC fragment remains unchanged, but in the amido complex **338** a proton has been transferred from the ammonia ligand to the former carbene position of the NHC to create a secondary alkyl ligand.

11. Dissociation reactions

NHCs are generally thought to form strong M–C bonds that provide enhanced stability. However, there are several examples of NHC ligands dissociating from metal centers or undergoing ligand exchange reactions that show that the M–C bond is more labile in some cases. The ligand exchange of NHCs with phosphine ligands was first examined in detail by Caddick, Cloke and co-workers.[176] They observed the dissociation of NHC ligands when Pd complexes **339** were treated with phosphines to provide complexes **340** (Scheme 61A). When Co complex **341** was treated with triphenylphosphine an equilibrium between complexes **341** and **342** was observed, and equilibrium constant and thermodynamic data were calculated (Scheme 61B).[177] It was suggested that this equilibrium was made possible by the combination of the Cp ligand and bulky NHC ligand. Complete and irreversible substitution of IMes by triphenylphosphine or tri(p-tolyl)phosphine was observed in halogenated solvents by Crudden and co-workers (Scheme 61C).[178] An unexpected product **345** was observed when 1,2-dichloroethane (DCE) was used as solvent. However, the authors suggested that dissociation of IMes occurs prior to reaction with DCE. Complete substitution was also observed when Rh complex **346** was treated with diphosphines to deliver complexes **347** (Scheme 61D).[179] Ligand exchange of a protic NHC and CO by a diphosphine within a W complex **348** was observed by Edwards, Hahn and co-workers (Scheme 61E).[180] This was unexpected as the reverse reaction typically takes place. The bulky complex Cp*Co(IDipp) (**351**) undergoes ligand exchange reactions when treated with excess trimethylphosphine or ethylene (Scheme 61F).[181] It was suggested that the bulk imparted by both Cp* and the NHC contributed to labilizing the M–C bond.

Pd–NHC complexes were treated with N-heterocyclic silylenes (NHSi) and complete substitution of the NHCs for the NHSis was observed

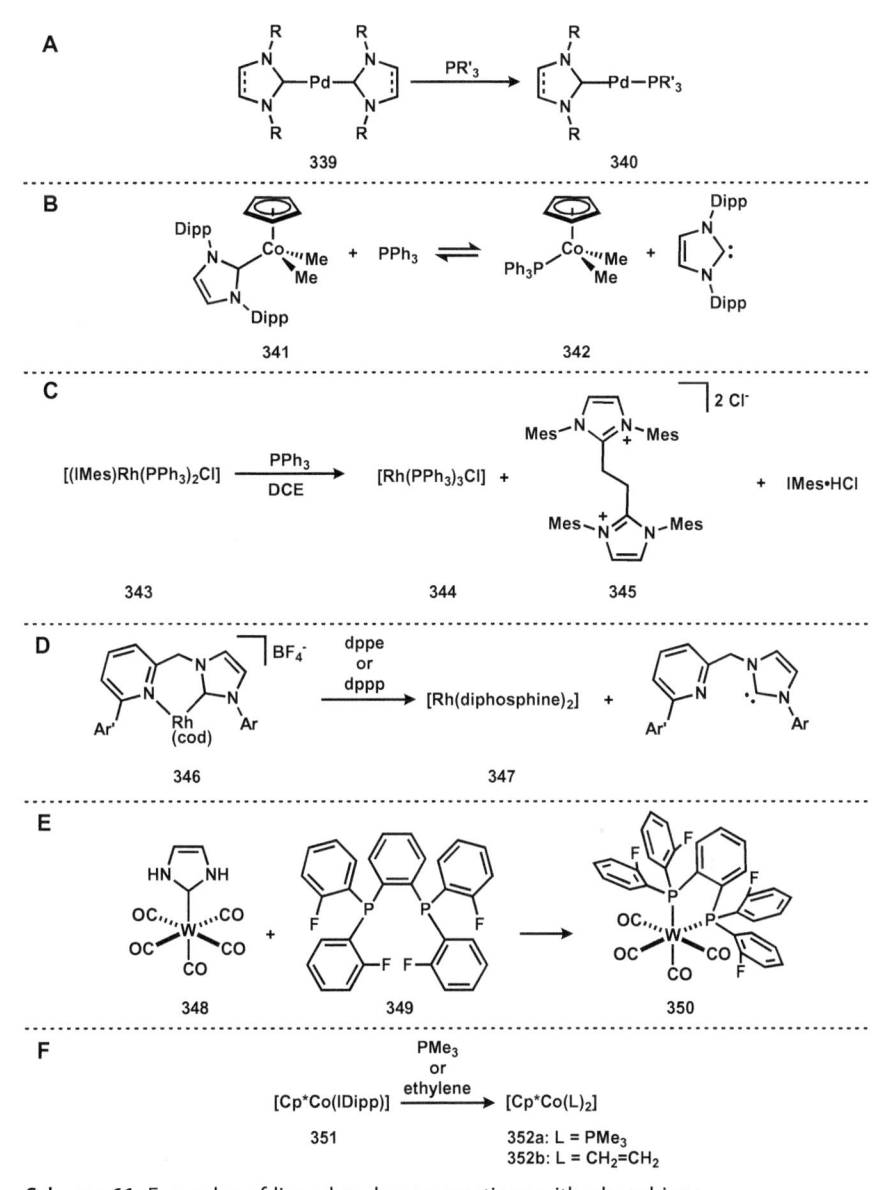

Scheme 61 Examples of ligand exchange reactions with phosphines.

(Scheme 62A).[182] While NHSis are commonly thought to produce a weaker bond with the metal than NHCs, in this case the large steric bulk of the ^{t}Bu or adamantyl *N*-substituents results in weaker M-C$_{NHC}$ bonds and allows ligand exchange. Displacement of ^{Me}I^{i}Pr within Ge complex **356** by MeLi was reported by Baines and co-workers (Scheme 62B).[183]

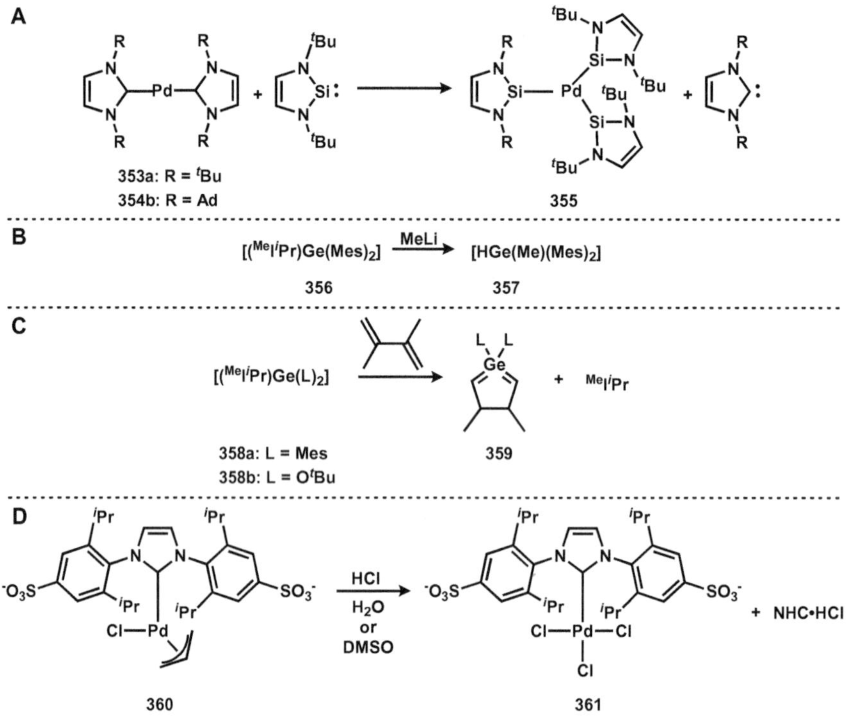

Scheme 62 Examples of NHC dissociation reactions.

This allowed access to anionic Ge complexes. Displacement of the same NHC (^{Me}I^{i}Pr) occurred when Ge complexes **358** were treated with dimethylbutadiene (Scheme 62C).[184] Interestingly, if the Mes or ^{t}Bu ligands were replaced with chloride the reaction did not occur. Protolysis of an aqueous soluble NHC complex **360** was investigated by de Jesús and co-workers (Scheme 62D).[185] Significant amounts of imidazolium salt were observed when treated with hydrochloric acid in D$_2$O, but this could be suppressed in DMSO to produce complex **361**.

12. Summary

NHCs are ubiquitous in organometallic chemistry and catalysis and have also been found to coordinate to several other elements across the whole periodic table. While they often act as spectator ligands, NHCs can become involved in reactivity, either at the carbene carbon or within the ligand architecture. This reactivity can be affected by the steric bulk and/or electronic properties of the ligand, in addition to external factors such as solvent, temperature or metal precursor. A full mechanistic understanding of these diverse processes can be exploited to take full advantage of NHCs, from promoting or enhancing catalysis through to the synthesis of new molecules.

Acknowledgment

We acknowledge financial support from EPSRC (EP/R009406/1) and the University of Leeds (UK).

References

1. Hine J. Carbon dichloride as an intermediate in the basic hydrolysis of chloroform. A mechanism for substitution reactions at a saturated carbon atom. *J Am Chem Soc.* 1950;72(6):2438–2445.
2. Hitchcock PB, Lappert MF, Pye PL. Spontaneous N-aryl (rather than P-aryl) orthometallation in the system [RuCl$_2$,(PPh$_3$)$_3$]-[=CN(Ar)(CH$_2$)$_2$NAr]$_2$ (Ar = C$_6$H$_4$Me-4); X-ray crystal and molecular structure of [RuCl(PEt$_3$)$_2${CN(Ar)(CH$_2$)$_2$N(C$_6$H$_3$Me-4)}], a stereochemically rigid 5-co-ordinate Ru^{II} complex, with a short (2.2 Å) Ru....H contact. *J Chem Soc Chem Commun.* 1977;196–198.
3. Hitchcock PB, Lappert MF, Pye PL, Thomas S. Carbene complexes. Part 16. Synthesis and properties of NN'N''N'''-tetra-aryl-substituted electron-rich olefin-derived carbeneruthenium(II) complexes containing a spontaneously formed ortho-metallated-N-Arylcarbene ligand; the crystal and molecular structures of [RuCl{CN(C6H4Me-4)CH2CH2NC6H3Me-4}(PEt3)2] (6) and [Ru(CO)Cl{CN(C6H4Me-)CH2CH2N6H3Me-4}(PEt3)2] (16). *J Chem Soc Dalton Trans.* 1979;1929–1942.
4. Hitchcock PB, Lappert MF, Terreros P. Synthesis of homoleptic tris(organochelate)iridium(III) complexes by spontaneous ortho-metallation of electronrich olefin-derived N,N'-diarylcarbene ligands and the X-ray structures of fac-[Ir{CN(C$_6$H$_4$Me-)(CH$_2$)$_2$NC$_6$H$_3$Me-p}$_3$] and mer-Ir{CN(C$_6$H$_4$Me-)(CH$_2$)$_2$NC$_6$H$_3$Me-p}$_2$ {CN(C$_6$H$_4$Me-)(CH$_2$)$_2$NC$_6$H$_4$Me-p}Cl (a product of HCl cleavage). *J Organomet Chem.* 1982;239:C26–C30.
5. Owen MA, Pye PL, Piggott B, Capparelli MV. The synthesis and characterisation of an ortho-metallated carberuthenium(II) complex, containing a metallated N-benzyl substituent with a 6-membered metallate nucleus, crystal structure of [RuCl(CN(CH$_2$C$_6$H$_5$)CH$_2$CH$_2$NCH$_2$C$_6$H$_4$)(PPh$_3$)$_2$]. *J Organomet Chem.* 1992;434:351–362.
6. Huang J, Stevens ED, Nolan SP. Intramolecular C–H activation involving a rhodium–imidazol-2-ylidene complex and its reaction with H$_2$ and CO. *Organometallics.* 2000;19:1194–1197.
7. Abdur-Rashid K, Fedorkiw T, Lough AJ, Morris RH. Coordinatively unsaturated hydridoruthenium(II) complexes of N-heterocyclic carbenes. *Organometallics.* 2004;23:86–94.
8. Torres O, Martın M, Sola E. Labile N-heterocyclic carbene complexes of iridium. *Organometallics.* 2009;28:863–870.
9. Jazzar RFR, Macgregor SA, Mahon MF, Richards SP, Whittlesey MK. C-C and C-H bond activation reactions in N-heterocyclic carbene complexes of ruthenium. *J Am Chem Soc.* 2002;124:4944–4955.

10. Diggle RA, Macgregor SA, Whittlesey MK. Computational study of C-C activation of 1,3-dimesitylimidazol-2-ylidene (IMes) at ruthenium: the role of ligand bulk in accessing reactive intermediates. *Organometallics*. 2008;27:617–625.
11. Dorta R, Stevens ED, Nolan SP. Double C−H activation in a Rh−NHC complex leading to the isolation of a 14-electron Rh(III) complex. *J Am Chem Soc*. 2004; 126:5054–5055.
12. Scott NM, Dorta R, Stevens ED, Correa A, Cavallo L, Nolan SP. Interaction of a bulky N-heterocyclic carbene ligand with Rh(I) and Ir(I). Double C−H activation and isolation of bare 14-Electron Rh(III) and Ir(III) complexes. *J Am Chem Soc*. 2005;127:3516–3526.
13. Scott NM, Pons V, Stevens ED, Heinekey DM, Nolan SP. An electron-deficient iridium(III) dihydride complex capable of intramolecular C- -H activation. *Angew Chem Int Ed Engl*. 2005;44(17):2512–2515.
14. Prinz M, Grosche M, Herdtweck E, Herrmann WA. Unsymmetrically substituted iridium(III)−carbene complexes by a CH-activation process. *Organometallics*. 2000;19: 1692–1694.
15. Tang CY, Smith W, Vidovic D, Thompson AL, Chaplin AB, Aldridge S. Sterically encumbered iridium bis(N-heterocyclic carbene) systems: multiple C−H activation processes and isomeric normal/abnormal carbene complexes. *Organometallics*. 2009;28: 3059–3066.
16. Danopoulos AA, Winston S, Hursthouse MB. C-H activation with N-heterocyclic carbene complexes of iridium and rhodium. *J Chem Soc Dalton Trans*. 2002;(16): 3090–3091.
17. Hanasaka F, Tanabe Y, Fujita K-I, Yamaguchi R. Synthesis of new iridium N-heterocyclic carbene complexes and facile intramolecular alkyl C−H bond activation reactions of the carbene ligand. *Organometallics*. 2006;25:826–831.
18. Tanabe Y, Hanasaka F, Fujita K-I, Yamaguchi R. Scope and mechanistic studies of intramolecular aliphatic C−H bond activation of N-heterocyclic carbene iridium complexes. *Organometallics*. 2007;26:4618–4626.
19. Semwal S, Ghorai D, Choudhury J. Wingtip-dictated cyclometalation of N-heterocyclic carbene ligand framework and its implication toward tunable catalytic activity. *Organometallics*. 2014;33(24):7118–7124.
20. Corberan R, Sanau M, Peris E. Aliphatic and aromatic intramolecular C−H activation on Cp*Ir(NHC) complexes. *Organometallics*. 2006;25:4002–4008.
21. Zhang C, Zhao Y, Li B, Song H, Xu S, Wang B. The intramolecular sp^2 and sp^3 C-H bond activation of (p-cymene)ruthenium(II) N-heterocyclic carbene complexes. *Dalton Trans*. 2009;26:5182–5189.
22. Segarra C, Mas-Marzá E, Mata JA, Peris E. Rhodium and iridium complexes with chelating C–C′-imidazolylidene–pyridylidene ligands: systematic approach to normal, abnormal, and remote coordination modes. *Organometallics*. 2012;31(14):5169–5176.
23. Danopoulos AA, Pugh D, Wright JA. "Pincer" pyridine-dicarbene-iridium complexes: facile C-H activation and unexpected η^2-imidazol-2-ylidene coordination. *Angew Chem Int Ed Engl*. 2008;47(50):9765–9767.
24. Raynal M, Pattacini R, Cazin CSJ, Vallee C, Olivier-Bourbigou H, Braunstein P. Reaction intermediates in the synthesis of new hydrido, N-heterocyclic dicarbene iridium(III) pincer complexes. *Organometallics*. 2009;28:4028–4047.
25. Boom MEvd, Gozin M, Ben-David Y, et al. Formation and X-ray structures of PCP ligand based platinum(II) and palladium(II) macrocycles. *Inorg Chem*. 1996;35: 7068–7073.
26. Danopoulos AA, Pugh D, Smith H, Sassmannshausen J. Structural and reactivity studies of "pincer" pyridine dicarbene complexes of Fe^0: experimental and computational comparison of the phosphine and NHC donors. *Chem Eur J*. 2009;15(22):5491–5502.

27. Poater A, Bahri-Laleh N, Cavallo L. Rationalizing current strategies to protect N-heterocyclic carbene-based ruthenium catalysts active in olefin metathesis from C-H (de)activation. *Chem Commun.* 2011;47(23):6674–6676.
28. Hong SH, Chlenov A, Day MW, Grubbs RH. Double C-H activation of an N-heterocyclic carbene ligand in a ruthenium olefin metathesis catalyst. *Angew Chem Int Ed Engl.* 2007;46(27):5148–5151.
29. Mathew J, Koga N, Suresh CH. C−H bond activation through σ-bond metathesis and agostic interactions: deactivation pathway of a Grubbs second-generation catalyst. *Organometallics.* 2008;27:4666–4670.
30. Hong SH, Wenzel AG, Salguero TT, Day MW, Grubbs RH. Decomposition of ruthenium olefin metathesis catalysts. *J Am Chem Soc.* 2007;129:7961–7968.
31. Leitao EM, Dubberley SR, Piers WE, Wu Q, McDonald R. Thermal decomposition modes for four-coordinate ruthenium phosphonium alkylidene olefin metathesis catalysts. *Chem Eur J.* 2008;14(36):11565–11572.
32. Endo K, Grubbs RH. Chelated ruthenium catalysts for Z-selective olefin metathesis. *J Am Chem Soc.* 2011;133(22):8525–8527.
33. Dumas A, Tarrieu R, Vives T, et al. A versatile and highly Z-selective olefin metathesis ruthenium catalyst based on a readily accessible N-heterocyclic carbene. *ACS Catal.* 2018;8(4):3257–3262.
34. Burling S, Paine BM, Nama D, et al. C-H activation reactions of ruthenium N-heterocyclic carbene complexes: application in a catalytic tandem reaction involving C-C bond formation from alcohols. *J Am Chem Soc.* 2007;129:1987–1995.
35. Burling S, Mas-Marzá E, Valpuesta JEV, Mahon MF, Whittlesey MK. Coordination, agostic stabilization, and C−H bond activation of N-alkyl heterocyclic carbenes by coordinatively unsaturated ruthenium hydride chloride complexes. *Organometallics.* 2009;28(23):6676–6686.
36. Choi G, Tsurugi H, Mashima K. Hemilabile N-xylyl-N'-methylperimidine carbene iridium complexes as catalysts for C-H activation and dehydrogenative silylation: dual role of N-xylyl moiety for ortho-C-H bond activation and reductive bond cleavage. *J Am Chem Soc.* 2013;135(35):13149–13161.
37. Tang CY, Smith W, Thompson AL, Vidovic D, Aldridge S. Iridium-mediated borylation of benzylic C-H bonds by borohydride. *Angew Chem Int Ed Engl.* 2011; 50(6):1359–1362.
38. Corberan R, Sanau M, Peris E. Highly stable Cp*−Ir(III) complexes with N-heterocyclic carbene ligands as C−H activation catalysts for the deuteration of organic molecules. *J Am Chem Soc.* 2006;128:3974–3979.
39. Giunta D, Hölscher M, Lehmann CW, Mynott R, Wirtz C, Leitner W. Room temperature activation of aromatic C-H bonds by non-classical ruthenium hydride complexes containing carbene ligands. *Adv Synth Catal.* 2003;345(910): 1139–1145.
40. Stylianides N, Danopoulos AA, Pugh D, Hancock F, Zanotti-Gerosa A. Cyclometalated and alkoxyphenyl-substituted palladium imidazolin-2-ylidene complexes. synthetic, structural, and catalytic studies. *Organometallics.* 2007;26:5627–5635.
41. Dible BR, Sigman MS, Arif AM. Oxygen-induced ligand dehydrogenation of a planar bis-μ-chloronickel(I) dimer featuring an NHC ligand. *Inorg Chem.* 2005;44: 3774–3776.
42. Caddick S, Cloke FG, Hitchcock PB, de Lewis AKK. Unusual reactivity of a nickel N-heterocyclic carbene complex: tert-butyl group cleavage and silicone grease activation. *Angew Chem Int Ed Engl.* 2004;43(43):5824–5827.
43. Burling S, Mahon MF, Powell RE, Whittlesey MK, Williams JMJ. Ruthenium induced C-N bond activation of an N-heterocyclic carbene: isolation of C- and N-bound tautomers. *J Am Chem Soc.* 2006;128:13702–13703.

44. Hu Y-C, Tsai C-C, Shih W-C, Yap GPA, Ong T-G. The zirconium benzyl mediated C−N bond cleavage of an amino-linked N-heterocyclic carbene. *Organometallics.* 2010;29(3):516–518.
45. Zhang L, Liu Y, Deng L. Three-coordinate cobalt(IV) and cobalt(V) imido complexes with N-heterocyclic carbene ligation: synthesis, structure, and their distinct reactivity in C-H bond amination. *J Am Chem Soc.* 2014;136(44):15525–15528.
46. Spencer LP, Beddie C, Hall MB, Fryzuk MD. Synthesis, reactivity, and DFT studies of tantalum complexes incorporating diamido-N-heterocyclic carbene ligands. Facile endocyclic C−H bond activation. *J Am Chem Soc.* 2006;128:12531–12543.
47. Corberan R, Lillo V, Mata JA, Fernandez E, Peris E. Enantioselective preparation of a chiral-at-metal Cp*Ir(NHC) complex and its application in the catalytic diboration of olefins. *Organometallics.* 2007;26:4350–4353.
48. Avello MG, de la Torre MC, Sierra MA, Gornitzka H, Hemmert C. Central (S) to central (M=Ir, Rh) to planar (metallocene, M=Fe, Ru) chirality transfer using sulfoxide-substituted mesoionic carbene ligands: synthesis of bimetallic planar chiral metallocenes. *Chem Eur J.* 2019;25(58):13344–13353.
49. Chernyshev VM, Denisova EA, Eremin DB, Ananikov VP. The key role of R–NHC coupling (R = C, H, heteroatom) and M–NHC bond cleavage in the evolution of M/NHC complexes and formation of catalytically active species. *Chem Sci.* 2020;11 (27):6957–6977.
50. Tan KL, Park S, Ellman JA, Bergman RG. Intermolecular coupling of alkenes to heterocycles via C−H bond activation. *J Org Chem.* 2004;69(21):7329–7335.
51. Lewis JC, Bergman RG, Ellman JA. Direct functionalization of nitrogen heterocycles via Rh-catalyzed C−H bond activation. *Acc Chem Res.* 2008;41(8): 1013–1025.
52. Cavell KJ, McGuinness DS. Redox processes involving hydrocarbylmetal (N-heterocyclic carbene) complexes and associated imidazolium salts: ramifications for catalysis. *Coord Chem Rev.* 2004;248(7):671–681.
53. McGuinness DS, Cavell KJ, Skelton BW, White AH. Zerovalent palladium and nickel complexes of heterocyclic carbenes: oxidative addition of organic halides, carbon−carbon coupling processes, and the heck reaction. *Organometallics.* 1999;18(9): 1596–1605.
54. Magill AM, Yates BF, Cavell KJ, Skelton BW, White AH. Synthesis of N-heterocyclic carbene palladium(ii) bis-phosphine complexes and their decomposition in the presence of aryl halides. *Dalton Trans.* 2007;31:3398–3406.
55. McGuinness DS, Green MJ, Cavell KJ, Skelton BW, White AH. Synthesis and reaction chemistry of mixed ligand methylpalladium–carbene complexes. *J Organomet Chem.* 1998;565(1):165–178.
56. McGuinness DS, Saendig N, Yates BF, Cavell KJ. Kinetic and density functional studies on alkyl-carbene elimination from PdII heterocyclic carbene complexes: a new type of reductive elimination with clear implications for catalysis. *J Am Chem Soc.* 2001; 123(17):4029–4040.
57. McGuinness DS, Mueller W, Wasserscheid P, et al. Nickel(II) heterocyclic carbene complexes as catalysts for olefin dimerization in an imidazolium chloroaluminate ionic liquid. *Organometallics.* 2002;21(1):175–181.
58. Clement ND, Cavell KJ. Transition-metal-catalyzed reactions involving imidazolium salt/N-heterocyclic carbene couples as substrates. *Angew Chem Int Ed Engl.* 2004; 43(29):3845–3847.
59. Normand AT, Hawkes KJ, Clement ND, Cavell KJ, Yates BF. Atom-efficient catalytic coupling of imidazolium salts with ethylene involving Ni−NHC complexes as intermediates: a combined experimental and DFT study. *Organometallics.* 2007;26(22):5352–5363.

60. Steinke T, Shaw BK, Jong H, Patrick BO, Fryzuk MD, Green JC. Noninnocent behavior of ancillary ligands: apparent trans coupling of a saturated N-heterocyclic carbene unit with an ethyl ligand mediated by nickel. *J Am Chem Soc.* 2009; 131(30):10461–10466.
61. Normand AT, Yen SK, Huynh HV, Hor TSA, Cavell KJ. Catalytic annulation of heterocycles via a novel redox process involving the imidazolium salt N-heterocyclic carbene couple. *Organometallics.* 2008;27(13):3153–3160.
62. McGuinness DS, Cavell KJ. Reaction of CO with a methylpalladium heterocyclic carbene complex: product decomposition routes implications for catalytic carbonylation processes. *Organometallics.* 2000;19(23):4918–4920.
63. Warsink S, de Boer SY, Jongens LM, et al. Synthesis and characterization of Pd^{II}-methyl complexes with N-heterocyclic carbene-amine ligands. *Dalton Trans.* 2009;35:7080–7086.
64. Magill AM, McGuinness DS, Cavell KJ, et al. Palladium(II) complexes containing mono-, bi- and tridentate carbene ligands. Synthesis, characterisation and application as catalysts in C-C coupling reactions. *J Organomet Chem.* 2001;617–618: 546–560.
65. Zhou X, Jordan RF. Synthesis, cis/trans isomerization, and reactivity of palladium alkyl complexes that contain a chelating N-heterocyclic-carbene sulfonate ligand. *Organometallics.* 2011;30(17):4632–4642.
66. Nielsen DJ, Cavell KJ, Skelton BW, White AH. A pyridine bridged dicarbene ligand and its silver(I) and palladium(II) complexes: synthesis, structures, and catalytic applications. *Inorg Chim Acta.* 2002;327(1):116–125.
67. Graham DC, Cavell KJ, Yates BF. Influence of geometry on reductive elimination of hydrocarbyl–palladium–carbene complexes. *Dalton Trans.* 2005;6:1093–1100.
68. Graham DC, Cavell KJ, Yates BF. The influence of N-substitution on the reductive elimination behaviour of hydrocarbyl–palladium–carbene complexes—a DFT study. *Dalton Trans.* 2006;14:1768–1775.
69. Astakhov AV, Soliev SB, Gordeev EG, Chernyshev VM, Ananikov VP. Relative stabilities of M/NHC complexes (M = Ni, Pd, Pt) against R–NHC, X–NHC and X–X couplings in M(0)/M(II) and M(II)/M(IV) catalytic cycles: a theoretical study. *Dalton Trans.* 2019;48(45):17052–17062.
70. Cavell K. N-heterocyclic carbenes/imidazolium salts as substrates in catalysis: the catalytic 2-substitution and annulation of heterocyclic compounds. *Dalton Trans.* 2008;47:6676–6685.
71. Caddick S, Cloke FGN, Hitchcock PB, et al. The first example of simple oxidative addition of an aryl chloride to a discrete palladium N-heterocyclic carbene amination precatalyst. *Organometallics.* 2002;21(21):4318–4319.
72. Lewis AKdK, Caddick S, Cloke FGN, Billingham NC, Hitchcock PB, Leonard J. Synthetic, structural, and mechanistic studies on the oxidative addition of aromatic chlorides to a palladium (N-heterocyclic carbene) complex: relevance to catalytic amination. *J Am Chem Soc.* 2003;125(33):10066–10073.
73. Albéniz AC, Espinet P, Manrique R, Pérez-Mateo A. Aryl palladium carbene complexes and carbene–aryl coupling reactions. *Chem Eur J.* 2005;11(5):1565–1573.
74. Campeau L-C, Thansandote P, Fagnou K. High-yielding intramolecular direct arylation reactions with aryl chlorides. *Org Lett.* 2005;7(9):1857–1860.
75. Lewis JC, Wiedemann SH, Bergman RG, Ellman JA. Arylation of heterocycles via rhodium-catalyzed C−H bond functionalization. *Org Lett.* 2004;6(1):35–38.
76. Lewis JC, Berman AM, Bergman RG, Ellman JA. Rh(I)-catalyzed arylation of heterocycles via C−H bond activation: expanded scope through mechanistic insight. *J Am Chem Soc.* 2008;130(8):2493–2500.

77. Lake BRM, Ariafard A, Willans CE. Mechanistic insights into the oxidative coupling of N-heterocyclic carbenes within the coordination sphere of copper complexes. *Chem Eur J*. 2014;20(40):12729–12733.
78. Williams TJ, Bray JTW, Lake BRM, et al. Mechanistic elucidation of the arylation of non-spectator N-heterocyclic carbenes at copper using a combined experimental and computational approach. *Organometallics*. 2015;34(14):3497–3507.
79. Couzijn EPA, Zocher E, Bach A, Chen P. Gas-phase energetics of reductive elimination from a palladium(II) N-heterocyclic carbene complex. *Chem Eur J*. 2010;16(18):5408–5415.
80. Normand AT, Nechaev MS, Cavell KJ. Mechanisms in the reaction of palladium(II)–π-allyl complexes with aryl halides: evidence for NHC exchange between two palladium complexes. *Chem Eur J*. 2009;15(29):7063–7073.
81. Gordeev EG, Eremin DB, Chernyshev VM, Ananikov VP. Influence of R–NHC coupling on the outcome of R–X oxidative addition to Pd/NHC complexes (R = Me, Ph, vinyl, ethynyl). *Organometallics*. 2018;37(5):787–796.
82. Gordeev EG, Ananikov VP. Switching the nature of catalytic centers in Pd/NHC systems by solvent effect driven non-classical R-NHC coupling. *J Comput Chem*. 2019;40(1):191–199.
83. Marshall WJ, Grushin VV. Synthesis, structure, and reductive elimination reactions of the first (σ-aryl)palladium complex stabilized by IPr N-heterocyclic carbene. *Organometallics*. 2003;22(8):1591–1593.
84. Eckert P, Organ MG. A path to more sustainable catalysis: the critical role of LiBr in avoiding catalyst death and its impact on cross-coupling. *Chem Eur J*. 2020;26(21):4861–4865.
85. Tan KL, Bergman RG, Ellman JA. Annulation of alkenyl-substituted heterocycles via rhodium-catalyzed intramolecular C−H activated coupling reactions. *J Am Chem Soc*. 2001;123(11):2685–2686.
86. Tan KL, Bergman RG, Ellman JA. Intermediacy of an N-heterocyclic carbene complex in the catalytic C−H activation of a substituted benzimidazole. *J Am Chem Soc*. 2002;124(13):3202–3203.
87. Hawkes KJ, Cavell KJ, Yates BF. Rhodium-catalyzed C−C coupling reactions: mechanistic considerations. *Organometallics*. 2008;27(18):4758–4771.
88. Downing SP, Danopoulos AA. Indenyl- and fluorenyl-functionalized N-heterocyclic carbene complexes of titanium and vanadium. *Organometallics*. 2006;25(6):1337–1340.
89. Astakhov AV, Khazipov OV, Chernenko AY, et al. A new mode of operation of Pd-NHC systems studied in a catalytic mizoroki–heck reaction. *Organometallics*. 2017;36(10):1981–1992.
90. Kostyukovich AY, Tsedilin AM, Sushchenko ED, et al. In situ transformations of Pd/NHC complexes with N-heterocyclic carbene ligands of different nature into colloidal Pd nanoparticles. *Inorg Chem Front*. 2019;6(2):482–492.
91. Haslinger S, Kück JW, Anneser MR, Cokoja M, Pöthig A, Kühn FE. Formation of highly strained N-heterocycles via decomposition of iron N-heterocyclic carbene complexes: the value of labile Fe-C bonds. *Chem Eur J*. 2015;21(49):17860–17869.
92. Heckenroth M, Neels A, Garnier MG, Aebi P, Ehlers AW, Albrecht M. On the electronic impact of abnormal C4-bonding in N-heterocyclic carbene complexes. *Chem Eur J*. 2009;15(37):9375–9386.
93. Wierenga TS, Vanston CR, Ariafard A, Gardiner MG, Ho CC. Accessing chelating extended linker Bis(NHC) palladium(II) complexes: sterically triggered divergent reaction pathways. *Organometallics*. 2019;38(15):3032–3038.
94. Vigalok A. Metal-mediated formation of carbon–halogen bonds. *Chem Eur J*. 2008;14(17):5102–5108.
95. Sheppard TD. Metal-catalysed halogen exchange reactions of aryl halides. *Org Biomol Chem*. 2009;7(6):1043–1052.

96. Lappert MF, Pye PL. Carbene complexes. Part 10. Electron-rich olefin-based mono-, bis-, and tris-carbene-tungsten(0) complexes and some derived six and seven-coordinate mono- and bis-carbene-dihalogenotungsten(II) and related molybdenum(II) species. *J Chem Soc Dalton Trans*. 1977;(13):1283–1291.
97. Liu S-T, Ku R-Z, Liu C-Y, Kiang F-M. Oxidative cleavage of metal carbene complexes by iodine. *J Organomet Chem*. 1997;543(1):249–250.
98. Fooladi E, Dalhus B, Tilset M. Synthesis and characterization of half-sandwich N-heterocyclic carbene complexes of cobalt and rhodium. *Dalton Trans*. 2004;22:3909–3917.
99. Fu C-F, Lee C-C, Liu Y-H, et al. Biscarbene palladium(II) complexes. Reactivity of saturated versus unsaturated N-heterocyclic carbenes. *Inorg Chem*. 2010;49(6):3011–3018.
100. Lin B-L, Kang P, Stack TDP. Unexpected $C_{carbene}-X$ (X: I, Br, cl) reductive elimination from N-heterocyclic carbene copper halide complexes under oxidative conditions. *Organometallics*. 2010;29(17):3683–3685.
101. Hirsch-Weil D, Snead DR, Inagaki S, Seo H, Abboud KA, Hong S. In situ generation of novel acyclic diaminocarbene–copper complex. *Chem Commun*. 2009;18:2475–2477.
102. Kolychev EL, Shuntikov VV, Khrustalev VN, Bush AA, Nechaev MS. Dual reactivity of N-heterocyclic carbenes towards copper(ii) salts. *Dalton Trans*. 2011;40(12):3074–3076.
103. Lake BRM, Willans CE. Remarkable stability of copper(II)–N-heterocyclic carbene complexes void of an anionic tether. *Organometallics*. 2014;33(8):2027–2038.
104. Younesi Y, Nasiri B, Baba-Ahmadi R, Willans CE, Fairlamb IJS, Ariafard A. Theoretical rationalisation for the mechanism of N-heterocyclic carbene-halide reductive elimination at Cu^{III}, Ag^{III} and Au^{III}. *Chem Commun*. 2016;52(28):5057–5060.
105. Fauché K, Nauton L, Jouffret L, Cisnetti F, Gautier A. A catalytic intramolecular nitrene insertion into a copper(I)–N-heterocyclic carbene bond yielding fused nitrogen heterocycles. *Chem Commun*. 2017;53(15):2402–2405.
106. Chernyshev VM, Khazipov OV, Shevchenko MA, et al. Revealing the unusual role of bases in activation/deactivation of catalytic systems: O–NHC coupling in M/NHC catalysis. *Chem Sci*. 2018;9(25):5564–5577.
107. Li D, Ollevier T. Mechanism studies of oxidation and hydrolysis of Cu(I)–NHC and Ag–NHC in solution under air. *J Organomet Chem*. 2020;906:121025.
108. Li D, Ollevier T. Synthesis of imidazolidinone, imidazolone, and benzimidazolone derivatives through oxidation using copper and air. *Org Lett*. 2019;21(10):3572–3575.
109. Saker O, Mahon MF, Warren JE, Whittlesey MK. Sequential formation of $[Ru(IPr)_2(CO)H(OH_2)]^+$ and $[Ru(IPr)(\eta^6-C_6H_6)(CO)H]^+$ upon protonation of $Ru(IPr)_2(CO)H(OH)$ (IPr = 1,3-bis(2,6-diisopropylphenyl)imidazol-2-ylidene). *Organometallics*. 2009;28(6):1976–1979.
110. Wu F, Dioumaev VK, Szalda DJ, Hanson J, Bullock RM. A tungsten complex with a bidentate, hemilabile N-heterocyclic carbene ligand, facile displacement of the weakly bound W−(CC) bond, and the vulnerability of the NHC ligand toward catalyst deactivation during ketone hydrogenation. *Organometallics*. 2007;26(20):5079–5090.
111. Khazipov OV, Shevchenko MA, Chernenko AY, et al. Fast and slow release of catalytically active species in metal/NHC systems induced by aliphatic amines. *Organometallics*. 2018;37(9):1483–1492.
112. Wang C-Y, Liu Y-H, Peng S-M, Chen J-T, Liu S-T. Palladium(II) complexes containing a bulky pyridinyl N-heterocyclic carbene ligand: preparation and reactivity. *J Organomet Chem*. 2007;692(18):3976–3983.
113. Asensio JM, Tricard S, Coppel Y, Andrés R, Chaudret B, de Jesús E. Synthesis of water-soluble palladium nanoparticles stabilized by sulfonated N-heterocyclic carbenes. *Chem Eur J*. 2017;23(54):13435–13444.

114. van Rensburg H, Tooze RP, Foster DF, Slawin AMZ. The synthesis and X-ray structure of the first cobalt carbonyl–NHC dimer. Implications for the use of NHCs in hydroformylation catalysis. *Inorg Chem*. 2004;43(8):2468–2470.
115. van Rensburg H, Tooze RP, Foster DF, Otto S. Synthesis and characterization of a novel cobalt carbonyl N-heterocyclic carbene salt. Crystal structure of [Co(CO)$_3$(IMes)$_2$]$^+$[Co(CO)$_4$]. *Inorg Chem*. 2007;46(6):1963–1965.
116. Gandolfi C, Heckenroth M, Neels A, Laurenczy G, Albrecht M. Chelating NHC ruthenium(II) complexes as robust homogeneous hydrogenation catalysts. *Organometallics*. 2009;28(17):5112–5121.
117. Jeletic MS, Jan MT, Ghiviriga I, Abboud KA, Veige AS. New iridium and rhodium chiral di-N-heterocyclic carbene (NHC) complexes and their application in enantioselective catalysis. *Dalton Trans*. 2009;(15):2764–2776.
118. Bacciu D, Cavell KJ, Fallis IA, Ooi L-L. Platinum-mediated oxidative addition and reductive elimination of imidazolium salts at C4 and C5. *Angew Chem Int Ed Engl*. 2005;44(33):5282–5284.
119. Danopoulos AA, Tsoureas N, Green JC, Hursthouse MB. Migratory insertion in N-heterocyclic carbene complexes of palladium; an experimental and DFT study. *Chem Commun*. 2003;6:756–757.
120. Pugh D, Boyle A, Danopoulos AA. Pincer' pyridine dicarbene complexes of nickel and their derivatives. Unusual ring opening of a coordinated imidazol-2-ylidene. *Dalton Trans*. 2008;(8):1087–1094.
121. Becker E, Stingl V, Dazinger G, Puchberger M, Mereiter K, Kirchner K. Facile migratory insertion of a N-heterocyclic carbene into a ruthenium–carbon double bond: a new type of reaction of a NHC ligand. *J Am Chem Soc*. 2006;128(20):6572–6573.
122. Becker E, Stingl V, Dazinger G, Mereiter K, Kirchner K. Migratory insertion of acetylene in N-heterocyclic carbene complexes of ruthenium: formation of (Ruthenocenylmethyl)imidazolium salts. *Organometallics*. 2007;26(6):1531–1535.
123. Fernández FE, Puerta MdC, Valerga P. Functionalized N-heterocyclic carbene nonspectator ligands upon internal alkyne activation reactions. *Inorg Chem*. 2013;52(11):6502–6509.
124. Fantasia S, Jacobsen H, Cavallo L, Nolan SP. Insertion of a N-heterocyclic carbene (NHC) into a platinum–olefin bond. *Organometallics*. 2007;26(14):3286–3288.
125. Ingleson MJ, Layfield RA. N-heterocyclic carbene chemistry of iron: fundamentals and applications. *Chem Commun*. 2012;48(30):3579–3589.
126. Riener K, Haslinger S, Raba A, et al. Chemistry of Iron N-heterocyclic carbene complexes: syntheses, structures, reactivities, and catalytic applications. *Chem Rev*. 2014;114(10):5215–5272.
127. Wang X, Zhang J, Wang L, Deng L. High-spin Iron(II) alkynyl complexes with N-heterocyclic carbene ligation: synthesis, characterization, and reactivity study. *Organometallics*. 2015;34(12):2775–2782.
128. Keaton RJ, Blacquiere JM, Baker RT. Base metal catalyzed dehydrogenation of ammonia–borane for chemical hydrogen storage. *J Am Chem Soc*. 2007;129(7):1844–1845.
129. Yang X, Hall MB. The catalytic dehydrogenation of ammonia-borane involving an unexpected hydrogen transfer to ligated carbene and subsequent carbon–hydrogen activation. *J Am Chem Soc*. 2008;130(6):1798–1799.
130. Yang X, Hall MB. Density functional theory study of the mechanism for Ni(NHC)$_2$ catalyzed dehydrogenation of ammonia–borane for chemical hydrogen storage. *J Organomet Chem*. 2009;694(17):2831–2838.
131. Romain C, Miqueu K, Sotiropoulos J-M, Bellemin-Laponnaz S, Dagorne S. Non-innocent behavior of a tridentate NHC chelating ligand coordinated onto a zirconium(IV) center. *Angew Chem Int Ed Engl*. 2010;49(12):2198–2201.

132. Holloczki O, Terleczky P, Szieberth D, Mourgas G, Gudat D, Nyulaszi L. Hydrolysis of imidazole-2-ylidenes. *J Am Chem Soc*. 2011;133(4):780–789.
133. Pelegrí AS, Elsegood MRJ, McKee V, Weaver GW. Unexpected ring expansion of an enantiopure imidazoline carbene ligand. *Org Lett*. 2006;8:3049–3051.
134. Zuo W, Braunstein P. Stepwise synthesis of a hydrido, N-heterocyclic dicarbene iridium(III) pincer complex featuring mixed NHC/abnormal NHC ligands. *Dalton Trans*. 2012;41(2):636–643.
135. Segarra C, Mas-Marza E, Benitez M, Mata JA, Peris E. Unconventional reactivity of imidazolylidene pyridylidene ligands in iridium(III) and rhodium(III) complexes. *Angew Chem Int Ed Engl*. 2012;51(43):10841–10845.
136. Hildebrandt B, Ganter C. Reactivity of a cationic N-heterocyclic carbene and its corresponding dicationic precursor. *J Organomet Chem*. 2012;717:83–87.
137. Gupta SK, Ghorai D, Choudhury J. A new type of palladium-pincer complexes generated via hydrolytic ring-opening of Imidazole-2-ylidenes. *Organometallics*. 2014; 33(12):3215–3218.
138. Collins LR, Riddlestone IM, Mahon MF, Whittlesey MK. A comparison of the stability and reactivity of diamido- and diaminocarbene copper alkoxide and hydride complexes. *Chem Eur J*. 2015;21(40):14075–14084.
139. Tao S, Guo C, Liu N, Dai B. Counteranion-controlled Ag_2O-mediated benzimidazolium ring opening and its application in the synthesis of palladium pincer-type complexes. *Organometallics*. 2017;36(22):4432–4442.
140. Pandolfi F, Mattiello L, Zane D, Feroci M. Electrochemical behaviour of 9-methylcaffeinium iodide and in situ electrochemical synthesis of hymeniacidin. *Electrochim Acta*. 2018;280:71–76.
141. Iversen KJ, Wilson DJ, Dutton JL. Activation of the C-N bond of N-heterocyclic carbenes by inorganic elements. *Dalton Trans*. 2014;43(34):12820–12823.
142. Wurtemberger-Pietsch S, Radius U, Marder TB. 25 years of N-heterocyclic carbenes: activation of both main-group element-element bonds and NHCs themselves. *Dalton Trans*. 2016;45(14):5880–5895.
143. Waltman AW, Ritter T, Grubbs RH. Rearrangement of N-heterocyclic carbenes involving heterocycle cleavage. *Organometallics*. 2006;25:4238–4239.
144. Arrowsmith M, Hill MS, Kociok-Kohn G, MacDougall DJ, Mahon MF. Berylliuminduced C–N bond activation and ring opening of an N-heterocyclic carbene. *Angew Chem Int Ed Engl*. 2012;51(9):2098–2100.
145. Al-Rafia SM, McDonald R, Ferguson MJ, Rivard E. Preparation of stable lowoxidation-state group 14 element amidohydrides and hydride-mediated ring-expansion chemistry of N-heterocyclic carbenes. *Chem Eur J*. 2012;18(43):13810–13820.
146. Schmidt D, Berthel JH, Pietsch S, Radius U. C-N bond cleavage and ring expansion of N-heterocyclic carbenes using hydrosilanes. *Angew Chem Int Ed Engl*. 2012; 51(35):8881–8885.
147. Iversen KJ, Wilson DJD, Dutton JL. A theoretical study on the ring expansion of NHCs by silanes. *Dalton Trans*. 2013;42(31):11035–11038.
148. Iversen KJ, Wilson DJD, Dutton JL. Comparison of the mechanism of borane, silane, and beryllium hydride ring insertion into N-heterocyclic carbene C–N bonds: a computational study. *Organometallics*. 2013;32(21):6209–6217.
149. Momeni MR, Rivard E, Brown A. Carbene-bound borane and silane adducts: a comprehensive DFT study on their stability and propensity for hydride-mediated ring expansion. *Organometallics*. 2013;32(21):6201–6208.
150. Su MD. Mechanistic investigations on E-N bond-breaking and ring expansion for N-heterocyclic carbene analogues containing the group 14 elements (E). *Inorg Chem*. 2014;53(10):5080–5087.
151. Iversen KJ, Wilson DJD, Dutton JL. Effects of the electronic structure of fivemembered N-heterocyclic carbenes on insertion of silanes and boranes into the NHC C–N bond. *Dalton Trans*. 2015;44(7):3318–3325.

152. Arrowsmith M, Hill MS, Kociok-Köhn G. Activation of N-heterocyclic carbenes by {BeH$_2$} and {Be(H)(Me)} fragments. *Organometallics*. 2015;34(3):653–662.
153. Franz D, Inoue S. Systematic investigation of the ring-expansion reaction of N-heterocyclic carbenes with an iminoborane dihydride. *Chem Asian J*. 2014;9(8):2083–2087.
154. Wang T, Stephan DW. Carbene-9-BBN ring expansions as a route to intramolecular frustrated Lewis pairs for CO$_2$ reduction. *Chem Eur J*. 2014;20(11):3036–3039.
155. Bose SK, Fucke K, Liu L, Steel PG, Marder TB. Zinc-catalyzed borylation of primary, secondary and tertiary alkyl halides with alkoxy diboron reagents at room temperature. *Angew Chem Int Ed*. 2014;53(7):1799–1803.
156. Pietsch S, Paul U, Cade IA, Ingleson MJ, Radius U, Marder TB. Room temperature ring expansion of N-heterocyclic carbenes and B-B bond cleavage of diboron(4) compounds. *Chem Eur J*. 2015;21(25):9018–9021.
157. Wurtemberger-Pietsch S, Schneider H, Marder TB, Radius U. Adduct formation, B-H activation and ring expansion at room temperature from reactions of HBcat with NHCs. *Chem Eur J*. 2016;22(37):13032–13036.
158. Schneider H, Hock A, Bertermann R, Radius U. Reactivity of NHC alane adducts towards N-heterocyclic carbenes and cyclic (alkyl)(amino)carbenes: ring expansion, ring opening, and Al-H bond activation. *Chem Eur J*. 2017;23(50):12387–12398.
159. Bruckner T, Arrowsmith M, Hess M, Hammond K, Muller M, Braunschweig H. Synthesis of fused B,N-heterocycles by alkyne cleavage, NHC ring-expansion and C-H activation at a diboryne. *Chem Commun*. 2019;55(47):6700–6703.
160. Phillips N, Tirfoin R, Aldridge S. Probing the limits of ligand steric bulk: backbone C-H activation in a saturated N-heterocyclic carbene. *Chem Eur J*. 2014;20(13):3825–3830.
161. Hernan-Gomez A, Kennedy AR, Hevia E. C-N bond activation and ring opening of a saturated N-heterocyclic carbene by lateral alkali-metal-mediated metalation. *Angew Chem Int Ed Engl*. 2017;56(23):6632–6635.
162. Prema D, Mathota Arachchige YL, Murray RE, Slaughter LM. "Decarbonization" of an imino-N-heterocyclic carbene ligand via triple benzyl migration from hafnium. *Chem Commun*. 2015;51(31):6753–6756.
163. Hatanaka T, Ohki Y, Tatsumi K. Coupling of an N-heterocyclic carbene on iron with alkynes to form η^5-cyclopentadienyl-diimine ligands. *Angew Chem Int Ed Engl*. 2014;53(10):2727–2729.
164. Lubitz K, Radius U. The coupling of N-heterocyclic carbenes to terminal alkynes at half sandwich cobalt NHC complexes. *Organometallics*. 2019;38(12):2558–2572.
165. Despagnet-Ayoub E, Henling LM, Labinger JA, Bercaw JE. Group 4 transition-metal complexes of an aniline–carbene–phenol ligand. *Organometallics*. 2013;32(10):2934–2938.
166. Yang C-F, Lu T, Chen X-T, Xue Z-L. Unusual rearrangement of an N-heterocyclic carbene via a ring-opening and ring-closing process. *Chem Commun*. 2018;54(56):7830–7833.
167. Frey GD, Masuda JD, Donnadieu B, Bertrand G. Activation of Si-H, B-H, and P-H bonds at a single nonmetal center. *Angew Chem Int Ed Engl*. 2010;49(49):9444–9447.
168. Liu H-J, Ziegler MS, Tilley TD. Ring-opening and double-metallation reactions of the N-heterocyclic carbene ligand in Cp*(IXy)Ru (IXy = 1,3-bis(2,6-dimethylphenyl) imidazol-2-ylidene) complexes. Access to an anionic Fischer-type carbene complex of ruthenium. *Polyhedron*. 2014;84:203–208.
169. Danopoulos AA, Massard A, Frison G, Braunstein P. Iron and cobalt metallotropism in remote-substituted NHC ligands: metalation to abnormal NHC complexes or NHC ring opening. *Angew Chem Int Ed Engl*. 2018;57(44):14550–14554.
170. Galan BR, Gembicky M, Dominiak PM, Keister JB, Diver ST. Carbon monoxide-promoted carbene insertion into the aryl substituent of an N-heterocyclic carbene ligand: Buchner reaction in a ruthenium carbene complex. *J Am Chem Soc*. 2005;127:15702–15703.

171. Poater A, Ragone F, Correa A, Cavallo L. Exploring the reactivity of Ru-based metathesis catalysts with a π-acid ligand trans to the Ru-ylidene bond. *J Am Chem Soc*. 2009;131:9000–9006.

172. Galan BR, Pitak M, Gembicky M, Keister JB, Diver ST. Ligand-promoted carbene insertion into the aryl substituent of an N-heterocyclic carbene ligand in ruthenium-based metathesis catalysts. *J Am Chem Soc*. 2009;131:6822–6832.

173. Zhang D, Kawaguchi H. Deprotonation attempts on imidazolium salt tethered by substituted phenol and construction of its magnesium complex by transmetalation. *Organometallics*. 2006;25:5506–5509.

174. Miranda-Soto V, Grotjahn DB, DiPasquale AG, Rheingold AL. Imidazol-2-yl complexes of Cp*Ir as bifunctional ambident reactants. *J Am Chem Soc*. 2008;130: 13200–13201.

175. Brown RM, Borau Garcia J, Valjus J, et al. Ammonia activation by a nickel NCN-pincer complex featuring a non-innocent N-heterocyclic carbene: ammine and amido complexes in equilibrium. *Angew Chem Int Ed Engl*. 2015;54(21):6274–6277.

176. Titcomb LR, Caddick S, Cloke FGN, Wilson DJ, McKerrecher D. Unexpected reactivity of two-coordinate palladium–carbene complexes; synthetic and catalytic implications. *Chem Commun*. 2001;15:1388–1389.

177. Simms RW, Drewitt MJ, Baird MC. Equilibration between a phosphine-cobalt complex and an analogous complex containing an N-heterocyclic carbene: the thermodynamics of a phosphine-carbene exchange reaction. *Organometallics*. 2002;21: 2958–2963.

178. Allen DP, Crudden CM, Calhoun LA, Wang R. Irreversible cleavage of a carbene–rhodium bond in Rh-N-heterocyclic carbene complexes: implications for catalysis. *J Organomet Chem*. 2004;689(20):3203–3209.

179. Wang C-Y, Liu Y-H, Peng S-M, Liu S-T. Rhodium(I) complexes containing a bulky pyridinyl N-heterocyclic carbene ligand: preparation and reactivity. *J Organomet Chem*. 2006;691(19):4012–4020.

180. Kaufhold O, Stasch A, Pape T, et al. Metal template controlled formation of [11]ane-P$_2$C^{NHC} macrocycles. *J Am Chem Soc*. 2009;131:306–317.

181. Andjaba JM, Tye JW, Yu P, Pappas I, Bradley CA. Cp*Co(IPr): synthesis and reactivity of an unsaturated Co(I) complex. *Chem Commun*. 2016;52(12):2469–2472.

182. Zeller A, Bielert F, Haerter P, Herrmann WA, Strassner T. Replacement of N-heterocyclic carbenes by N-heterocyclic silylenes at a Pd(0) center: experiment and theory. *J Organomet Chem*. 2005;690(13):3292–3299.

183. Hurni KL, Baines KM. Stabilization of a transient diorganogermylene by an N-heterocyclic carbene. *Organometallics*. 2007;26(10):4109–4111.

184. Rupar PA, Staroverov VN, Baines KM. Reactivity studies of N-heterocyclic carbene complexes of germanium(II). *Organometallics*. 2010;29(21):4871–4881.

185. Asensio JM, Andrés R, Gómez-Sal P, de Jesús E. Aqueous-phase chemistry of η^3-allylpalladium(II) complexes with sulfonated N-heterocyclic carbene ligands: solvent effects in the protolysis of Pd–C bonds and Suzuki–Miyaura reactions. *Organometallics*. 2017;36(21):4191–4201.